Liquid Crystal Display Drivers

Liquid Crystal Display Drivers

Techniques and Circuits

by

David J.R. Cristaldi
DIEES - University of Catania, Italy

Salvatore Pennisi
DIEES - University of Catania, Italy

and

Francesco Pulvirenti
STMicroelectronics, Catania, Italy

Springer

Dr. David J.R. Cristaldi
Università Catania
Dipto. Ingegneria Elettrica
Elettronica e dei Sistemi (DIEES)
Viale Andrea Doria, 6
95125 Catania
Italy
dcristaldi@diees.unict.it

Dr. Francesco Pulvirenti
STMicroelectronics
Stradale Primosole, 50
95121 Catania
Italy
francesco.pulvirenti@st.com

Prof. Salvatore Pennisi
Università Catania
Dipto. Ingegneria Elettrica
Elettronica e dei Sistemi (DIEES)
Viale Andrea Doria, 6
95125 Catania
Italy
spennisi@diees.unict.it

ISBN 978-90-481-2254-7 e-ISBN 978-90-481-2255-4

DOI 10.1007/978-90-481-2255-4

Library of Congress Control Number: 2009920981

Printed on acid-free paper

9 8 7 6 5 4 3 2 1

springer.com

The amateur photo in the cover was taken by S. Pennisi during a summer thunderstorm in front of Taormina, Sicily, where he lives. The idea of the cover was suggested by his friend Eng. Giuseppe Cavallaro.

Contents

Forewords

The origins of this book can be found in the scientific collaboration (officially started in March 2003) between Francesco Pulvirenti and me. Besides its obvious technical aim, this collaboration has produced eighteen theses for second-level laurea degree in electronic engineering and fifteen theses for first level laurea degree in electronic engineering. It has given many students the possibility to work within a skilled and experienced design team in a friendly environment, while measuring their strengths against advanced and exciting microelectronic issues. Many of these students are now appreciated designers working all over the world. One of these graduate students was David Joseph Roger Cristaldi, who has just completed his Ph.D. course program.

This book has been written also to leave the marks of this hard work, commitment, respect, and friendship.

Salvatore Pennisi

The activity on Display Drivers started in 1999 when the former CEO of STMicroelectronics, Pasquale Pistorio, and the General Manager of TECDIS (a company of passive displays based in Chatillon, Italy), Ettore Morezzi, decided to create a joint-venture, called DORA, with the purpose to design display drivers. At the beginning this activity was managed by Roberto Gariboldi, today site manager of DORA design center, now based in Aosta (Italy) and by Giuseppe Collé, today quality assurance manager in EM Microelectronics based in Marin (Switzerland).

Starting from January 2001, Pietro Menniti, General Manager of the Industrial and Power Supply Division, now called Industrial and Power Conversion Division, involved me in the design activity. At the beginning I had a team of about 15 people, shared between the DORA design center, originally based in Chatillon, and the STMicroelectronics Catania (Italy) site.

This new challenging adventure started with the contribution and thrust of all the main actors: Roberto, Giuseppe and Pietro. A huge support to my activity came also from Maria Rosa Borghi, Design Director of the Industrial and Power Supply Division, based in Castelletto (Italy) and Mario Paparo, Director of the TPA (Telecommunication, Peripheral and Automotive) R&D team based in Catania. Today Maria Rosa is still covering the same position, while Mario is managing the Automotive R&D team in Shenzen (China).

My traveling from Catania to Agrate (Italy) and Chatillon became more and more frequent, and I used to spend at least one week a month in Chatillon. During my staying in Chatillon I felt very comfortable there thanks to Giuseppe support and his frequent invitations to dinner with his family.

Even the members of the design team Alberto Iorio, Roberto Aletti, Mara Concolato, Riccardo Lavorerio, Ivo Pannizzo and Leonardo Sala became soon my friends sharing all efforts and satisfactions. We used to spend our evenings in a restaurant eating grilled steaks, smoking cigars and drinking some good whisky.

The first product developed under my responsibility was the STE2002, an 81-row×128-column driver for monochrome displays designed in 0.5-μm lithography. We sold several millions units for many years, but the best seller among the monochrome display drivers was the STE6596, a 65×96 driver for Nokia Mobile Phones, designed in the same lithography.

At that time, the business of those products was under the Industrial Business Unit, managed by Alberto De Marco, now director of the Health Care Business Unit in the MAHRS Division. The marketing responsible was Luca Rodeschini.

After the reorganization occurred in ST at the end of 2003, all the activity moved into the new Display Division, which grouped three main businesses: CRTs, large size and small size flat panel displays. The Division was managed by Philippe Berger, who was based in Grenoble with most of the divisional staff members. Luca Rodeschini became the responsible of the small size display Business Unit. He was based in Agrate together with the BU program managers and most of the marketing and application teams.

In this new context I had the responsibility of the design activity for small size display drivers. The design team suddenly grew to 45 people, based in Grenoble (France), Aosta and Catania; the Aosta team had moved from Chatillon.

The team in Grenoble was technically prepared and autonomous. Before my arrival it was managed by Olivier Le-Briz, who took the responsibility of the divisional R&D team continuing to guarantee a great support to the design activity. I was very lucky as I found skilled people with consolidated know-how in OLED design and application. Celine Mas had the leadership of most of the projects developed in Grenoble and the responsibility of the analog team. Olivier Scouarnec had the responsibility of the digital design and the related team. Christian Michon had the responsibility of the back-end activity and the related team.

In Catania, most of the team members had worked with me since before the beginning of the activity on display drivers, thus we were well tuned as a diapason. Gregorio Bontempo was my deputy and one of the first pioneer in the display driver design; he left the company in 2004 to be closer to his family and now he is a high school teacher of electronics. Salvo Pappalardo and Santo Ilardo became respectively the responsible of the Analog and Digital design teams. Tiziana Signorelli managed alternatively the Memory blocks for all devices and the Trocadero project. At the end of 2003 other colleagues joined my team: in particular, Bruno Cavallaro and Amedeo La Scala were respectively the project leaders of the Caravelle and Videophone projects. This is of course only a rough and limited description of the activity and role of each member of my design teams.

In the Display Division the design activity suddenly grew, leading to the development of drivers for OLED and Active Matrix LCD technologies. We developed several drivers for the above mentioned technologies, and two of them are today still in production: the STE2028, a 162×132 RGB driver, 262k colors, for AMLCDs developed in 0.18-μm lithography, nicknamed Caravelle mounted in a display for Nokia 6070 model mobile phone, and the STE2102, a 240×320 RGB driver, 262k colors, for AMLCDs developed in 0.18-μm lithography, nicknamed Videophone.

Since the activity was spread in many sites, it was perceived as an additional incentive and motivation. Everybody intensified traveling in order to keep solid and fresh contacts with the other teams. I remember the past years as a wonderful period of my professional life. I had the opportunity to meet and work with important people in the display industry such as Marshall J. Bell, staff engineer of Flat Panel Display Group in National Semiconductor. Ken Foo, Manager of Platform Display Module and Driver ICs development in Motorola. Atsunari Tsuda, Manager of Platform Display Module development in Sanyo-Epson.

The design activity was strictly linked to the development of new technologies and the interaction with the technology department was intensified. Many skilled people worked for the purpose, among them the key leaders were: Enrico Laurin, responsible of the development of the HCMOS5 technology, in 0.5-μm lithography, and Giuseppe Croce, responsible of the development of the HVG6 and HVG8 technologies, respectively in 0.35-μm and 0.18-μm lithography.

Among the sale offices visited during my travels, the one in Japan remains for me an indelible experience for the human qualities of the people met there: Kamiya Akikatsu, responsible of the Photovoltaic team, Eric Benoit, marketing manager and Matsuo Takahiro, application manager. When the activity was slowing down Eric went back to Grenoble and, one year later, Matsuo left the company.

During these years I have had a friend and companion of adventure, who has supported me in all technical issues, with whom I have shared many moments of my private and professional life: prof. Pennisi, with whom, in conclusion of the activity on the display drivers, I accepted to write this book.

I am proud to have worked with all these persons. For the sake of conciseness, I have been obliged to shrink the wide list of people really involved and contacted during the intense activity on display drivers. However, I cannot omit to report my collaborators not mentioned yet, most of them still working with me: Michele Battista, Mirko Dondini, Sergio Fa', Laurent Ferchaud, Cinzia Ferrara, Salvatore Di Fazio, Basilio Filocamo, Angela Gambina, Marco Giuffrida, Corinne Ianigro, Rene Krawiec, Pascal Lelievre, Massimiliano Licciardello, Igor Lisciandra, Antonio Maimone, Rosario Murabito, Roberto Nicolosi, Nicola Nigido, Rosalba Pani, Maurizio Patane', Danika Perrin, Vincenzo Petitto, Salvatore Privitera, Salvatore Puglisi, Domenico Ragonese, Massimiliano Ragusa, Regis Rousset, Francesco Rua', Vincenzo Sambataro, Massimo Sorbera, Angelo Scuderi, Remi Tores.

I am too emotionally involved to tell my opinion on this story, but I leave the reader to judge the results. In less than seven years we developed 3 new technologies and more than 20 display drivers, most of them industrialized and produced with revenue of hundreds million dollars.

In June 2007 I joint back the Industrial and Power Conversion Division, with most of the design team based in Catania. Then Pietro Menniti gave me a new challenging adventure: developing new products for photovoltaic applications.

Few weeks before my moving back in the I&PC Division, the activity on display drivers was stopped for strategic reasons.

Francesco Pulvirenti

Preface

Liquid Crystals (LCs) represent a paradigm studied and exploited in several different contexts. For example, the membrane of a biological cell is made up of *lyotropic* type LCs which are therefore broadly investigated in the fields of biochemistry, biophysics, and bionics. LCs are also studied in the area of pattern-forming mechanisms and in the bifurcation theory [C1991, BK1996, TMBC2001], to understand and describe nature structures: snowflakes, honeycomb, and animal coat markings are just some examples. They are preferred to standard isotropic fluids, because they provide more stable and larger regions with regular patterns, and are therefore more easily observable and controllable (by means of electric and magnetic fields). LCs are also used in the field of image processing and analysis. The diffusion of LCs is so wide all over the world that studies on their reuse, recycling, and recovery, as well as on their toxicity, have been performed. As a consequence, LC producers agreed to market no severely toxic, carcinogenic or mutagenic LC substances [MSB2004].

Nevertheless, LCs are certainly better known by most of us for their use in the realization of electronic displays. Flat Panel Displays (FPDs) of palm-top computers, notebook-PCs, Personal Digital Assistants (PDAs), mobile/smart phones, video games, desktop monitors, TV-sets, projectors, view finders for digital cameras/camcorders, CD/MP3/MP4 players, GPS navigation systems, airplane cockpits, e-Books, electronic labels (for supermarket shelves), photocopy and medical equipments, are the most noteworthy applications. FPDs are the principal medium actually linking people and information: they have enabled applications otherwise impossible[1] which are limited only by the designer fantasy.

The multi-billion-dollar Liquid Crystal Display (LCD) industry has been rapidly growing since 1971. Technological progress has enabled wide-ranging-size commercial products with diagonal size below 1 in. to 40 in. and more. This is nowadays the mainstream display technology playing an important role in the economy of many Asian countries. LCDs dominate the scene in portable electronic equipments and notebook computers. LCD monitors for desktop computers have almost completely replaced CRTs (cathode-ray-tubes). Furthermore, LCDs have recently surpassed CRTs even for television technology, outselling Plasma Displays by more than 7 to 1.

---

[1]Not only compact hand-held equipments have taken advantage of FPDs, but also large screens. For instance, high definition TV (HDTV) has received widespread acceptance only with the availability of FPDs larger than 40 in. Indeed, large-screen CRT-based displays, both direct view and projection, are far too bulky for the average home.

This explosion is the consequence of concurrent and decisive research efforts in:

1. Materials (LC mixtures, polarizer films, and color pigments);
2. Processing technologies. The size of the glass substrates increased from 320 mm × 400 mm in 1989 (Generation 1) to 2160 mm × 2460 mm in 2006 (Generation 8). Corning Incorporated,[2] the world leader in specialty glass and ceramics, and the dominant supplier to the LCD industry, is going to begin the mass production of the Generation 10 glass substrates (2850 mm × 3050 mm) by 2010.
3. Micro-optics and precision machinery designs.
4. Electronic designs (driver ICs, ASIC ICs, image processing, color management);

While the above key elements are still continuously improving, other relevant research issues are emerging. Better electrical power utilization, better material utilization and environmental designs to save natural resources, are only some of the new topics. Besides, the LCD panels are migrating from *simple* output devices to *two-way* communication devices with many functions as human-machine interface (*integrated* touch panels are under development with signal-sensing pixels without an additional touch-pad attached on the LCD panel) [L2007].

Despite many alternative display technologies have been developed or may emerge in the near future, their chances of success are rather weak because of the huge investments in massive worldwide LCD business infrastructures made by the few dominating enterprises. OLED (Organic Light Emitting Diodes) is the only technology that can compete in picture quality, viewing angle issues and low power, but, to be competitive with LCDs, further reduction in cost, increase in throughput, increase in longevity, and scaling to larger sizes will be needed [V2008].

The sudden increase of interest in Liquid Crystals has been accompanied also by the publication of numerous books related to the topic. The reader may thus wonder why writing another one. The answer is that most of these books are focused on the LC materials and/or LCD technologies and many are written from the physics or chemistry point of view, or even from an historical perspective. This book, in contrast, deals with LCDs from the *electronic engineering* point of view and is specifically focused on those reliable techniques, architectures, and design solutions amenable to efficiently design driving circuits for such systems. To this aim the book, after providing an introduction to the physical-chemical properties of LC substances, their evolution and application to LCDs, converges to the examination and in-depth explanation of addressing techniques and suitable driving schemes for passive-matrix and active-matrix LCDs.

---

[2]Since 2000, the Corning glass size has doubled about every 1.5 years by following Moore's law also for LCDs. A number of twelve 30-in. panels can be simultaneously derived from a single Generation-8 motherglass.

LCDs are intrinsically "passive" devices, they are simple *light valves*. The managing and control of the data to be displayed is relegated to one or more circuits commonly denoted as LCD drivers. The semiconductor market of LCD driver ICs is very competitive, similar to the DRAM aggressive market. Reduction of costs and area is vital, especially for small-size panels. Considering hand-held equipments for consumer electronics, the product life time is very short (1–3 years) and, due to the fast IC price erosion, the major revenues occur only in the first 6 months. These considerations mandate for very short development cycles, leaving no possibility of failure. Having a deep comprehension of the technology, the existing architectures and design techniques is fundamental for the implementation of next generation two-way display devices and to gain insight into the future of the industry.

In this rapidly evolving scenario, *Liquid Crystal Display Drivers* is focused on the presentation of practical approaches and circuit solutions regularly adopted for mass production, but also unconventional and emerging techniques are mentioned. Even if the attention is mainly paid on LCDs, the topics treated have more general validity and find application also in alternative display technologies (OLEDs, Electrophoretic Displays, etc.).

The book is not only a reference for engineers and system integrators who work in the field of displays with their future development, but it is written also for scientific researchers, educators and students. It is a valuable resource for advanced undergraduate and graduate students attending display systems courses, and it may prove interesting even for a non-expert in the field, as it is reasonably simply written without referring to unnecessary mathematics and by privileging the intuitive description. Besides, we have tried to put in evidence the men and the ideas behind the things they produced. To improve clarity and for the sake of completeness we also included 22 Photos, 20 Tables and more than 200 Figures and 400 bibliographic references.

Writing a book of this nature has also required extensive discussions with colleagues, designers and technologists. Our thanks to: Giuseppe Musumarra, Giuseppe Falci, Gino Sorbello, professors at the University of Catania. A special thank is due to Santo Ilardo and Amedeo La Scala at STMicroelectronics, to Giuseppe Collè at EM Microelectronics, and to Prof. Ingo Dierking for his photographs, as well as the people at Merck, Sharp and Corning, who have kindly and positively replied to our several requests. We are also indebted to professors Gaetano Palumbo and Domenico Pappalardo for their contributed chapter on charge pumps. Finally, we would gratefully acknowledge the financial support of STMicroelectronics.

> *There are no limits to the creativity of the display developers or to the tools available for discovery and development. The challenge is to find the right combination of display performance and manufacturing cost to make FPDs attractive and affordable to billions of people, which in turn enrich their lives through technology* [M2002].

Catania, Italy David J.R. Cristaldi

Catania, Italy Salvatore Pennisi

Catania, Italy Francesco Pulvirenti

References

[BK1996] A. Buka and L. Kramer, *Pattern Formation in Liquid Crystals,* Springer, Berlin, 1996.

[C1991] P.E. Cladis, *Pattern Formation in Complex Dissipative Systems,* World Scientific, Singapore, 1991.

[L2007] C.T. Liu, "Revolution of the TFT LCD Technology," *IEEE Journal of Display Technology*, Vol. 3, No. 4, pp. 342–350, 2007.

[M2002] D.E. Mentley, "State of Flat-Panel Display Technology and Future Trends," *Proceedings of the IEEE*, Vol. 90, No. 4, pp. 453–459, 2002.

[MSB2004] R. Martin, B. Simon-Hettich, and W. Becker, "Safety of Liquid Crystals (LCs) and Recycling of Liquid Crystal Displays (LCDs) in Compliance with WEEE," *Merck Liquid Crystal Newsletter*, No. 19, pp. 27–30, 2004.

[TBM2001] S.J. Tavener, G.I. Blake, T. Mullin, K.A. Cliffe, "A Numerical Bifurcation Study of Electrohydrodynamic Convection in Nematic Liquid Crystals," Phys. Rev. E, Vol. 63, No. 1, pp. 1-12, 2001.

[V2008] R.J. Visser, "The Business of Displays: OLED Displays on the Verge of Commercial Breakthrough," *Journal of the SID*, Vol. 5, No. 4, p. 8, pp. 102–103, 2008.

Chapter 1

Liquid Crystals

Discovery, Classification, and Physics

This chapter begins with the early history of liquid crystals (LCs). Subsequently, a brief classification and introduction to the physics of LCs and their electro-optic properties is discussed, particularly, for those materials suitable for display applications.

It will emerge that the initial studies on LCs involve botanists, chemists, physicists (and even mineralogists and mathematicians) almost exclusively from Europe. The interested reader is referred to [SDS2004] for a thorough report concerning the history of LCs and for an excellent collection and translation in English of the fundamental documents in the field (most of which were originally written in German and French). Other important sources are, ordered by publication year, [K1973, H1984, K1988, KK1989, S1989, V1992, K2000, K2002, C2005].

Complete treatment of the physical LCs properties can be found in [GP1995] (a standard), another source is [DFL2001], a good review of basic properties is presented in [SS1974], optical and electro-optical properties are specifically discussed in [KW1993, BC1994, L2001].

1.1 Early History

Solid, liquid and gaseous states were the only known states of matter before the discovery of LCs. A solid can be either crystalline or amorphous, being the former characterized by a three-dimensional lattice with both long-range positional and orientational atomic order and the latter having only short-range order. When a solid is heated above its melting point, it turns into an isotropic liquid having neither positional nor orientational order, which, after cooling, turns back into a solid.

In 1888,[1] a young Austrian botanist, **Friedrich Reinitzer**, was working at the University of Prague (Prague is now the capital of the Czech Republic, but was then the capital of the province of Bohemia in the Austro-Hungarian empire). In his

---

[1]Nine years later, in 1897, the German scientist Karl F. Braun introduced the first cathode ray tube scanning device.

D.J.R. Cristaldi et al., *Liquid Crystal Display Drivers*, DOI 10.1007/978-90-481-2255-4_1,

experiments he was extracting cholesterol from carrots in order to determine its exact chemical formula, and to find whether it was the same compound found in the cells of many animals. At this purpose he isolated from human gallstones two substances we now know as *cholesteryl benzoate*, and *cholesteryl acetate*. Both organic compounds are solid crystalline at room temperature, and he wanted to measure their melting point. When heating the compounds, he noticed the reflection colors of their melts, as had others before him for other cholesterol derivatives. However, he also noted two distinct melting points. The *cholesteryl benzoate* melted at 145.5°C, becoming cloudy and viscous, but rising the temperature to 178.5°C it became isotropic and clear, typical of the liquid state (for this reason this point was lately called **clearing point**; the **melting point,** in which a solid enters the liquid crystalline state, is sometimes called **freezing point**). The phenomenon was reversible. The other substance, the cholesteryl acetate, showed a melting point at 94.8°C and a clearing point at 114.3°C, in this case however, the appearance of colours was observed only once on cooling [SDS2004].

Reinitzer attributed the two-melting-point *anomaly* to the presence of impurities, but after several attempts and specimen refinements, he observed always the same behavior. Unable to explain the phenomenon, he communicated with **Otto Lehmann** a young German physicist.

Lehman studied physics at Strassburg University (contemporary Strasbourg is in France, but at that time, like the rest of the disputed provinces of Alsace and Lorraine, was in Germany). He was fascinated with microscopes and microscopy (his father, schoolmaster, had been an amateur microscopist before him), and to complement his research, Lehmann spent a large amount of time and energy developing and improving (in a series of implementations) his invention, the heating stage microscope, that he called the *crystallization microscope*, Fig. 1.1. This instrument is still standard equipment in liquid crystal research laboratories today.

Fig. 1.1 Lehmann's original hot-stage microscope and some Merck LC substances from 1904. Picture source: Merck KGaA, Corporate History

An epistolary communication between the two scientists produced the first actual investigation of liquid crystals, leading to a fundamental insight into the nature of this new phase of matter.

On 14 March 1888, Reinitzer wrote to Lehmann his, now famous, first 16-page-long letter describing the two melting points:

...just below the second melting temperature, 178.5 °C, ... violet and blue colors appear, which rapidly vanish as the sample becomes more milk-like turbid, but still fluid. On further cooling, the blue and violet colors reappear, to disappear again as the substance solidifies to form a white crystalline mass.

Along with the letter, Reinitzer sent samples of the two materials.

In 1888, Lehmann was 33. One year later he would become professor of physics at the Technical University of Karlsruhe, as the successor to Heinrich Hertz, who had lately demonstrated experimentally Maxwell's theory of electromagnetism.

Lehmann soon confirmed the observations of the Austrian botanist. At this purpose, the *crystallization microscope* revealed itself extremely useful to understand the behaviour of cholesteryl benzoate, because not only it allowed observations in polarized light, but also enabled *in situ* high temperature observations thanks to its hot stage.

In one of the many correspondences with Reinitzer, Lehmann wrote:

My results confirm your previous observations, which the substance consists of a crystal very soft ... It is completely homogenous and does not contain another liquid, unlike you suggested... It is truly of great interest for physicists that a crystal so soft that it can be called liquid exists.

Lehmann realized that he was dealing with a new state of the matter, between a solid crystal and an isotropic liquid, which appears fluid and turbid and that only certain materials exhibit. He also ascertained that this state shows a double refraction (typical of crystals, see Section 1.3.7). Because the shared properties of both liquids and solids, in a celebrated article of 1889 [L1889], he called these materials, *Fliessende Kristalle* (flowing crystals) or *Schleimig flussige Kristalle* (slimy liquid crystals). Lehmann found materials that even exhibited *three* melting points, he named this additional phase *Kristalline Flüssigkeit* (crystalline fluid) or *Tropfbar flüssige Kristalle* (liquid crystals which form drops). All this researches were organized in an enormous tome, simply entitled "Liquid Crystals" [L1904].

At first, many scientists underestimated the importance of the discovery while others were skeptic and thought that the newly-discovered state was just a mixture of solid and liquid components.

It is interesting to observe that Lehmann asked the support of the **E. Merck**[2] company in Darmstadt, Germany (see again Fig. 1.1), in order to obtain highly pure substances:

---

[2]Since 1904, Merck produced and sold for experimental purposes chemical preparations exhibiting LC properties.

My work in the area of liquid crystals ... also among the experts in this field [remains] misunderstood and unknown. Nobody is able to procure the necessary materials and instruments.

Nevertheless, after the seminal works of Lehmann, new LC substances were continuously discovered. In 1890, **Ludwig Gattermann** (at that time 30-year-old Assistant Professor at the University of Heidelberg, and that later became Full Professor at the University of Freiburg and a famous organic chemist[3]) published the *first* report of the complete synthesis of one of these substances [GR1890]. The article describes the synthesis of **para-azoxyanisole** (**PAA**, a liquid crystal at temperature between 116°C to 134°C). The method of synthesis was well defined and relatively easy, and the temperature range was more accessible than in the case of cholesteryl benzoate. These favourable features caused this material to become that of choice for the successive studies in liquid crystal properties. In the subsequent years, the physical chemist **Rudolf Schenck** of Marburg, recorded new 24 LC compounds. **Daniel Vorländer** of the University of Halle and his students synthesized hundreds of LC compounds and the first *thermotropic smectic* compound. Even in 1908 he had enough material to be organized in a book [V1908].

In 1909, Lehmann (then 54 and with a recognized reputation) visited Geneva and Paris, giving seminars and experimental demonstrations. This stimulated the formation of a French school of LC science that became important up today. Among those influenced by Lehmann's visit were **George Friedel** and **Charles-Victor Mauguin**.

Friedel, was a crystallographer who formulated basic laws concerning the external morphology and internal structure of crystals. He clarified that LCs have three types of molecular organization: *smectic*, *nematic*, and *cholesteric* (that will be discussed in Section 1.2) leaving a mark on the science of LCs. Friedel gave Lehmann the responsibility of the initial skepticism of the scientific community, and to avoid misunderstanding he proposed to adopt the term **mesophase** to describe the LC state, or phase. After this work, we now use **mesomorphic** as the attribute related to both features and properties peculiar to this state of matter, and **mesogen** to designate an organic compound able to generate mesophases. Nevertheless, the term liquid crystal continued to be used.

In the introduction of his famous article published in 1922, Friedel wrote [F1922]:

I use the term mesomorphic to designate those states of matter observed by Lehmann in the years following 1889, and for which he conceived the terms liquid crystal and crystalline fluid. Lehmann had the great merit of drawing attention to these materials, but he erred greatly in naming them. The unfortunate names have been repeated again and again over the last 30 years. As a result many people suppose that these substances are merely crystalline materials, albeit rather more

[3]Gattermann's textbook *Die Praxis des Organischen Chemikers* (The practice of Organic Chemist) was adopted by generations of students.

fluid than those hitherto known. The exact opposite is the case. Indeed, these materials are infinitely more interesting than they would be if they were simply crystals exhibiting some unexpected degree of fluidity.

On the other side, C. Mauguin, of the École Normale Supérieure, studied the behavior of LC thin layers (of thickness between 10 and 150 μm) confined between plates. In his experiments he used also Gattermann's PAA. His achievements are fundamental for the subsequent evolution of the subject. In particular, he examined "birefringent liquid films with a helicoidal structure" and found that an incident linearly polarized beam exits the sample elliptically polarized. Moreover, under certain circumstances the polarization of incident light is twisted (this is the result of what we would now call a twisted nematic cell). He demonstrated that if the ratio of the twist pitch to the wavelength of light is large, the polarization follows the twisting birefringence [M1911a]. Even more significantly, Mauguin established that (magnetic) fields orient liquid crystals [M1911b].

Between 1922 and the World War II, **Carl W. Oseen**,[4] of the University of Uppsala, Sweden, and **Hans Zöcher** of the University of Prague developed a macroscopic mathematical model for the study of liquid crystals and introduced the *order parameter* (see Section 1.3.1) [O1933, Z1933]. Besides, the aligning properties of LCs due to magnetic fields was theoretically analyzed by **Vsevolod Freédericksz** and his group at the Physico-Technical Institute of Leningrad [FZ1929, FZ1933]. After these fundamental works, the scientific community gradually lost interest in LCs, as it was believed that all their relevant features had been discovered.

In the 1950s the works by **Glenn Brown**,[5] from the University of Cincinnati, **George W. Gray**,[6] from the University of Hull, and **Charles Frank**[7] from the University of Bristol led to a renewed interest in liquid crystals [F1958], while **Wilhelm Maier** and **Alfred Saupe**[8] formulated a molecular theory of LCs [MS1958]. It is also worth mentioning that in 1956, a young physicist named **James L. Fergason**, decided that LCs would be the lifetime subject of his research. He

---

[4]C.W. Oseen was professor of mechanics and mathematical physics. He was also influential (and despotic) member of the committee for the Nobel Prize from 1921 to 1944, judging nominations for theoretical physics and in particular atomic physics. He awarded eminent personalities such as Albert Einstein, Niels Bohr, Werner Heisenberg, Erwin Schrödinger and Paul Dirac [F2002].

[5]G. Brown learnt about LCs while looking for a research topic for his graduate students. Subsequently, he wrote a review of the existing literature in *Chemical Reviews* [BS1957]. The article was widely read and encouraged Brown to a lifetime study on the LC materials.

[6]The work of G.W. Gray was fundamental for the successive development of new high-performance LC materials, see Section 2.3.

[7]Sir C. Frank was a physicist interested in crystallography. He wrote only four papers on LCs, and one of them (the 1958 referenced paper) gave him long term recognition. Actually, he recapitulated in part the two previous works of Oseen and Zöcher (who can perhaps be regarded as the father of the Frank-Oseen theory [SDS2004], see Section 1.3.2).

[8]A. Saupe was a German physicist who worked also at Kent State University. The Maier-Saupe theory is originated from his diploma thesis elaborated together with his advisor prof. W. Maier in Karlsruhe University.

would become one of the foremost pioneers in this field, and one of the American most prolific living inventors, as we will see in the next chapter.

From the 1960's the *commercial* interest in LCs started to grow, as reported in the next chapter, but the theoretical studies on LCs did not paused, culminating eventually in the 1991 Nobel Prize to **Pierre-Gilles de Gennes**, of College de France, Paris, with the following motivation:

...for discovering that methods developed for studying order phenomena in simple systems can be generalized to more complex forms of matter, in particular to liquid crystals and polymer.

In fact, de Gennes expanded the phase transitions theory of Lev D. Landau [LL1984] into liquid crystals, providing a phenomenological description of molecular order in the LC phases, as well as elastic and hydrodynamic properties. He showed that phase transitions in such seemingly widely-differing physical systems as magnets, liquid crystals, polymer solutions and superconductors can be described analytically in a general unified manner.

To close the section, we include in Fig. 1.2 the picture of some of the scientists involved in the history of LCs and mentioned before.

Fig. 1.2 (**a**) Friedrich Reinitzer (1857–1927), (**b**) Otto Lehmann (1855–1922), (**c**) Georges Friedel (1865–1933), (**d**) Carl W. Oseen (1878–1944), (**e**) Hans Zöcher (1893–1969), (**f**) George W. Gray (1926), (**g**) James L. Fergason (1934), (**h**) Pierre-Gilles de Gennes (1932–2007). [K1989, K2002]

1.2 Classification

A vast variety of chemical compounds are now known to exhibit one or several liquid crystalline phases. These molecules show significant differences in chemical composition, but they share some common features in chemical and physical properties.

Many concepts discussed in the following are due to Friedel, who, as already mentioned, suggested the liquid crystal classification scheme which is used today.

1.2.1 Cause of Formation

A first classification makes a distinction between **thermotropic** and **lyotropic** LCs, according to the mechanism that initiates the transition to the liquid crystal phase. Most known liquid crystals are produced by varying the temperature of various substances (in particular, by raising the temperature of certain solids or lowering the temperature of certain liquids) and are commonly termed thermotropic. Below the melting point temperature, T_{mp}, thermotropic liquid crystals are solid, crystalline and anisotropic; as the temperature is raised above the clearing point $T_{cp}>T_{mp}$, they are a clear isotropic liquid. Between T_{mp} and T_{cp} they are in the liquid crystal mesophase,[9] as illustrated in Fig. 1.3. The process is usually reversible by lowering the temperature, though there may be a small temperature hysteresis (for example, T_{mp} when reducing temperature may be slightly less then T_{mp} when increasing temperature). The liquid crystals Reinitzer and the early researchers discovered were all thermotropic. Among the thermotropic liquid crystals there are two fundamental classes of substances, those that are **enantiotropic**, which are able to enter the LC state both by cooling a liquid and heating a solid, and those that are **monotropic**, which can enter the LC state via one or the other of those methods, but not both.

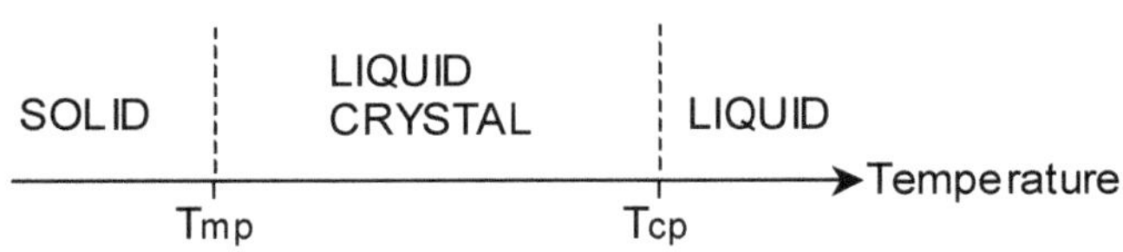

Fig. 1.3 The temperature range of a thermotropic LC is comprised between the melting and clearing points

The other fundamental way to make a substance enter a liquid crystalline phase is by the action of a solvent, and the concentration of the solution is primarily responsible for the occurring phase. More specifically, concentration (and secondarily temperature) is the most important controllable parameter for the lyotropic phase, while temperature (and secondarily pressure) is the most important

[9] An example of a typical phase sequence is Cr 20 N 60 I, meaning that the LC phase (in this case *nematic*) is located between 20°C and 60°C. Cr (or sometimes letter K) stands for the crystalline phase and I for the isotropic phase.

controllable parameter for the thermotropic phase. Lyotropic LCs are broadly present in nature. For instance, by adding water to soap (which is not a LC) above a minimum concentration, we cause the soap molecules to form lyotropic LCs. The term “lyo” comes from the Greek word *liein* which means “to dissolve” (just like soap does in the water). Steroids, glicolipids, phospholipids, liposomes, to cite some examples, all form lyotropic mesophases. Finally, the membrane of a biological cell is made up of lyotropic LCs. As a consequence, this type of LCs is primarily investigated in the fields of biochemistry, biophysics, and bionics.

1.2.2 Shape and Molecular Structure

As far as the shape of LC molecules is concerned, we distinguish two main types (though there are less usual but more complex shapes): in the first the molecules appear like a rod (rod-like, cigar-like or **calamitic** LCs) as sketched in Fig. 1.4a, in the second the molecules are shaped like a disc (disc-like or **discotic**[10] LCs), as sketched in Fig. 1.4b.

(a) (b)

Fig. 1.4 Shape of liquid crystal molecules: calamitic (**a**) and discotic (**b**)

We represented the molecules in Fig. 1.4 as made up of a rigid central part (core) plus some flexible ramifications (lateral, or side chains) whose chemical structures are generally different each other. Hereafter, the chains will not be plotted for the sake of simplicity, as they are not essential to the discussion. Moreover, since calamitic and thermotropic liquid crystals are the most important for display applications, we shall treat only these liquid crystals.

It is noteworthy that a careful balance between rigid and flexible parts is essential to produce the LC phase. The molecule must not be completely flexible, as it will not have orientational order, but if completely rigid it will directly pass from solid to liquid, with increasing temperature.

The characteristics of a calamitic LC are obtained by the interconnection of two (or more) rigid cyclic units, or rings, causing the elongated shape.

The planar conformation is ensured by the linking groups such as -(CH=N)-, -N=N-, $-(CH=CH)_n-$, -CH=N-N=CH-, etc., that with their multiple bonds reduce the freedom of rotation. Some representative examples of these LC molecules and the

[10]Discotic LCs were discovered by the group of **Sivaramakrishna Chandrasekhar** in Bangalore, India, only in 1977. The paper announcing the discovery was published in an Indian physics journal [CSS1977]. Discotic LCs found a limited application in electronic displays only in the late 1990s. They are used to make a sheet of film that expands the viewing angle of a twisted-nematic (TN) display [K2002]. Recently, other bizarre and more complex LC shapes have been added, like the banana, or boomerang, one [NSW1996].

temperature range in which they exhibit nematic mesophase are reported in Table 1.1.

Table 1.1 Some Liquid Crystal compounds

Acronym	Molecular structure and full name	Nematic range (°C)
PAA	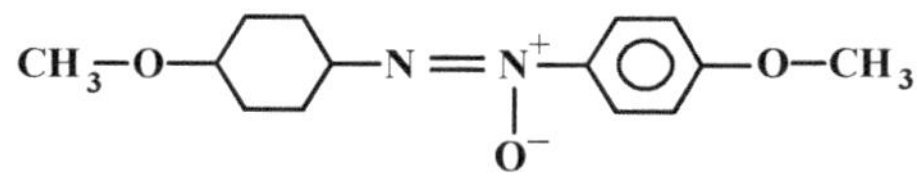 para-azoxyanisole	116–134
MBBA	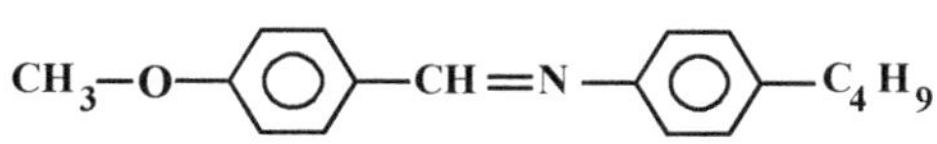 N-(4′-methoxybenzylidene)-4-butylaniline	22–47
EBBA	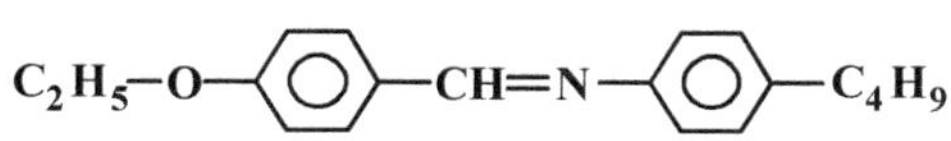 N-(4′-ethoxybenzylidene)-4-butylaniline	35–77
5CB	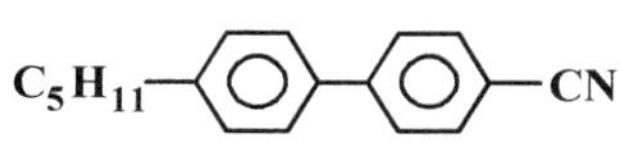 4′-(n-pentyl)-4-cyanobiphenyl	22–35
6CB	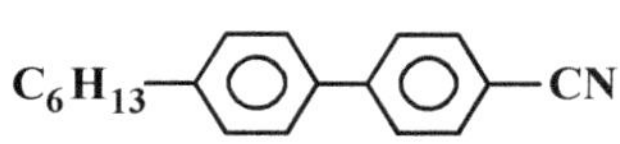 4′-(n-hexyl)-4-cyanobiphenyl	15–29
PCH	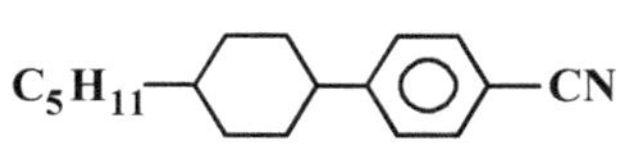 4′-(trans-4-n-pentylcyclohexyl) benzonitrile	30–55

All the structures in Table 1.1, except the last one, contain two benzene rings.[11] We recall from the organic chemistry that when the benzene ring is attached to some other group of atoms in a molecule it is called a *phenyl group* and is represented in several ways, as it is illustrated in Fig. 1.5a. Moreover, when two substituents of the hydrogen atoms are present, their relative positions are usually indicated by the prefixes *ortho*, *meta* and *para*, (or alternatively by 2, 3, and 4, respectively) as illustrated in Fig. 1.5b. In LC molecules, the linking groups are attached in the *para* positions, ensuring the most elongated shape (usually, in the range of a few nanometers). In general, a typical LC possesses a linear structure with a central core that contains several rings, a linkage and two terminal chains. It was seen that short chains helped the formation of the nematic state, as well as the combination of one

[11]Benzene, C_6H_6, is conventionally and concisely depicted in several ways. We adopt the hexagon with a circle in the middle, to indicate that benzene is a resonance hybrid.

short alkyl chain on one ring and a polar substituent on the other to maximize the degree of anisotropic polarizability of electron density.

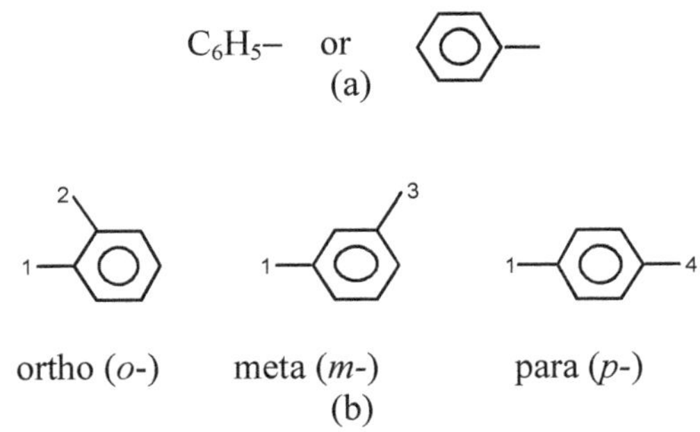

Fig. 1.5 (**a**) Symbols of the phenyl group. (**b**) Relative positions and nomenclature of substituents in a benzene ring

The first compound in Table 1.1 is the Gattermann's PAA which exhibits nematic state at temperatures greater than 116°C. The following entries, MBBA and EBBA, belong to a class of compounds called **Schiff's bases**, whose general formula is

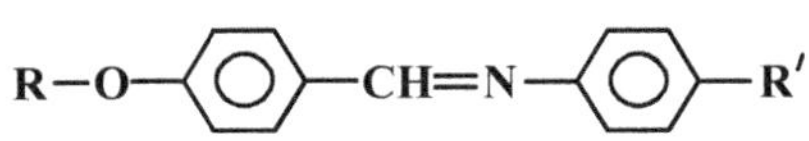

where the symbols **R** and **R′** at the left and right represent the terminal groups. **R** is of the type C_nH_{2n+1}, with *n* integer. **R′** includes C_4H_9, CH_3COO, and CN. They exhibit nematic state nearly at room temperature, thus making their use much easier for experimental purposes. Unfortunately they have poor chemical stability, as they are easily hydrolyzed by water. To avoid this problem, chemically stable substances, like for instance Cyanobiphenils (5CB and 6CB are two examples) and their derivatives were synthesized. They are characterized by two phenyl groups attached directly (**biphenyl**). The last entry in Table 1.1 shows PCH, as an example of a class of compounds in which one aromatic ring is substituted by a *cyclohexane* (C_6H_{12}) ring. Further LC compounds and mixtures utilized for display applications will be discussed throughout the Chapter 2 of this book.

1.2.3 Liquid Crystal Phases

Unlike liquids, whose molecules possess no type of order and are free to move in all the three dimensions (Fig. 1.6), LCs possess a certain degree of orientational order and positional relationships between the molecules. Different **phases** for thermotropic LCs can be defined in a sequence given by increasing temperature (namely, the *smectic* and either *nematic* or *cholesteric* phases).

The first phase above T_m is the **smectic** phase, which is viscous and fluid. Smectic is derived from the Greek word for soap, *smectos*. The slipperiness of soap is indicative of the ease with which these layers shift about. The smectic liquid crystals are closer to solids because they exhibit both (two-dimensional) positional and orientational order. The molecules are arranged side by side in a series of layers and their movement is mainly limited inside the layers, which are free to slide over each other.

A unit vector **n**, known as the director, can be defined, parallel to the average direction of the long axis of the molecules in the immediate neighborhood. This vector is not constant throughout the whole medium, but is a function of space. To put in evidence this last aspect, **n**(r) is often used to designate the director.[12]

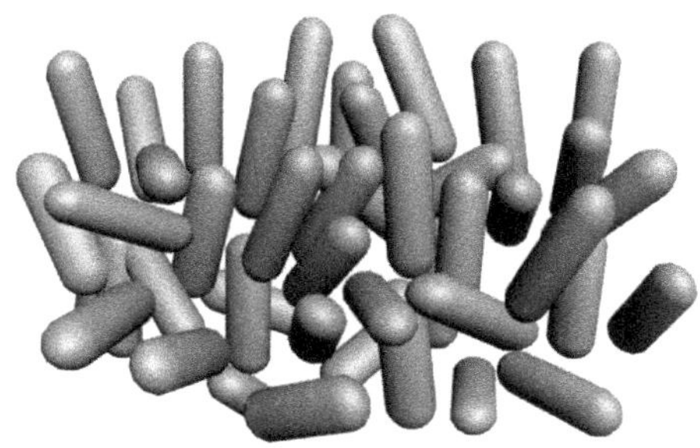

Fig. 1.6 Molecular disorder in liquids

There are three main smectics sub-categories. The two best known of these are *smectic A*, in which the (calamitic) molecules align their long axes, with random deviations, perpendicularly to the layer planes, as shown in Fig. 1.7a, and *smectic C*, where the director is tilted from the layers normal by a fixed angle, as shown in Fig. 1.7b. The transition to smectic C occurs by cooling the smectic A phase.

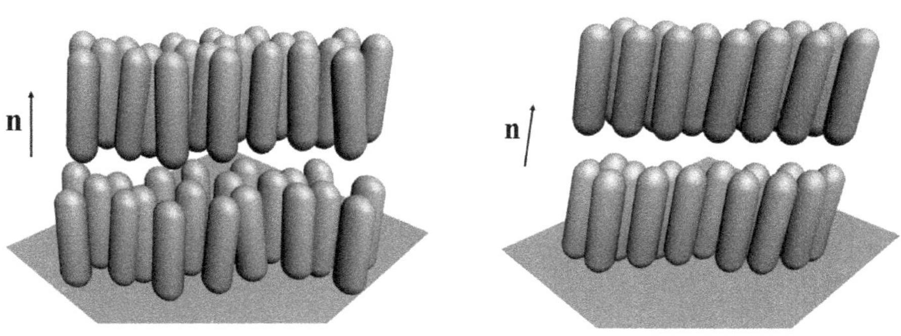

(a) (b)

Fig. 1.7 Molecular arrangement in (**a**) *smectic A* LCs and (**b**) *smectic C* LCs

[12]Liquid crystals tend to form microdroplets. Within each droplet, molecules have the same orientation (director), but the director can be different for adjacent droplets. This explains the milky appearance of some LCs, as scattering of light occurs due to different director orientation. At the nematic-isotropic transition temperature (clearing point), the fluid no longer is made up of microdroplets and light scattering does not occur.

If materials which are intrinsically chiral or have added chiral components[13] are used, the smectic C phase changes into the *smectic C** (or *chiral smectic C* phase). This phase possess a layered smectic structure in which the parallel long axes of the molecules are rotated from layer to layer on the surface of a cone, resulting in a helix, as shown in Fig. 1.8. The helical structure, as discussed in the followings, is responsible for optical rotation, and other peculiar properties.

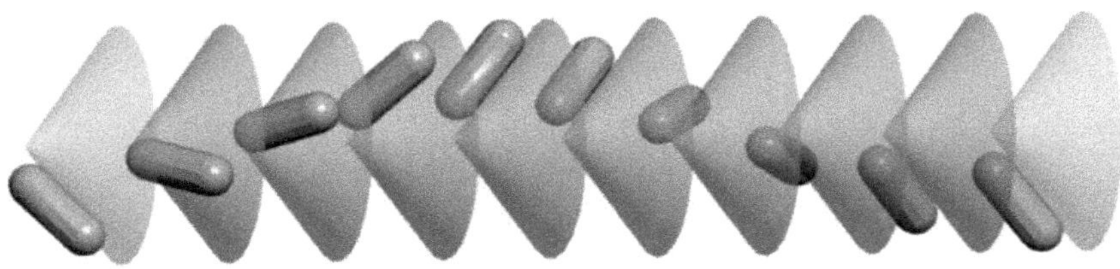

Fig. 1.8 The helix structure of the smectic C* phase

Next to the clearing point, increasing the temperature, the **nematic** phase appears. Nematic liquid crystals are the most widely used liquid crystals in display applications.

The name nematic is derived from a Greek word *nematos* meaning thread-like. In the nematic phase all molecules are aligned approximately parallel to each other, with only a one-dimensional (orientational) order and without a positional order. Molecules can translate in all the three directions and can rotate freely along the long molecular axes. An illustration of nematic phase is given in Fig. 1.9, along with an arbitrary system of coordinated axis included for reference. Some have compared the organization of the molecules in nematics to toothpicks in a box, which are free to translate in all directions and to rotate along the long (z) axis, but maintain their original orientation, since they are not allowed to rotate along the y and x axis by their neighbors.

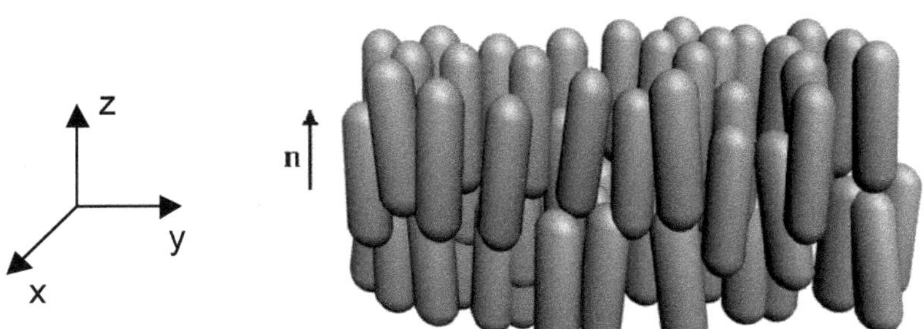

Fig. 1.9 Molecular arrangement in *nematic* liquid crystals. Molecules are aligned approximately parallel to each other. The local average preferred direction is described by the unity vector **n** (the director)

[13]The term chiral is used to designate an object that cannot be superimposed to its mirror image. The classic example is constituted by the human hands. The left hand is a non-superimposable mirror image of the right hand.

It should be noted that on optical inspection of a nematic, one rarely sees the idealized equilibrium configuration. Some very prominent structural perturbation appear as threads, as illustrated in the photographs in Fig. 1.10 [D2003], and from which nematics take their name.

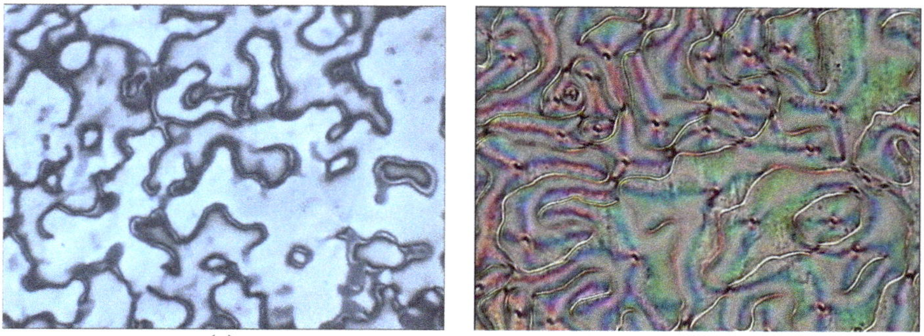

(a) (b)

Fig. 1.10 Thread-like textures of nematic materials. Photo courtesy of Ingo Dierking, Wiley-VCH and Taylor and Francis [D2003]

A schematic representation of the phase transitions with temperature described above is illustrated in Fig. 1.11.

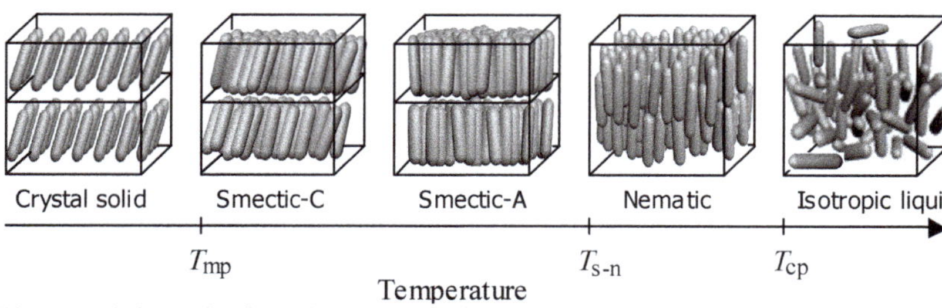

Fig. 1.11 Schematization of phase transitions as a function of temperature for rod-like molecules. T_{mp}, $T_{s\text{-}n}$ and T_{cp} are respectively the temperatures at the melting point, smectic-to-nematic transition and clearing point

If chiral compounds such as cholesterol esters are added, the nematic phase changes into the **chiral-nematic** or **cholesteric** phase. The molecules in cholesteric liquid crystals are again arranged in layers (see Fig. 1.12). Layers are very thin and, within each layer, molecules are aligned in parallel with their long axes averagely parallel to the plane of the layers, similar to those in nematic liquid crystals. The main difference is that the director in each layer is displaced slightly from the corresponding director of the adjacent layers. The director is hence rotated (twisted) from layer to layer tracing out a helical path whose *pitch* is of the order of the wavelengths of visible light. Observe that the pitch is a function of temperature, and in particular it decreases as temperature increases.

Because of the helical structure, cholesteric LCs exhibit optical rotation, and selective reflection.[14] Moreover, cholesteric liquid crystals are also used as additives in the Twisted Nematic and Super Twisted Nematic technologies (that will be treated in the next chapter).

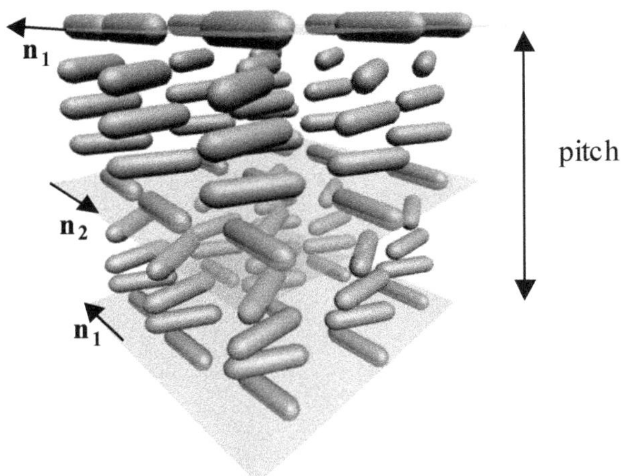

Fig. 1.12 Layers in cholesteric liquid crystals

1.3 Physical Characteristics

In this section some fundamental physical properties of liquid crystals limited, for the purposes of this book, to the nematic liquid crystal phase will be discussed. It will be shown that nematic LCs, being not as rigid as solids, are easily reoriented, realigned, or deformed by applying mechanical stresses, electric and magnetic fields, and by the proximity actions with surfaces that have been properly prepared.

Because of their specific molecular shape and alignment, nematic LCs exhibit anisotropic physical characteristics. Their dielectric susceptibility, electrical conductivity, magnetic permeability, refractive index, and viscosity measured in the direction of the long axis are different from those measured in the plane normal to the long axis.

[14] The phenomenon of selective reflection was exploited by J. L. Fergason in the first LC commercial application; well before liquid crystals were used for displays (see Sections 1.3.7 and 2.1.1).

1.3.1 Molecular Order

In the nematic phase, the molecules alignment is described by two parameters:

(a) The already introduced director, **n**. It describes the macroscopic preferred direction of molecules in a volume, which is small compared to the total LC system, but which is large compared to one LC molecule.

(b) The **order parameter**, *S*. It describes the distribution of molecules around the director direction within the volume. Defining as ϑ the polar angle (i.e., the angle comprised between the director and the longitudinal axis of the molecule), *S* is expressed by

$$S = \frac{1}{2}\left\langle 3\cos^2\vartheta - 1\right\rangle \tag{1.1}$$

where the symbol $\langle\cdot\rangle$ indicates an ensemble average. Observe that averaging this function, instead of just ϑ alone, gives a value between 0 and 1 for the amount of orientational order. In a perfectly ordered state, $\vartheta = 0$, hence $S = 1$. A completely unordered phase has $S = 0$. In a typical nematic phase, *S* ranges between 0.4 and 0.7, indicating that the molecules have a certain degree of disorder [Z1933, O1933, F1958, L1999].

1.3.2 Elastic Properties

The lowest energy state for bulk nematic LCs corresponds to a single director orientation throughout the material. Boundaries, mechanical stress and external fields deform the LC molecular and director alignment; this induces reaction (elastic) forces. As a consequence of the orientational elasticity there exists always a local restoring torque (*elastic torque*) on the director which opposites to director variations.

The deformation of nematic liquid crystals can be considered for three elementary cases, illustrated in Fig. 1.13.

The first is a "splay", where molecules are spread by external stress (Fig. 1.13a), the second is a "twist" where molecules are twisted by an external stress (Fig. 1.13b), and the third is a "bend" where molecules are bent by an external stress (Fig. 1.13c). General deformations are a combination of these three types.

The relationship between the deformation and the restoring torques opposing to the director deformation is expressed by the splay, twist and bend **elastic moduli** (also known as Frank elastic constants), k_{11}, k_{22}, k_{33}, whose dimension is *energy/length* and hence N in SI units.

The *elastic* increment of the volume *free energy* density (per cm^3), f_k, due to splay, twist and bend is, according to the elastic theory for (non compressible) LCs, quadratic in the director gradients

$$f_\text{k} = \frac{1}{2}\left[k_{11}\left|\nabla\bullet\mathbf{n}\right|^2 + k_{22}\left|\mathbf{n}\bullet\nabla\times\mathbf{n}\right|^2 + k_{33}\left|\mathbf{n}\times\nabla\times\mathbf{n}\right|^2\right] \tag{1.2}$$

This is the well-known Frank-Oseen elastic free energy density for nematics (and also cholesteric) LCs [F1958, C1992, GP1995].

The three elastic moduli are of the same order of magnitude (10^{-11} N), with $k_{22} < k_{11} < k_{33}$ for most nematics. The value of these moduli is much lower than that of ordinary elastic material; this facilitates alignment modification of liquid crystals by the application of external mechanical stress, electric field and magnetic field.

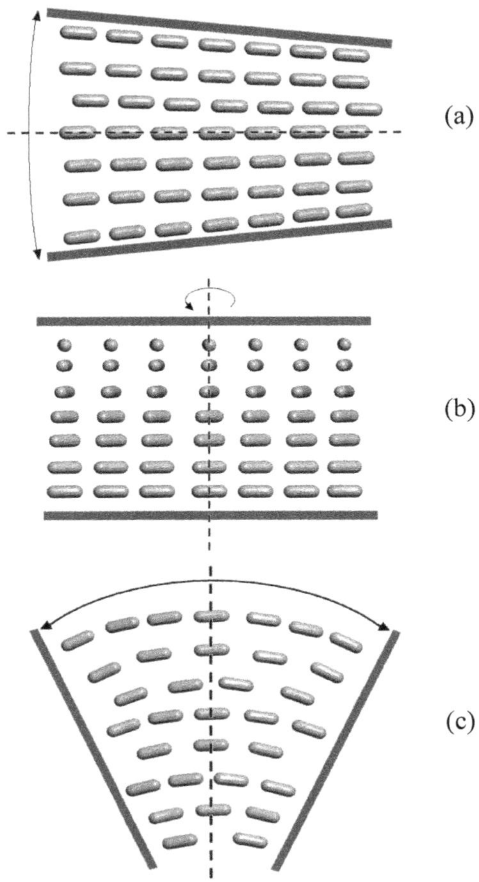

Fig. 1.13 Molecular alignments of nematic liquid crystals under external mechanical stress: (**a**) splay, (**b**) twist, (**c**) bend [F1958]

1.3.3 Surface Aligning Properties

In most experiments and applications, a thin nematic layer (of 2–10 μm, for display applications) is sandwiched between two substrates, usually glass plates. Near the substrate surface the LC molecules can exhibit aligning effects, that can be accentuated by special surface coatings (through certain organic or inorganic films) and/or treatments to allow the director *alignment* to be controlled. The so called orienting (or alignment) layers force the director to a preferred orientation near their surface. Often, the interaction between the liquid crystal and the surface is strong enough not to allow a change of the direction of **n** at the boundaries (strong *anchoring*) even in presence of director gradients within the bulk. The surface treatments combined with the elastic torques originating from (1.2) ensure the initial homogeneous director alignment of liquid crystal cells (i.e., without spatial variations in the plane of the layer).

Two basic geometries exist: the **planar**[15] (or **homogeneous**) one, where **n** is parallel to the surface, and the **homeotropic** one, where **n** is normal to the surface, as shown in Fig. 1.14.

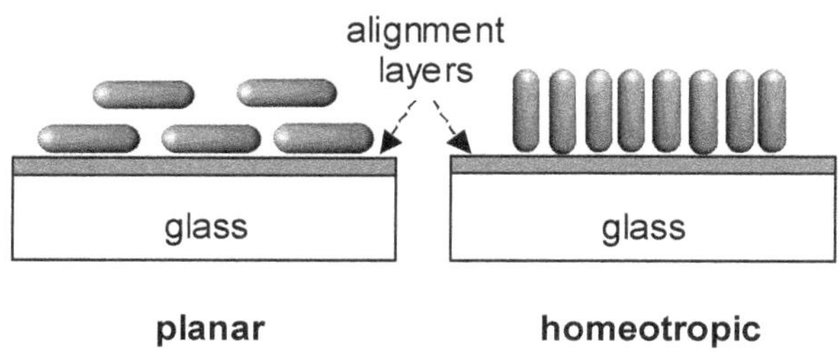

Fig. 1.14 Alignment of nematic liquid near a solid plate (typically glass) trough suitable alignment layers. In the planar case the director is parallel to the substrate, whereas it is perpendicular in the homeotropic case

A simple and widely used process to achieve planar alignment is **rubbing**. A coating polymer layer (e.g. polyimide, nylon or polyvinylalcohol) is deposited on the glass surface and rubbed repeatedly in the same direction (about 100 times) with a soft tissue (cotton cloth) [C1941]. In this manner, microscopic grooves are created in the surface, which align the director of the near LC molecules parallel to the direction of rubbing,[16] as depicted in Fig 1.15.

This aligned LC monolayer induces long range alignment through the LC cell. The technique is related to an earlier method dated back at least to 1925 [Z1925, Z1932]. It is simple, but constitutes a source of yield loss causing non-uniform black areas in the LCD. New methods have been proposed to replace the rubbing approach

---

[15]This case, important from the technology of LCDs, will be discussed also in the following chapter (see Section 2.4).

[16]Actually, the underlying physical mechanism is not completely understood. It was originally supposed that *only* the morphology of the grooves were responsible of the alignment of LC molecules in the groove direction. However, observations made with the aid of an atomic force microscope, seem to indicate that this is not the case [LSU1992]. Indeed, the widths of the grooves (in the order of 600 Å) are much wider than the length of the long axis of nematic LCs (typically 30 Å). In other words, the LC molecules could align in any direction in the wide-pitch grooves. Therefore the alignment and anchoring effect are now attributed to the *stresses* in the surface.

such as oblique evaporation,[17] Langmuir-Blodgett formation, unidirectional stretching and groove formation by stamping, but without real success.

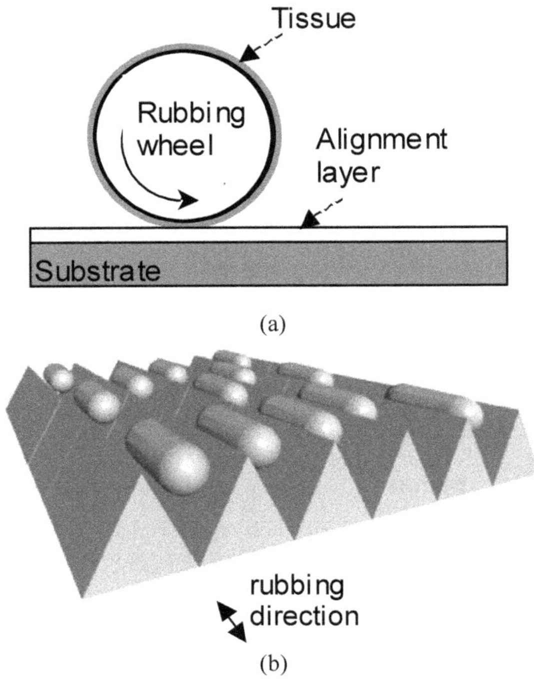

Fig. 1.15 (**a**) Roller used for rubbing. (**b**) In proximity of a surface where grooves are rubbed, LC molecules align along the direction of the grooves

It should be noted that rubbing gives a **pretilt** to the LC molecules that is, one end of the molecule is slightly lifted of some degrees. This speeds up and homogenizes the LC realignment under an electric field (avoiding the occurrence of reverse tilt and reverse twist creating different domains).

Some LCD modes, such as the Vertical Alignment discussed in Section 2.4.4 do not require rubbing (and this is considered an advantage).

In the homeotropic alignment the molecules are arranged perpendicular to the surface. This is coated with a surfactant like octadecyloxysilane (ODSE) that attaches to the surface with its long octadecyl chains projecting away in a direction approximately normal to the same surface.

[17]An important problem of early LCDs was their limited life, caused by moisture as displays were not sealed hermetically. The development of techniques to align the crystals without rubbing and also allowing to seal displays hermetically was a major step toward the large-scale manufacture of LCDs. However, in 1977 new moisture resistant LCs were synthesized, so manufacturers abandoned hermetic sealing and went back to rubbing (or aligning the cells with epoxy stamped with tiny grooves).

1.3.4 Viscosity

Viscosity has remarkable effects on the dynamic behavior of LCs. Indeed, the increase of viscosity at low temperatures reduces the possibility of molecules to move, and is one of the most limiting factors in the application of LCs.

Cinematic and dynamic viscosity can be defined, they are respectively referred to as ν and η. The relationship that relates these two parameters is

$$\nu = \frac{\eta}{\delta} \tag{1.3}$$

Where δ is the material density. However, since the major part of the LC materials is characterized by a value of δ around 1 N×s^2/mm^4, a distinction between ν and η is often avoided.

Four **viscosity coefficients** are required to characterize completely a nematic LC. Three of these are of the translational type η_1, η_2, η_3 (also termed Miesowicz viscosities [M1936, M1946]) and one is rotational, γ.

Figure 1.16 exemplifies how the four viscosity coefficients can be defined.

– η_1: the director is perpendicular to the flux and parallel to the gradient of velocity.
– η_2: the director is parallel to the flux and parallel to the gradient of velocity.
– η_3: the director is perpendicular to both the flux and the gradient of velocity.
– γ: the molecules rotate around an axis that is orthogonal to the director.

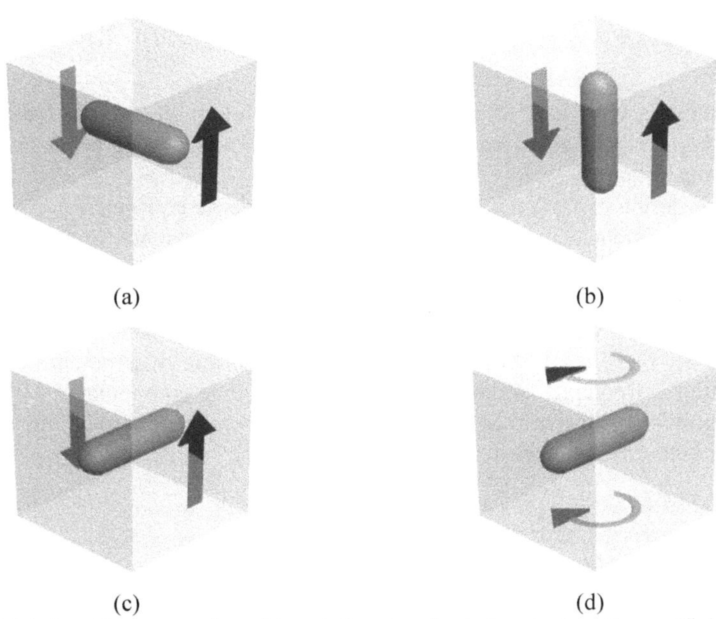

Fig. 1.16 Graphical examples of translational and rotational viscosity coefficients in LCs: **(a)** η_1, **(b)** η_2, **(c)** η_3 and **(d)** γ

The last parameter, the rotational viscosity, is of great importance for the switching properties of LCs within a display, when the nematic molecules reorientate under the effect of an electric field. Indeed, in LCDs the response time[18] is proportional to g times d^2, where d is the thickness of the LC layer. Typical values of dynamic rotational viscosities of LCs are in the range of 0.02–0.5 Pa×s.

1.3.5 Electromagnetic Properties

Owing to their orientational order, nematics are **anisotropic** substances. Therefore, in contrast to isotropic fluids, many physical quantities should be described by tensors. However, nematic LCs are free to rotate along the director axis. This fact produces an averaging action for the components of a given electromagnetic parameter in the plane perpendicular to the director. Thus, the dielectric susceptibility ε, magnetic susceptibility μ and the electrical conductivity σ are tensors, each having only two different components in their principal-axis system: $\varepsilon_{\|}$, $\sigma_{\|}$, $\chi_{\|}$ and $\varepsilon_{\perp}$, $\sigma_{\perp}$, $\chi_{\perp}$, respectively,[19] where subscripts $\|$ and $\perp$ indicate a direction parallel and perpendicular to the director, respectively.

As an important consequence, consider the behavior of a nematic LC under an external electric field **E**. Let us decompose the electric displacement vector, **D,** into its three components along the $\hat{\mathbf{x}}$, $\hat{\mathbf{y}}$ and $\hat{\mathbf{z}}$ axis of a local right-handed Cartesian coordinate system, in which, without loss generality, **n** is parallel to $\hat{\mathbf{z}}$

$$\begin{aligned}\mathbf{D} &= \varepsilon_x E_x\hat{\mathbf{x}} + \varepsilon_y E_y\hat{\mathbf{y}} + \varepsilon_z E_z\hat{\mathbf{z}} = \varepsilon_\perp E_x\hat{\mathbf{x}} + \varepsilon_\perp E_y\hat{\mathbf{y}} + \varepsilon_\| E_z\mathbf{n} \\ &= \varepsilon_\perp\left(E_x\hat{\mathbf{x}} + E_y\hat{\mathbf{y}} + E_z\mathbf{n}\right) + \left(\varepsilon_\| - \varepsilon_\perp\right)E_z\mathbf{n} = \varepsilon_\perp\mathbf{E} + \Delta\varepsilon(\mathbf{n}\cdot\mathbf{E})\mathbf{n}\end{aligned} \tag{1.4}$$

LC materials with $\varepsilon_{\|} > \varepsilon_{\perp}$ are said to be **positive** or of ***p* type**. Otherwise, if $\varepsilon_{\|} < \varepsilon_{\perp}$, they are said **negative** or of ***n* type**. Both *p*- and *n*- type substances can be found among nematics.

We also define in (1.4) the anisotropy of the dielectric constant $\Delta\varepsilon = \varepsilon_{\|} - \varepsilon_{\perp}$. In practice, an efficient way to obtain large $\Delta\varepsilon$ values is to attach at one end of the molecule a strongly polar group, aligned to the long axis, such as a CN group (see 5CB, 6CB and PCH in Table 1.1). PAA with its N^+–O^- group almost perpendicular to the long axis and with a strong permanent dipole moment is another example of *n*-type material. The Schiff's bases are also of *n* type. Note that the sign of $\Delta\varepsilon$ may change with the frequency and/or the temperature in some compounds.

Analogous constitutive relations can be derived between the magnetic vectors **B** and **H** and between the electric current density **j** and **E**.

---

[18]Response time (the time a pixel takes to turn from ON to OFF or vice versa) for a twisted nematic LC cell is discussed in Section 2.2.4.

[19]In other words, parameters $\varepsilon_{\perp}$, $\sigma_{\perp}$, $\chi_{\perp}$ can be thought as average values of their components in the plane perpendicular to the director. For instance, the dielectric susceptibility in the plane (**xy**) perpendicular to the director **n** can be approximated to $\varepsilon_{\perp}=(\varepsilon_x+\varepsilon_y)/2$.

The anisotropy of the magnetic susceptibility $\Delta\mu = \mu_{\parallel} - \mu_{\perp}$ is positive for the major part of nematics due to aromatic rings of the constituent molecules, exceptions with $\Delta\mu < 0$ are caused by the use of exclusively non-aromatic (e.g. cyclohexane) rings. In most cases anisotropy of the conductivity $\Delta\sigma = \sigma_{\parallel} - \sigma_{\perp}$ is positive (i.e., charges are easier transported parallel to the director rather than perpendicular). In the (layered) smectic phases, on the contrary, $\Delta\sigma < 0$ is typically observed. The sign of the anisotropies $\Delta\varepsilon$ or $\Delta\mu$ determines the behavior of the liquid crystal in an electric (**E**) or magnetic (**H**) field via an *electromagnetic* contribution, f_{em}, to the *free energy* density:

$$f_{\text{em}} = -\int_0^E \mathbf{D} \cdot \text{d}\mathbf{E} - \int_0^H \mathbf{B} \cdot \text{d}\mathbf{H} = \\ = -\frac{1}{2}\varepsilon_{\perp} E^2 - \frac{1}{2}\Delta\varepsilon(\mathbf{n} \cdot \mathbf{E})^2 - \frac{1}{2}\mu_{\perp} H^2 - \frac{1}{2}\Delta\mu(\mathbf{n} \cdot \mathbf{H})^2 \tag{1.5a}$$

Albeit (1.5a) indicates a similarity between the behavior in electric and magnetic fields, the magnetic contribution is usually negligible ($\Delta\varepsilon$ and $\Delta\chi$ are respectively in the order of the unity and 10^{-7}, in SI units). Hence, (1.5a) is rewritten as

$$f_{\text{em}} \approx f_{\text{e}} = -\frac{1}{2}\varepsilon_{\perp} E^2 - \frac{1}{2}\Delta\varepsilon(\mathbf{n} \cdot \mathbf{E})^2 \tag{1.5b}$$

that is made up of two terms in which only the latter depends on the direction of $\mathbf{n}$.

An important effect of (1.5b) is that the molecules of *p*-type LCs ($\Delta\varepsilon$ usually ranging from 2 to 20) align with the director parallel to an applied electric field **E** (greater than a certain critical value E_c) because this condition *minimizes* the electric energy f_{e}. Conversely, in *n*-type materials ($\Delta\varepsilon$ usually from –0.8 to –6), the molecules align with the director orthogonal to the electric field because the electric energy is minimized when **n** is orthogonal to **E**. The alignment properties under an electric field of *p*- and *n*-type LCs are exemplified in Fig. 1.17.

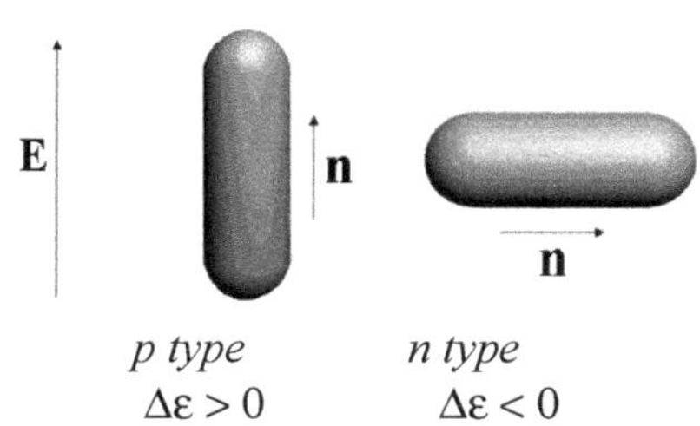

Fig. 1.17 Relation between an applied electric field **E** and the director in *p*-type ($\Delta\varepsilon > 0$) and *n*-type ($\Delta\varepsilon < 0$) liquid crystals

We have hence devised a simple and practical mean to reorient the LC molecules. If the field is strong enough and at sufficiently low frequency,[20] the LC molecules will reorient to follow its direction. Owing to the related studies by V. Freéedericksz on magnetic fields, the reorientation caused by an applied field is often called (simplifying the name) Frederick transition [FZ1929, FZ1933].

A fundamental case in applications occurs when nematic LCs are sandwiched by two parallel plates and the electric field is applied normally to the plates. In this case, the threshold field (E_c) at which the reorientation discussed above occurs, is expressed by

$$E_c = \left(\frac{\pi}{d}\right)\sqrt{\frac{k_{ii}}{\left|\Delta\varepsilon \cdot \varepsilon_0\right|}} \tag{1.6}$$

where d is the spacing of the plates and k_{ii} depends on molecular alignment. This expression results from the minimization of the total free energy F_{free}, of the liquid crystal volume (for the equilibrium configuration in terms of molecular alignment), which is expressed by the summation of electric energy and elastic energy (neglecting the interface energy related to the alignment of the director at the surfaces of the considered volume)

$$F_{\text{free}} = \int_V (f_{\mathrm{e}} + f_{\mathrm{k}})dV \tag{1.7}$$

When the liquid crystals are of positive type and the alignment is twisted, k_{ii} is equal to $k_{11} + (k_{33} - 2k_{22}) / 4$ and the **threshold voltage** for molecular alignment transition is

$$V_{th} = \pi\sqrt{\frac{k_{11} + \dfrac{k_{33} - 2k_{22}}{4}}{\left|\Delta\varepsilon \cdot \varepsilon_0\right|}} \tag{1.8}$$

The threshold is independent of the spacing of the plates.[21] It can be reduced if $k_{33}-2k_{22}$ is negative and decreases with increasing $\Delta\varepsilon$.

---

[20]An LC molecule cannot follow a rapidly changing filed due to its inertia. This fact is of fundamental importance in practice, since it allows the LC cell to be driven by zero mean AC voltages instead of DC ones, without causing any director variation. We will show in the next subsection that the alignment of a liquid crystal by an electric field is complicated by the presence of conducting impurities which mandates for the use of alternating electric fields.

[21]This property allows some tolerances in the cell spacing and high yields in manufacturing.

The LC dielectric anisotropy exhibits a strong dependency on the frequency range. The frequency at which the dielectric anisotropy changes its sign is called **crossover frequency**. Usually, this crossover frequency is quite high (greater than 50 kHz). At low frequency the dielectric anisotropy is almost constant. When approaching the crossover frequency the value of $\varepsilon_{\parallel}$ decreases monotonically as the frequency is increased. However, $\varepsilon_{\perp}$ does not change when the frequency is increased even up to several MHz (where the dipole moment of molecules cannot follow the change in electric field). As a result, the dielectric anisotropy changes from a positive to a negative value.

From (1.8) it is seen that V_{th} is inversely proportional to the square root of the dielectric anisotropy. Hence, it is always desirable to operate an LCD in the frequency spectrum where the dielectric anisotropy remains constant.

1.3.6 Ion Transport

Though liquid crystals are insulators, they usually contain some ionic impurities which are responsible for a limited electric conductivity.[22]

The fundamental equation describing the charge transport through a nematic medium under the influence of an applied electric field **E** is

$$\mathbf{j} = \sigma_{\perp}\mathbf{E} + \Delta\sigma\mathbf{n}(\mathbf{n}\cdot\mathbf{E}) \tag{1.9}$$

where **j** is the current density and the electrical conductivity σ has components $\sigma_{\parallel}$ and $\sigma_{\perp}$, respectively parallel and perpendicular to the director direction. As already mentioned, ions in liquid crystals are responsible for the electric conductivity. Some lyotropic LCs have high ion concentration (and conductivity), but this must be avoided in LCs for display applications. Ion impurities are generated during the production process, for instance contamination of the substrates or the polymer alignment layer can introduce ions into the liquid crystal bulk. Weak dissociations of the compounds and injected charges from the cell electrodes can also contribute to the conductivity. Typical ion concentrations in nematic liquid crystals are in the order of 10^{16}–10^{20} m^{-3} and an electric conductivity lower than 10^{-12} $(\Omega\cdot\text{cm})^{-1}$ is normally found. Ion concentrations above 10^{20} m^{-3} reduce the display life and give rise to **flicker**[23] and **image retention**, as discussed below.

---

[22]Ensuring a conductivity equal or lower than 10^{-11} $\Omega^{-1}\text{cm}^{-1}$ is particularly important for active matrix LCDs. As we will see, these types of display adopt a capacitor to store the data voltage between two subsequent frames. The LC conductivity should be high enough to avoid appreciable voltage discharge within a frame time interval.

[23]**Flicker** is an unwanted effect that is perceived by the human eye as a periodic variation in brightness of large display areas. Given the eye response to luminance variations and to its frequency, flicker is perceived when the luminance varies by more than a few percent at frequencies less than about 40 Hz. For this reason, refresh frequencies are equal or higher than 50 Hz.

Two **ion transport mechanisms,** of *short term* and *long term* type, can be distinguished in LCs. The first type is related to fast ions, whose transfer and influence on the internal electric field lead to effects as flicker in the optical transmission at relatively low frequency (10–40 Hz) and errors in the obtained grey level. The second type is related to slow effects and can originate from dc voltages in excess to 50 mV applied during several hours. Slow effects cause the generation and migration of ions. The ions gathered at the alignment layers can lead to a compensating voltage which persists even after the external dc voltage is removed, causing image retention (*image sticking* or *ghosting*), as illustrated in Fig. 1.18.

For this reason the use of dc voltages to drive LCDs is always avoided and alternating (with zero mean) driving voltage waveforms are invariantly adopted.

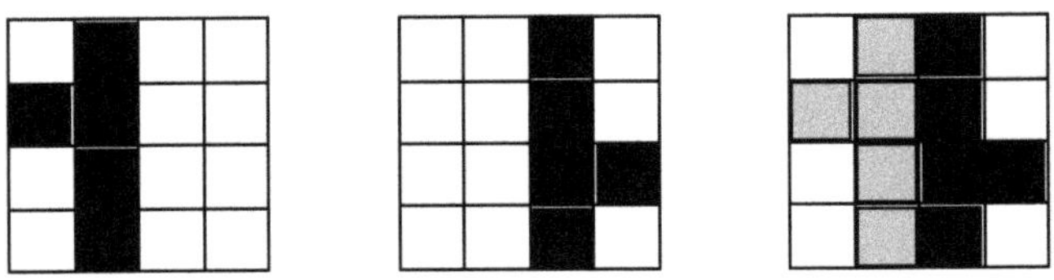

(a) (b) (c)

Fig. 1.18 First applied image (**a**), ideal second applied image (**b**), second image with ghosting (**c**)

1.3.7 Optical Properties and Birefringence

Optical waves involve electric fields, in principle they can reorient the liquid crystal director like the electrically applied fields. However, in a display this process can be neglected, since the intensity of the optical fields and its frequency are respectively much lower and much higher than those of the fields generated by the applied voltages. To make a distinction, the refractive index is given for optical waves instead of the dielectric susceptibility. The refractive index, n, is defined as the ratio of the light velocity in the vacuum and in the specific material. It is expressed in terms of the dielectric susceptibility as $n=\sqrt{\varepsilon}$. As a consequence of the LC dielectric anisotropy, we get also optical anisotropy $\Delta n = n_{//} - n_{\perp}$, where $n_{//}$ and $n_{\perp}$ are the refractive indexes along and perpendicular to the director. For typical *p*-type nematic LCs, $n_{//}$ is approximately 1.5 and the anisotropy, Δn, may range between 0.05 and 0.5. The direction where the refractive index is maximum is called the *optical axis*. As a result, one of the most evident characteristics that distinguish LCs from ordinary isotropic liquids is their **birefringence.** As we will see, birefringence allows to rotate the plane of oscillation of light and is therefore one of the fundamental properties exploited in the display application.

To understand birefringence and its effect on the propagation of light, light must be again represented by an electric field. For a plane wave propagating in a specific direction, the electric field vector describes an ellipse in the plane perpendicular to the propagation direction. This ellipse represents the polarization of the light. Some special cases are the linear polarization and the circular polarization where this ellipse is distorted to a straight line or a perfect circle. Each ellipsoidal polarization can be decomposed as a superposition of linearly polarized waves along two perpendicular axes with a well defined phase relationship.[24]

In an *isotropic medium*, these two linearly-polarized waves propagate with the same phase velocity c/n_{medium}. This case is exemplified in Fig. 1.19a, being curves (i) and (ii) the representation of the two polarizations. A simplified notation, useful for a pencil and paper description, is illustrated in Fig. 1.19b. Note that the free rotation in liquids averages out any asymmetry of molecular shape and renders them optically isotropic. Some crystals are optically isotropic.

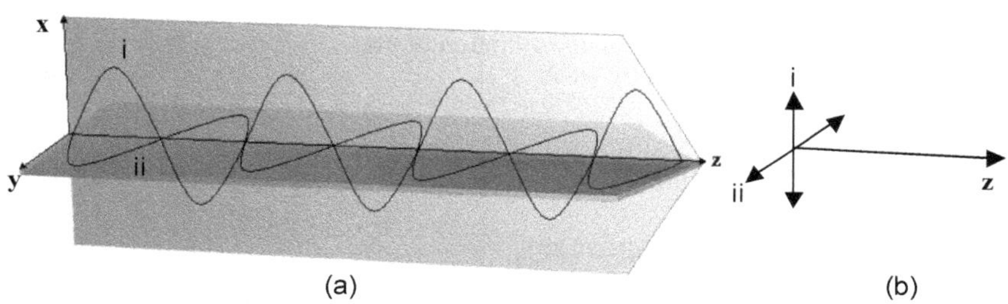

Fig. 1.19 Light propagation in an isotropic medium: (**a**) The two polarizations, curves i and ii, are subject to the same refractive index and have the same phase velocity, hence the two polarizations emerge with the same relative phase. (**b**) Simplified representation

In contrast, the optical properties of nematics are those of an *anisotropic medium* and correspond particularly to those of uniaxial crystals, in which an incident wave is subject to a different refractive index when it oscillates in the plane perpendicular to the director or along the director. Therefore, they propagate through the liquid crystal with a different phase velocity as illustrated in Fig. 1.20.

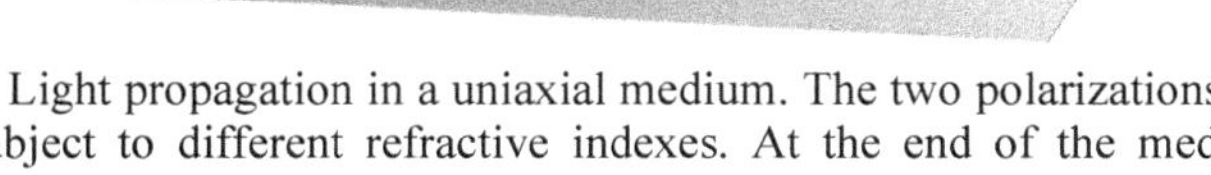

Fig. 1.20 Light propagation in a uniaxial medium. The two polarizations, curves i and ii, are subject to different refractive indexes. At the end of the medium, the two polarizations emerge with a different relative phase. Hence, in general, the polarization changes

[24]Alternatively, one can decompose a generic polarization into two circularly polarized components with opposite handedness. For example, if we consider linear polarization, it can be decomposed into two equal-amplitude right and left circular polarizations.

Let us better describe this effect. Assume that light incident forms an angle α with the director, as illustrated in Fig. 1.21.

We call the **ordinary wave** the one oscillating orthogonally to the plane formed by the director and the propagation direction. It is thereby subject to a refractive index, n_o, (refractive index for the ordinary wave) that is equal to $n_\perp$ and its velocity of propagation is equal to $c/n_\perp$, independent of α. On the other hand, the **extraordinary wave** oscillates in the plane formed by the director and the propagation direction and its refractive index, n_e, can vary between $n_{||}$ and $n_\perp$, depending on α.

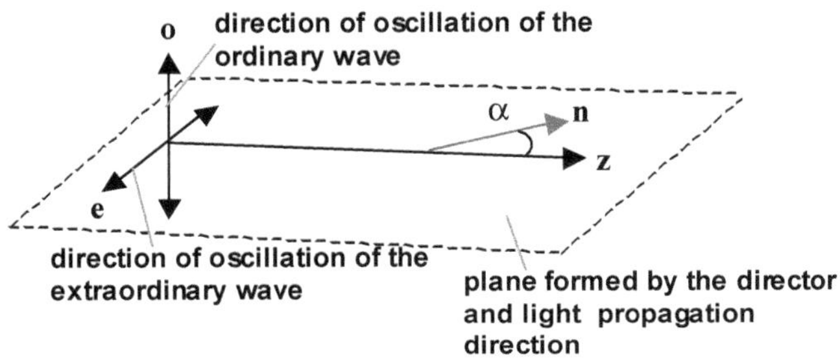

Fig. 1.21 Decomposition of light (propagating through z and forming an arbitrary angle α with the director) into ordinary and extraordinary component

As limit cases, the refractive index n_e is equal to $n_{||}$ for α=90°, whereas it is equal to $n_\perp$ for α=0°. These two cases are illustrated in Fig. 1.22a and 1.22b, respectively.

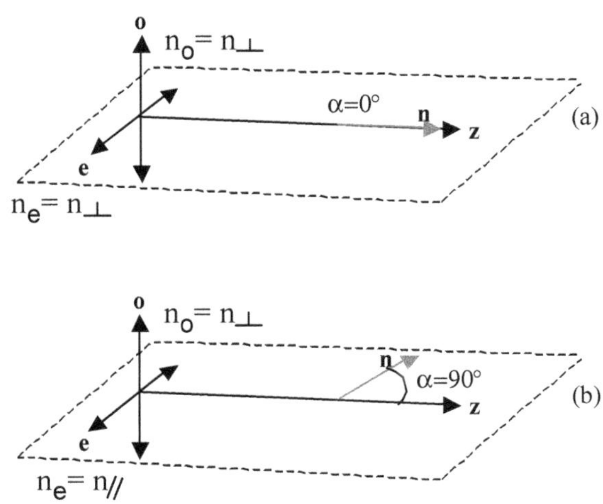

Fig. 1.22 Ordinary and extraordinary indexes in a birefringent medium for: (**a**) angle α=0° and (**b**) α=90°

As a result, in birefringent media, the different speed of the ordinary and extraordinary waves causes a phase difference, φ, between the two modes, that, at the end of the medium will result in a different polarization ellipse. The thickness, *d*, of the LC sample is in this context an important parameter, because the phase shift accumulates as long as the light propagates in the birefringent material. If the phase difference equals 2π, the wave returns to its original polarization state.[25]

Let us consider the important case of Fig. 1.22b, where **n** is perpendicular to **z**. A wave polarized along the ordinary axis will experience a phase delay $\varphi_o = (2\pi/\lambda)n_o d$, while the phase delay of a wave polarized along the extraordinary axis is $\varphi_e = (2\pi/\lambda)n_e d$, with λ is the free-space wavelength. The phase difference, $\Delta\varphi$, is

$$\Delta\varphi = \frac{2\pi}{\lambda}(n_e - n_o)d \tag{1.10}$$

Note that light of different colors (wavelength) will be subjected to different $\Delta\varphi$. For example, consider the direction of oscillation be either orthogonal or parallel to the director as illustrated in Fig. 1.23. In both cases the waves propagate through the medium and exit unchanged and linearly polarized and will experience different phase delays. In the case depicted in Fig. 1.23c, the oscillation direction forms a generic angle with the director.

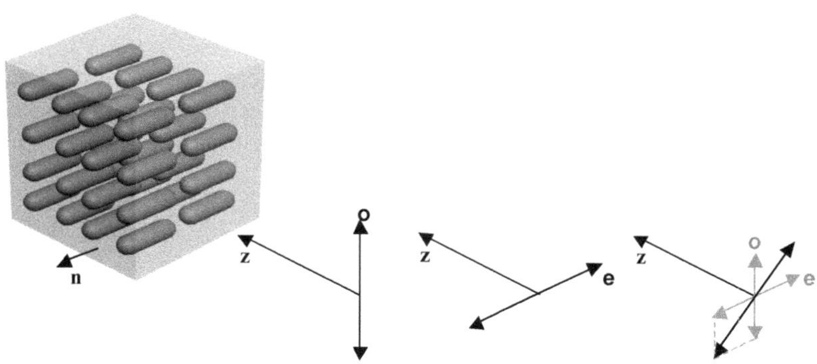

Fig. 1.23 Linearly polarized light entering an LC medium with α =90°: (**a**) Pure ordinary wave. (**b**) Pure extraordinary wave. (**c**) Generic wave made up of ordinary and extraordinary components

[25]In a typical LC, the birefringence and length are not constant over the entire sample. If the sample is observed between cross polarizers, under *monochromatic light*, some areas appear light and others appear dark. These light and dark areas denote regions of different director orientation, birefringence, and length. If the sample is observed under *polarized white light*, birefringence can lead also to multicolored images as shown in the microscope picture of a nematic liquid crystal, in Fig. 1.10b. Color patterns observed in the polarizing microscope are very useful in the study of liquid crystals in many situations, including the identification of textures, of liquid crystal phases and the observations of phase changes.

The ordinary and extraordinary modes emerging at the end of the medium have a phase difference $\Delta\varphi$ according to (1.10). As a consequence, the exiting wave changes its polarization from linear to elliptic. In particular, for $d=\lambda/[4(n_e-n_o)]$, φ results to be 90°. In this case, if the angle formed by the oscillation direction and director is 45° (equal ordinary and extraordinary wave amplitudes), the exiting polarization will be circular. Besides, if $d=\lambda/[2(n_e-n_o)]$ the phase difference $\Delta\varphi$ is 180°. In this case, for α=45° the exiting polarization is still linear but with the polarization plane rotated of 2α, that is 90°. This effect is fundamental in some recent LCD operating modes like the In Plane Switching and Vertical Alignment described in Section 2.4.4.

Chiral nematic LCs, owing to their helical molecular organization, are birefringent in a different way: they exhibit **circular birefringence**. Assume that the helical structure is aligned with the direction of propagation of the light. Then, circularly polarized light will propagate through the crystal at different phase velocities depending on whether it is right-circularly polarized or left-circularly polarized. In particular, if the light is linearly polarized (it can be viewed as the sum of left and right circular polarization, see note 24), the component of circularly polarized light that matches the chirality of the crystal structure will travel faster than the other component. Moreover, a *new* notable effect is the total reflection of an ideally precise wavelength (**selective reflection**) given by

$$\lambda = pitch\sqrt{2\left(n_{\parallel}^2 + n_{\perp}^2\right)} \tag{1.11}$$

A film of LC that reflects the light will appear hence colored. Since the pitch varies with temperature, the color of the substance will also vary. This effect was exploited in the first applications of LCs as thermometers (see Section 2.1.1).

1.3.8 Temperature Effects

All the physical parameters mentioned above are dependent on the temperature. Some general trends, characteristic for most nematics are the following. Increasing the temperature, the absolute values of the anisotropies usually decrease, until they drop to zero at the nematic-isotropic phase transition. The electrical conductivities increase with increasing temperature as well, while the viscosity coefficients, as already mentioned, decrease.

1.4 Concluding Remarks

In conclusion of this Chapter, it should be remarked that the performance of a liquid crystal display relies strongly on:

- Temperature of smectic-to-nematic transition and of clearing point, determining the so called operating temperature range. For typical applications, a nematic temperature between –40°C to 100°C is needed.

- Elastic constants and (rotational) viscosity. They are important for the response time and the former also for the threshold voltage.
- Dielectric anisotropy, determining the behavior (*p* or *n* type) under an electric field. A large ε_a decreases the threshold voltage.
- Optical anisotropy, determining the optical behavior.
- Threshold voltage, determining the operating voltage range towards low power consumption.

None of the single liquid crystal materials has all the correct physical properties that fulfil the specifications of even the simplest display. The substances that are presently used in liquid crystal displays are eutectic[26] mixtures of up to 20 chemically, photo-chemically and electro-chemically stable mesogens. Table 1.2 compares some selected performance parameters of the 5CB compound (already introduced in Table 1.1) and a typical mixture used in LCDs.

Table 1.2 Some physical parameters of 5CB and of a typical mixture for LCD applications. $T_{S\text{-}N}$ and T_{CP} are the smectic-to-nematic transition temperature and clearing point temperature

Parameter	**5CB**	**Typical mixture**
$T_{s\text{-}n}$	30°C	–40°C
T_{cp}	55°C	80°C
Optical anisotropy	$\Delta n = n_{\parallel} - n_{\perp}$ $= 1.617 - 1.492 = 0.125$	$\Delta n = n_{\parallel} - n_{\perp}$ $= 1.562 - 1.477 = 0.085$
Dielectric anisotropy	$\Delta\varepsilon = \varepsilon_{\parallel} - \varepsilon_{\perp}$ $= 17.5 - 4.8 = 12.7$	$\Delta\varepsilon = \varepsilon_{\parallel} - \varepsilon_{\perp}$ $= 10.5 - 3.5 = 7$
Elastic constants	$k_{11} \approx k_{22} \approx k_{33} \approx 10^{-11}$ N	$k_{11} \approx k_{22} \approx k_{33} \approx 10^{-11}$ N
Rotational viscosity	$\gamma = 150$ mPa·s	$\gamma = 100$ mPa·s

[26]An eutectic, or eutectic mixture, is a mixture of two or more phases at a composition that has the lowest melting point, and where the phases simultaneously crystallize from molten solution at this temperature. For example, the melting temperature of a mixture of water (H_2O) and salt (NaCl) is −10°C, while the melting temperature of the individual components is 0°C and 804°C, respectively.

References

[BC1994] L.M. Blinov and V.G. Chigrinov, *Electrooptic Effects in Liquid Crystal Materials*, Springer-Verlag, 1994.

[BS1957] G.H. Brown and W.G. Shaw, "The Mesomorphic State," *Chemical Reviews*, Vol. 57, No. 6, pp.1049–1157, 1957.

[C2005] J.A. Castellano, *Liquid Gold: The Story of Liquid Crystal Displays and the Creation of an Industry*, World Scientific Publishing, 2005.

[C1941] P. Chatelain, "Sur l'Orientation des Cristaux Liquides par les Surfaces Frottées (On the Orientation of Liquid Crystals by Rubbed Surfaces)," *Comptes rendus de l'Académie des Sciences*, Vol. 213, pp. 875–876, 1941.

[C1992] S. Chandrasekhar, *Liquid Crystals*, University Press, 1992.

[CSS1977] S. Chandrasekhar, B.K. Sadashiva, and K.A. Suresh, "Liquid Crystals of Disc-Like Molecules," *Pramana*, Vol. 9, No. 5, pp. 471–480, 1977.

[D2003] I. Dierking, *Textures of Liquid Crystals*, Wiley-VCH, 2003.

[DFL2001] D.A. Dunmur, A. Fukuda, and G.R. Luckhurst (editors), *Physical Properties of Liquid Crystals: Nematics*, Inspec, 2001.

[F1922] G. Friedel, "Les États Mésomorphes de la Matière (The Mesomorphic States of Matter)," *Annales de Physique*, Vol. 18, pp. 273–474, 1922.

[F1958] F.C. Frank, "On the Theory of Liquid Crystals," *Discussions of the Faraday Society*, Vol. 25, pp. 19–28, 1958.

[F1964] J.L. Fergason, "Liquid Crystals," *Scientific American*, Vol. 211, pp. 77–85, 1964.

[F2002] R.M. Friedman, "Quantum Theory and the Nobel Prize," *Physics World*, Vol. 15, pp. 33–38, 2002.

[FZ1929] V. Fréedericksz and V. Zolina, "On the Use of a Magnetic Field in the Measurement of the Forces Tending to Orient an Anisotropic Liquid in a Thin Homogeneous Layer," *Transactions of the American Electrochemical Society*, Vol. **55**, pp. 85–96, 1929.

[FZ1933] V. Freédericksz and V. Zolina, "Forces Causing the Orientation of an Anisotropic Liquid," *Transactions of Faraday Society*, Vol. 29, pp. 919–930, 1933.

[GP1995] P.G. de Gennes and J. Prost, *The Physics of Liquid Crystals*, Clarendon Press, second edition, 1995.

[GR1890] L. Gattermann and A. Ritschke, "Ueber Azoxyphenetoläther (On Azoxyphenol Ethers)," *Berichte der Deutschen Chemischen Gesellschaft*, Vol. 23, pp. 1738–1750, 1890.

[H1984] C. Hilsum, "The Anatomy of a Discovery – Biphenyl Liquid Crystals," *Technology of Chemicals and Materials for Electronics*, E.R. Howells (editor), Ellis Horwood Ltd, 1984.

[K1973] H. Kelker, "History of Liquid Crystals," *Molecular Crystals and Liquid Crystals*, Vol. 21, pp. 1–48, 1973.

[K1988] H. Kelker, "Survey of the Early History of Liquid Crystals," *Molecular Crystals and Liquid Crystals*, Vol. 165, pp. 1–43, 1988.

[K1989] H. Kelker and P.M. Knoll, "Some Pictures of the History of Liquid Crystals," *Liquid Crystals*, Vol. 5, No. 1, pp. 19–42, 1989.

[K2000] I. Korenic, "A Millennium of Liquid Crystals," *Optics and Photonics News*, Vol. 11, No. 17, pp. 16–22, 2000.

[K2002] H. Kawamoto, "The History of Liquid-Crystal Displays," *Proceedings of IEEE*, Vol. 90, No. 4, pp. 460–500, 2002.

[KW1993] I. Khoo and S. Wu, *Optics and Nonlinear Optics of Liquid Crystals*, World Scientific, 1993.

[L1889] O. Lehmann, "Über Fliessende Kristalle (On Flowing Crystals)," *Zeitschrift für Physikalische Chemie*, Vol. 4, pp. 462–472, 1889.

[L1904] O. Lehmann, *Flüssige Kristalle (Liquid Crystals)*, Wilhelm Engelmann, 1904.

[L1999] F.M. Leslie, in *Physical Properties of Liquid Crystals*, D. Demus, J.W. Goodby, G.W. Gray, H.-W. Spiess, and V. Vill (editors), p. 25, Wiley, 1999.

[L2001] E. Lueder, *Liquid Crystal Displays: Addressing Schemes and Electro-Optical Effects*, Wiley, 2001.

[LL1984] E.M. Lifshitz and L.D. Landau, *Statistical Physics*, Butterworth-Heinemann, third edition, 1984.

[LSU1992] E.S. Lee, Y. Saito, and T. Uchida, "Detailed Morphology of Rubbed Alignment Layers and Surface Anchoring of Liquid Crystals," *Proceedings of International Display Research Conference*, pp. 595–598, 1992.

[M1911a] C. Mauguin, "Sur Les Cristaux Liquides de Lehmann (On the Liquid Crystals of Lehmann)," *Bulletin de la Société Française de Minéralogie*, Vol. 34, pp. 71–117, 1911.

[M1911b] C. Mauguin, "Orientation des Cristaux Liquides par le Champ Magnétique (Orientation of Liquid Crystals by a Magnetic Field)," *Comptes rendus de l'Académie des Sciences*, Vol. 152, pp. 1680–1683, 1911.

[M1936] M. Mięsowicz, "Influence of a Magnetic Field on the Viscosity of Para-Azoxyanisole," *Nature*, Vol. 136, p. 261, 1936.

[M1946] M. Mięsowicz, "The Three Coefficients of Viscosity of Anisotropic Liquids," *Nature*, Vol. 158, p. 27, 1946.

[MS1958] W. Maier and A. Saupe, "Eine Einfache Molekulare Theorie des Nematischen Kristallinflüssigen Zustandes (A Simple Molecular Theory of the Nematic Liquid-Crystalline State)," *Zeitschrift Naturforschung, Vol.* **13a**, pp. *564–66, 1958.*

[NSW1996] T. Niori, T. Sekine, J. Watanabe, T. Furukawa, and H. Takezoe, "Distinct Ferroelectric Smectic Liquid Crystals Consisting of Banana-Shaped Achiral Molecules," *Material Chemistry Communication*, Vol. 6, pp.1231–1236, 1996.

[O1933] C.W. Oseen, "The Theory of Liquid Crystals," *Transactions of the Faraday Society*, Vol. 29, pp. 883–900, 1933.

[S1989] H. Sackmann, "*Smectic Liquid Crystals, a Historical Review*," *Liquid Crystals*, Vol. 5, pp. 43–55, 1989.

[SDS2004] T.J. Sluckin, D.A. Dunmur, and H. Stegemeyer, *Crystals That Flow: Classic Papers from the History of Liquid Crystals*, Taylor & Francis, 2004.

[SS1974] M.J. Stephen and J.P Straley, "Physics of Liquid Crystals," *Reviews of Modern Physics*, Vol. 46, pp.617–704, 1974.

[V1908] D. Vorländer, *Kristallinisch-flüssige Substanzen* (Liquid Crystalline Substances), Enke Verlag, 1908.

[V1992] V. Vill, "Early History of Liquid Crystalline Compounds," *Condensed Matter News*, Vol. 1, pp. 25–28, 1992.

[Z1925] H. Zöcher, "Optical Anisotropy of Selectively Absorbing Substances; Mechanical Production of Anisotropy," *Naturwissenschaften*, Vol. 13, pp. 1015–1021, 1925.

[Z1932] H. Zöcher, "Polarizers of Light and a Method of Preparation," *U.S. Patent #1 873 951*, issued 1932.

[Z1933] H. Zöcher, "The Effect of a Magnetic Field on the Nematic State," *Transactions of the Faraday Society*, Vol. 29, pp. 945–957, 1933.

Chapter 2

Liquid Crystal Displays

Materials, Operating Modes, and Applications

The modern history of liquid crystals is closely related to the development of electronic displays, whose main advances are summarized in this chapter. For a more enthusiastic report, dense of interesting details, the reader is referred to the already cited paper by Hirohisa Kawamoto, "who was part of the story from 1970" [K2002], and from which paper several data and information included in the chapter are taken:

... (it) is a story of the hard work, disappointments and successes of worldwide competition and cooperation that encompassed the U.S.A., Europe and Japan. Each industrial center contributed its particular strengths: in America, it was the quickness of forming new ideas and demonstrating their feasibility; in Europe, it was the fundamental science and synthesis of basic materials; and in Japan, it was the process of perfecting implementation and moving it to the production line

2.1 Towards The First LCD

During the first half of the 20th century, an accurate model of the LC phenomenon, based on the molecular basis, was developed. Nevertheless, very few and marginal applications were found.[1]

---

[1]The first patented application of LC technologies was the *Liquid Crystal Light Valve*, by the **Marconi Wireless Telegraph Company** and dates back to 1936 [LL1936]. In the 1940s, following the works of Zöcher, the American scientist **John F. Dreyer** refined the rubbing technique to obtain oriented lyotropic LC films of uniform thickness that could serve as base for polarizers, and with which he fabricated polarized lenses for sunglasses. He also made the first 3-D movie glasses [D1946–D1957].

D.J.R. Cristaldi et al., *Liquid Crystal Display Drivers*, DOI 10.1007/978-90-481-2255-4_2,

2.1.1 The First LC Application

In 1956, **Westinghouse Electric,** Pennsylvania, assigned to James L. Fergason to explore ways of using thermal imaging in making vacuum tubes for televisions. Fergason observed the optical properties of LCs for the first time in November 1957 and decided it would be the subject of his research. At Westinghouse, Fergason formed (with some difficulties) the first non-academic scientific group focused on LCs. At this purpose he said:

I had to get people to believe there was such a thing as liquid crystal, even though it's colorful and all that sort of thing, and get them to believe that it was more than just a lab curiosity. And it took a lot of measurements and a lot of work. Then I had to show them not only that liquid crystal was important but also that what you did with liquid crystal was important.

In 1958 he invented the first significant application, cholesteric liquid crystals that turned colors considerably as the temperature changed. We showed in Section 1.3.7 this effect. Liquid crystals allowed to determine temperature just by looking at the color of the thermometer. By mixing different compounds, a device for practically any temperature range could be built.

Since then, Fergason started to see many LC applications and worked even on the implementation of an optical-display system, but he was not supported toward product development at Westinghouse. It took from 1957 to 1964 for the potential of liquid crystal in products to be recognized for the first time. Flexible films and tapes that could be applied to the surface of objects to record temperatures were produced. They could be applied also to the skin of a human body to locate veins and arteries and to electronic circuit boards to locate trouble spots [F1964]. In the following years such thermal mapping application was further improved to screen for breast cancer and other medical conditions. However it was not well sold. [2]

The outstanding and prolific aptitude of Fergason (he now holds more than 150 patents in the U.S.A, and in 1998 was inducted into the U.S.A. National Inventors Hall of Fame) made him a protagonist in the evolution of the LCD industry.

2.1.2 The Pioneering Role of RCA

Still in the early 1960s only a few institutions were studying and performing research experiments on LC materials. The **RCA** (Radio Corporation of America) was one among these. At that time, RCA was the leader of the electronics market.

[2] In reality, there was an intrinsic inaccuracy to much of the medical imaging. For instance, Bayer Corporation developed equipment that used a plastic-enclosed liquid-crystal plate for breast exams, but this system affected the temperature of the skin when they pressed against it. Liquid crystal thermal imaging was eventually replaced with high-resolution digital infrared imaging, using remotely cameras and computers.

Innovations such as audio records, radio, and color television were all developed at RCA, under the guidance of its creative chairman David Sarnoff. Eloquently for this subject, one of the long-term Sarnoff dreams was a flat-panel TV that could hang on a wall like a picture.

In 1962, **Richard Williams** of RCA discovered some electrooptic characteristics of liquid crystals, which demonstrated their possible use in the implementation of display devices [W1963a, W1963b].

Williams used PAA, confined between two glass plates placed at a distance of 50 μm. Each plate was also covered in the inner face by a conductive and transparent oxide material, properly contacted to an external voltage source. After applying the voltage, an electric field perpendicular to the glass surfaces was generated. If the field was sufficiently high, it caused the molecules to organize themselves in a regular pattern similar to a fingerprint (see Fig. 2.1) consisting in an array of long parallel regions to which he referred as domains and lately called **Williams Domains** (see Section 2.1.3 for a physical explanation of the phenomenon). Both dc and ac voltages were used, but the latter showed better properties in terms of pattern stability and lower electrochemical deterioration.

The domain pattern appeared (disappeared) within about 2 ms (20 ms) after the electric field was applied (removed). The sample was mounted on the stage of a microscope and observed by unpolarized transmitted light. Considerable scattering of light was caused by the domain pattern and the transmitted light intensity changed during its formation to a stable value when the formation was complete. A similar change occurred when the electric field was removed and the domain pattern disappeared.

A representative behavior of the transmitted light intensity versus the intensity of the (slowly increasing) applied electric field, is shown in Fig. 2.2. It is seen that, for a certain threshold value (about equal to 1300 V/cm), there is a sharp change of light intensity followed by a slower change as the field increases further. The sharp change was caused by the formation of domains.

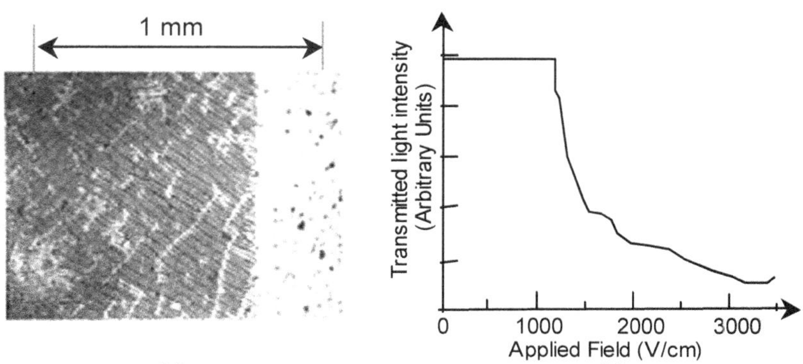

(a) (b)

Fig. 2.1 Williams experiment: (**a**) Williams domains. In the *left dark area,* a 1-kHz AC field of 2500 V/cm directed perpendicular to the plane is applied through transparent electrodes. Parallel regions (domains) are apparent. The *bright area at the right* is without the transparent conductive coating, therefore there is no field applied. Sample thickness is about 50 μm and temperature is 125°C. (**b**) Qualitative behaviour of transmitted light versus the electric field

The characteristics of the domains did not change replacing the conductive coating and did not depend on the wavelength of the monochromatic light. The domains were hence recognized as a property of the liquid crystal itself.

Williams's experiments attracted **George H. Heilmeier** of RCA, who in 1964 demonstrated the **guest-host** mode effect [HZ1968]. He doped a *p*-type liquid crystal (butoxy benzoic acid, liquid crystalline between 147°C and 161°C) with a strong dye (pleochroic) whose molecules have the same elongated shape as the LC molecules. The LC (called the *host*) and the dye (the *guest*) resulted aligned in their longitudinal axes, as shown in Fig. 2.2.

The dye was also chosen for its fundamental property of absorbing light when the direction of polarization of linearly-polarized light is the same of its long axis, while it transmits light when these two directions are different. Then, Heilmeier sandwiched the host-guest mixture between two glass slides coated with transparent tin oxide (SnO_2) electrodes, heated the mixture and applied a dc voltage. He observed that the cell changed color from red to colorless as a function of the applied field.

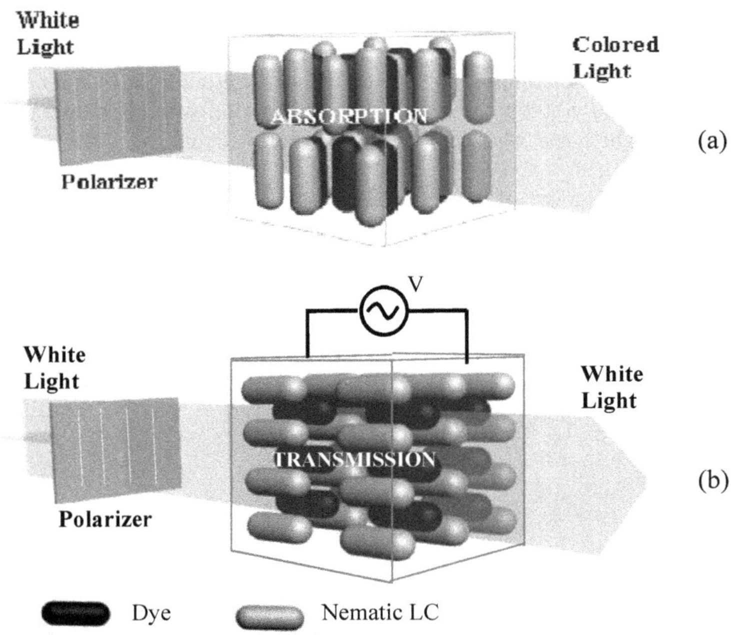

Fig. 2.2 The guest-host effect: (**a**) With no applied voltage, the direction of polarization of linearly-polarized light is parallel to the molecules long axis (i.e., it is vertical), some light frequencies are hence absorbed by the dye and the cell appears colored. (**b**) With an applied voltage, the direction of polarization of linearly-polarized light is parallel to the molecules long axis. Light is completely transmitted and the cell appears transparent

The effect was more apparent if a polarizer was used to filter the incident light (as in Fig. 2.2). Without an applied electrical field the incident light is absorbed (in certain wavelengths of the blue region) and consequently the cell appears red colored; with an applied electric field the guest molecules were reoriented following the host ones causing light to be transmitted and the cell to appear transparent.

The cell required less than 10 V to work. When compared to the more than 1000 V required by Cathode Ray Tubes (CRTs), this constituted a remarkable advantage towards the implementation of low-power and compact displays. However, as Heilmeier lately recognized [H1976]:

There were obvious problems with the guest-host effect. The dyes and their liquid crystal hosts were not stable over long periods of time in applied fields, the effect was sensitive to surface orientation effects, it required polarized light, it was viewed in transmission, it required heating to maintain the host in its nematic phase, uniformity was a problem, and so on.

In 1982, the guest-host mode combined with an Active Matrix drive would be used in a wristwatch television [S1982] (see also note 14 of this chapter).

2.1.3 The First LCD

The first LCD was developed at RCA Laboratories[3] and exploited another more interesting effect discovered by Heilmeier, **Louis A. Zanoni** and **Luke A. Barton** in the classes of *n*-type LCs [HZB1968]. They called it **dynamic scattering,** and dynamic scattering mode (**DSM**) the associated method to electronically control the transmission of light. To this purpose, the materials that yielded the best performance were members of organic compounds we have already introduced in 1.2.2 as Schiff's bases and which, when highly pure, are essentially transparent in the visible. For sufficiently high electric fields, Schiff's bases exhibited a marked turbulence that turned them from transparent to milky white, observable without the need of polarizer.

This property was observed in a sandwich cell consisting of a transparent front electrode and a specularly reflecting back electrode. In its quiescent state with no field applied, the liquid crystal is, as said, transparent. This means that if a specularly reflecting back electrode is faced into a black background, the cell appears black. When a dc field of the order of 5×10^3 V/cm is applied, the liquid becomes turbulent and scatters light. In this state the cell appears white. Increasing the field results in increased brightness so that a gray scale is obtainable.

[3]As we will see, most discoveries were carried out in the U.S.A., mainly at RCA, but also at Westinghouse. However, these firms never took advantage of their inventions that instead were commercially exploited by other (predominantly Japanese) companies. The Sarnoff dream of a wall-hanging TV set was not put in practice at RCA. But RCA and Westinghouse were not alone: between the late 1960–1980s, Japanese companies constantly capitalized on discoveries made by their U.S.A. competitors that were unable to bring them to market, as well described in [J1999].

The dynamic scattering is an electroconvection process (i.e., hydrodynamic instability driven by an electric field[4]) that now is also known as the **Carr-Helfrich** effect, from the name of the two researches **Edward F. Carr** of the University of Maine, U.S.A and **Wolfgang Helfrich** of Hoffmann-La Roche in Basel, Switzerland, that contributed to explain it in terms of a positive feedback mechanism [C1969, H1969], as illustrated in Fig. 2.3.[5]

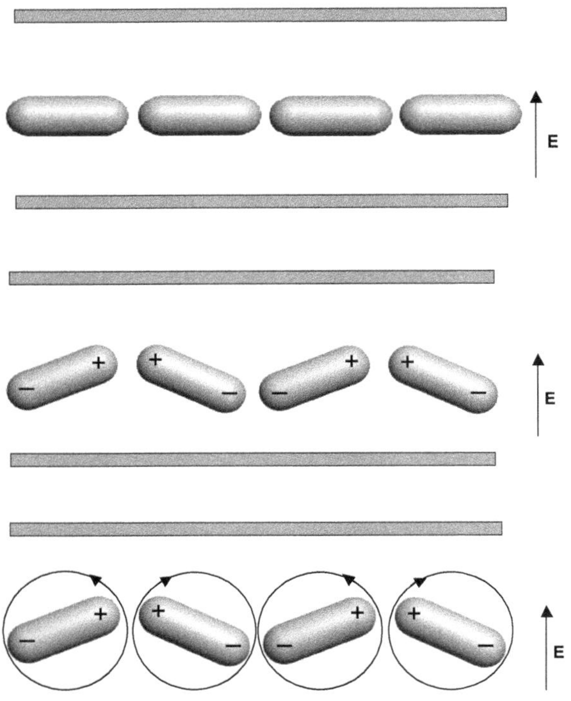

(a) Ideally aligned *n*-type nematic molecules under a vertical electric field.

(b) Fluctuations in the director alignments cause charge separation in the molecules, due to the higher conductivity along the director axis.

(c) The internal and external electric fields generate a torque that reinforces the initial director fluctuation and causes a circular motion of the molecules.

Fig. 2.3 Illustration of the Carr-Helfrich effect, which explains the Williams Domains and the Dynamic Scattering Mode

---

[4]Convection instabilities driven by *temperature* gradients are perhaps more common in nature. They are in fact important factors for the dynamics of our atmosphere.

[5]Helfrich considered a nematic liquid crystal placed in a uniform dc field and treated the problem in only one dimension. The calculation to an ac regime was later expanded in [DDG1971] and two dimensions and boundary conditions were taken into account in [PF1972]. The investigation of the threshold behaviour was carried out in [BZK1988].

In simple words, it is stated that any director fluctuation in an electric field leads to charge separation within a nematic LC (due to the fact that the electrical conductivity along the director is larger than in the short-axis directions). The induced field and external field give rise to a shear torque causing molecules to move circularly and reinforcing the initial director fluctuation (in a positive-feedback fashion). The mechanism is opposed by viscous damping of the flow and the elastic and dielectric torques, such that electroconvection appears only above a certain critical field. Above this value, the feedback mechanism overcomes the elastic restoring forces of the director field and a periodically deformed director and charge field is stabilized. In the simplest case a stationary state, characterized by convection rolls normal to the initial director alignment, is reached. This state corresponds to the Williams Domains. By further increasing the external field, the molecules go into a mechanically unstable state or turbulence. In modern parlance, this is an example of a bifurcation [TMB2000]. In this mode, the pattern becomes distorted and the positions of the domains fluctuate. The molecules now randomly scatter the light incident on them and appear milky white.

It is worth noting that to aid the formation of space charges and to reduce the operating temperature, ionic currents were required in the DSM mode. A small quantity of impurity was usually introduced in the LC material to control the value of these currents.

The DSM was immediately attractive for such applications as alphanumeric indicators. Indeed, it exhibited a rise time of 1–5 ms and a decay time lower than 30 ms together with dc operating voltages in the range of 10–100 V. Moreover, maximum contrast ratios in excess of 20 to 1 and maximum brightness of the order of 50% of $MgCO_3$ (the standard for white) were obtained. However, before fabrication of prototypes, room-temperature nematic materials were needed. **Joel E. Goldmacher** and **Joe A. Castellano** at RCA, solved the problem by developing a mixture of Schiff's base materials that were nematic at room temperatures [GC1970]. They found that by mixing similar kind liquid crystals, a melting temperature lower than the lowest one among the mixture's components could be achieved. For instance, the melting temperature of one of their mixture consisting of three Schiff's bases was 22°C, while the lowest of the melting temperatures of the three mixture components was 49°C.

With the availability of these materials Heilmeier, Zanoni, and other RCA engineers designed and fabricated prototype devices based on DSM. In May 1968 RCA *announced the discovery* of a new type of electronic display, totally different from traditional CRTs because it was lightweight, thin, with reduced power consumption. At its press conference, RCA distributed various photographs, see Fig. 2.4, showing the devices it had developed with the use of liquid crystals operating in DSM [RCA1968]. One was the first LCD digital clock.[6]

[6]The prototype nature of these devices is clearly described by the words of Bob Lohman, who posed in the picture: "*The clock lasted just long enough to take a picture.*"

(a) (b)

Fig. 2.4 RCA prototypes. (**a**) Heilmeier shows one of the first LCDs based on dynamic scattering. (**b**) The first LCD digital clock

Heilmeier and his team wanted to further refine and develop the LCD applications, and specifically towards an LCD television screen. However, this would have been impossible without yet-to-come achievements such as for instance: new liquid-crystal *p*-type materials and amorphous silicon technologies (both of which to be developed in Europe in the subsequent 25 years), Twisted Nematic (TN) Mode, and Active Matrix (AM) driving.

In any case, RCA top management and product division rejected the idea of LCDs because they represented a distraction or even a menace to the existing CRT business. Also caused by these frustrating interactions, the engineers and chemists working in the RCA liquid-crystal team started to drift away. **Nunzio Luce,** L. Zanoni, and others left RCA and joined Optel Corporation in Princeton, New Jersey. In 1970, the Optel team leaded by N. Luce succeeded in designing the first wrist watch having an LCD (operating in the DSM mode).

2.1.4 The Pocket Calculator

The first popular product incorporating an LCD was a pocket calculator made by **Sharp Corporation**.

At the beginning of 1960s calculators (that at that time were only of the desktop type) used Nixie[7] tubes or vacuum fluorescent[8] tubes as display elements. This caused the calculator to be large, heavy, expensive, and power consuming. In 1964 Sharp produced the first desktop calculator made only of transistors and diodes (priced 545 000 yen and weighted 25 kg). The key towards an impressive miniaturization was the use of complementary metal-oxide-semiconductor (CMOS)

---

[7] A Nixie (Numeric Indicator eXperimental) tube display is a tube filled with neon with a small amount of mercury. The anode is a transparent metal mesh winded around the front of the display and the 10 cathodes are in the shape of the numeral. The cathodes are stacked so that different numerals appear at different depths relatively to the viewer. When a DC voltage of 180–200 V is applied between the anode and any cathode, the digit emits light with an orange-red color.

[8] In vacuum fluorescent tubes, individual digits were housed in vacuum tubes like the Nixie tube. In this case however, specific phosphors allow the emission of red, yellow, and green as well as the more common blue-green color.

process. In 1969, with the use of ICs manufactured by Rockwell Corporation, Sharp reduced the calculator weight to 1.4 kg and could lower the price to 99 800 yen [K2002]. The subsequent step in miniaturization would be the use of a passive nonself-emitting display with a low-cost technology and compatible with the CMOS technology in order to place large-scale integrated (LSI) circuits on a glass substrate.

Tonio Wada, a chemical engineer at Sharp, was immediately attracted by RCA's 1968 announcement. He asked his management to verify the willingness of RCA to go into production of the LCDs for use in a pocket calculator. But RCA was not interested, as it believed the LCD's time response was too slow for applications different from watches. This difficulty confused the Sharp management, but Wada supported by his leader **Isamu Washizuka** insisted and in 1970 the researches of the company on LCDs started.

A special project team of 20 members was created by assembling engineers and scientists from related but different disciplines, which had the task of completing the commercialization in one to one-and-a-half year. The project was code-named S734: S stood for secret and 734 stood for completion by 1973 in the fourth month of the year.

The team had to face a number of issues. The first was to find a liquid crystal operating at room temperature and exhibiting a high contrast. They mixed over 3000 kinds of liquid crystals, synthesized over 500 mixtures and finally arrived at a mixture of Schiff's bases (MBBA, EBBA and BBBA), which exhibited nematic phase in the temperature range from –20°C to 63°C. It was *n* type and, in the DSM, it worked in the temperature range from 0°C to 40°C. The structure of BBA is shown in Fig. 2.5.

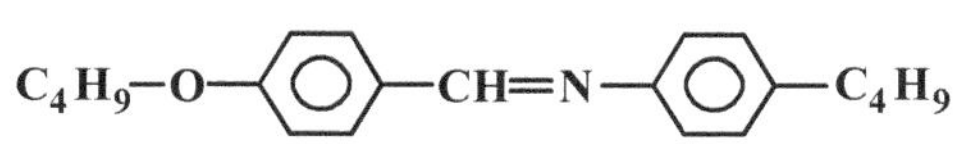

Fig. 2.5 Structure of BBBA

Then, they investigated which driving method, ac or dc, was the most suitable. The dc field generated bubbles in the mixture and degraded the material rather quickly. The ac field required a more complex electronic circuit, but it was the approach chosen.

They also searched for a transparent but electrically conductive plate. Wada found that the Laboratory of Ministry of International Trade and Industry (MITI) in Osaka had been working on indium tin oxide (**ITO**[9]). Sharp sent an engineer to the laboratory to learn the ITO technique.

Finally, they needed to find a company that would supply them with CMOS circuits for consumer use. RCA and Fairchild were involved in the commercialization of the CMOS circuits, but they were only interested in military applications. In that time, **Yasoji Suzuki** of the Toshiba Corporation was working

[9]Indium tin oxide is a mixture of indium oxide, In_2O_3 (typically 90%). and tin oxide, SnO_2 (10%). It is transparent and colorless in thin layers. ITO main feature is the combination of electrical conductivity and optical transparency. However, a compromise has to be reached during film deposition, as high concentration of charge carriers will increase the material's conductivity, but decrease its transparency.

on such CMOS circuits. He had developed a "clocked" CMOS circuit that allowed the impressive reduction of power consumption to less than 2 mW compared to 500 mW of conventional pMOS circuits [SHO1973]. Suzuki was asked to demonstrate its circuits to Sharp engineers and, ultimately, he supplied the clocked CMOS LSIs to Sharp.

Just before April 1973, Sharp completed the design of the pocket calculator and constructed a production line for its manufacturing. On May 1973, Sharp announced the first liquid-crystal pocket calculator (the Elsi Mate EL-805, shown in Fig. 2.6). It was a revolutionary calculator, with a DSM LCD and five ICs placed on a common glass substrate, more thin and lightweight than other electronic calculators available at that time. Its power consumption was 1/9000 of existing calculators! It was the first commercially successful product to use an LCD.

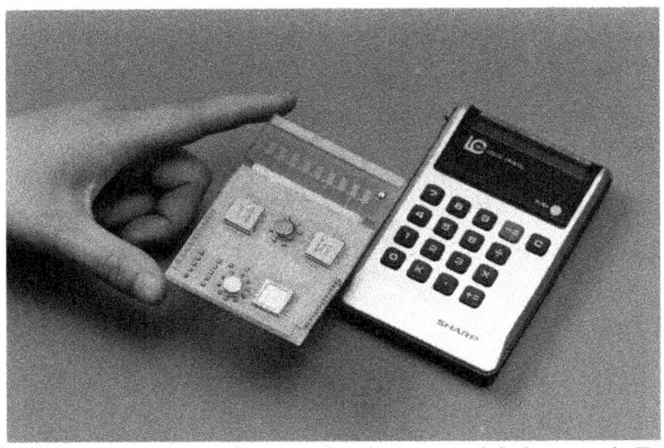

Fig. 2.6 The first liquid-crystal pocket calculator, which used DSM. Courtesy of Sharp Corporation

2.2 The Twisted Nematic Mode

The history of today's field-effect LCDs begins in 1970 when W. Helfrich and **Martin Schadt** of **Hoffmann-La Roche** in Basel, Switzerland (then the world's largest pharmaceutical firm), and independently J. L. Fergason developed the **Twisted Nematic** (**TN**) operation mode [HS1972, SH1971].

Heilfrich started his research activity on LCs at RCA, within the Liquid Crystal Group of Heilmeier. In 1969, he conceived and reported to Heilmeier a new idea, based on the molecular torsion induced on *p*-type LCs, in order to implement a different kind of display, but Heilmeier showed little interest also because two polarizers were required.

As many other colleagues, Heilfrich subsequently left RCA. In October 1970, he joined Hoffmann-La Roche where, with the collaboration of Schadt, demonstrated in a few weeks the totally new twisted nematic mode (see Fig. 2.7 illustrating the TN LCD prototype).

Fig. 2.7 TN LCD prototype presented by Helfrich and Schadt to the board of directors at Hoffmann-La Roche

The new effect required the orientation pattern to be twisted with no applied field, as described below.

2.2.1 TN Principle of Operation

We know that nematic LC molecules in their natural state are organized with the longitudinal axes almost parallel each other. Besides, if microgrooves are carved on a polyimide surface (deposited on the ITO, covering in turn the glass), LC molecules lying in the surface will align along the direction of the grooves.

Consider now an LC material sandwiched between two glass plates whose surfaces were grooved in orthogonal directions, as illustrated in Fig. 2.8. LC molecules near the plates must follow the direction of grooves and, as a consequence, the molecules in the bulk are forced to describe a 90° helical twist, from which the name twisted nematic for this operating mode.

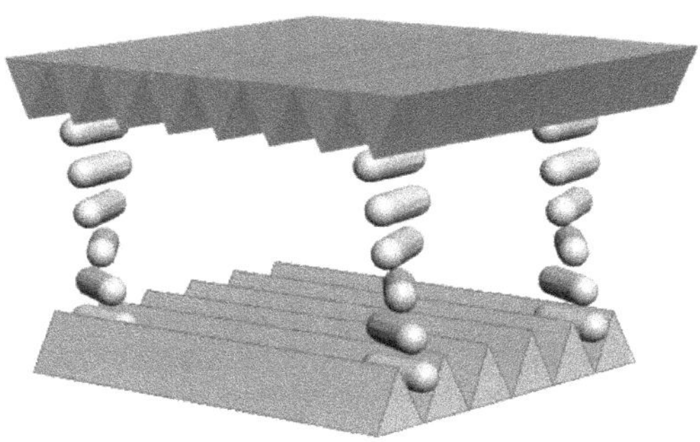

Fig. 2.8 Representation of the 90° helical rotation of LC molecules in Twisted Nematic cells

Starting from this structure and by adding two polarizing filters, better referred to as the **polarizer** (light in) and the **analyzer** (light out), on the outer surfaces of the two aligning layers, the basic TN cell can be devised, as illustrated in Fig. 2.9. Note that in an ideal polarizer, 50% of the incident energy of a beam of unpolarized light is absorbed and 50% is transmitted as a beam of polarized light. The transmission axis of the two polarizing filters must be parallel to the direction of the grooves. Hence, only the light with the polarization in axis with that of the polarizer can pass through.

It should be also noted, in passing, that depending on the type of light source, three types of LCDs have been devised: **transmissive**, **reflective** and **transflective**. A **backlight** module is required for transmissive LCDs which are better suited for indoor applications, but in very bright environments their images can be washed out. Reflective LCDs exploit the ambient light and do not require backlighting. They are hence lighter and consume less power (this behavior is appropriate for calculators and watches) but have poor indoor readability. In a transflective LCD, each pixel is divided into two subpixels one transmissive and the other reflective, whose percentage area can be tuned for the particular application.

Returning to the structure in Fig. 2.9, an important optical property is that *the plane of the polarization of light transmitted through the cell will rotate by following the director twist* (in a waveguiding fashion) provided that the Mauguin condition is fulfilled [M1911b]

$$\Delta n \cdot d \gg \frac{\phi \lambda}{\pi} \tag{2.2}$$

where Δn is the birefringence, ϕ is the total twist angle, d is the layer thickness and λ is the wavelength of incident light, the product $\Delta n\, d$ is called the *optical path*.

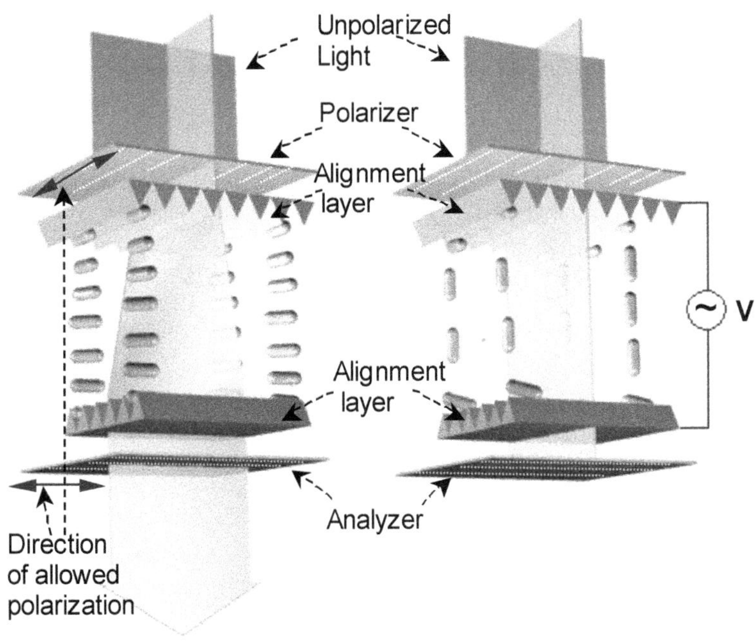

(a) (b)

Fig. 2.9 The Twisted Nematic effect in a normal white cell: (**a**) The direction of polarization of the light entering the liquid-crystal region rotates by 90° along the helix arrangement of the LC molecules and the screen appears *white*. (**b**) Light is blocked by the analyzer and the cell appears *black*

For a 90° twisted nematic cell the Mauguin condition reduces to

$$\Delta n \cdot d \gg \frac{\lambda}{2} \tag{2.3}$$

Therefore, the linearly polarized light passes through the analyzer (Fig. 2.9a) and the screen appears white. If a *p*-type nematic LC is considered, after applying a voltage across the two plates, the LC molecules tend to align along the direction of the electric field. In particular, when a voltage greater than the threshold V_{th} is applied, the molecules located at the center of the cell farthest away from the electrodes start to deviate from their quiescent state. With increasing voltage the twist is gradually removed and light polarization is no longer rotated since the LC is not birefringent in this orientation (Fig. 2.9b). The light is blocked at the bottom polarizing filter and the screen appears black.

The process described above is referred to as **normal white** contrast (or mode). If the analyzer is rotated by 90°, it results in parallel with the polarizer and light is blocked in absence of an applied voltage, this is the **normal black** mode.

2.2.2 Electro-Optic Transfer Function (EOTF)

In the liquid crystal cells actually used in TN displays, the Mauguin inequality is only approximately fulfilled, resulting in a reduction of the display brightness. More important, it was later demonstrated that the transmission depends on wavelength according to the Gooch-Tarry theory [GT1974, GT1975]

$$T_{(NW)} = 1 - \frac{\sin^2\left(\phi\sqrt{1+u^2}\right)}{1+u^2} \tag{2.4a}$$

where ϕ is the twist angle ($\phi = \pi/2$ in TN mode) and $u = \dfrac{\pi\, \Delta n\, d}{\lambda\, \phi}$ is called the retardation index. In the case of normal white contrast, transmission with zero voltage applied is

$$T_{(NB)} = \frac{\sin^2\left(\phi\sqrt{1+u^2}\right)}{1+u^2} \tag{2.4b}$$

Figure 2.10a shows T as a function of the *optical path* at $\lambda = 555$ nm. Transmission is maximum only for discrete values of u, resulting form $u = \sqrt{4n^2 - 1}$, with n natural number. The first two of these are known as the first ($u = \sqrt{3}$, $\Delta n\, d = 0.48$ µm) and second maximum ($u = \sqrt{15}$, $\Delta n\, d = 1.08$ µm) conditions respectively.

Similar discussion is possible for the normally black case, whose transmission versus $\Delta n\ d$ at $\lambda = 555$ nm is shown in Fig. 2.10b.

It is important to understand that achieving a good OFF state with the normally black mode is possible only for *monochromatic light* with a wavelength equal to one of the minima. Since LCDs operate with *white light*, luminance can be appreciable in the black state. Therefore, the normally white contrast is virtually exploited in all TN mode LCDs. It achieves a good black state for high applied voltage and a good contrast ratio exceeding 500:1, though for narrow viewing angles. With the advent of wide viewing angle displays, two other basic modes, namely the IPS and MVA were developed (see Section 2.4.4). They operate typically in normal black contrast.

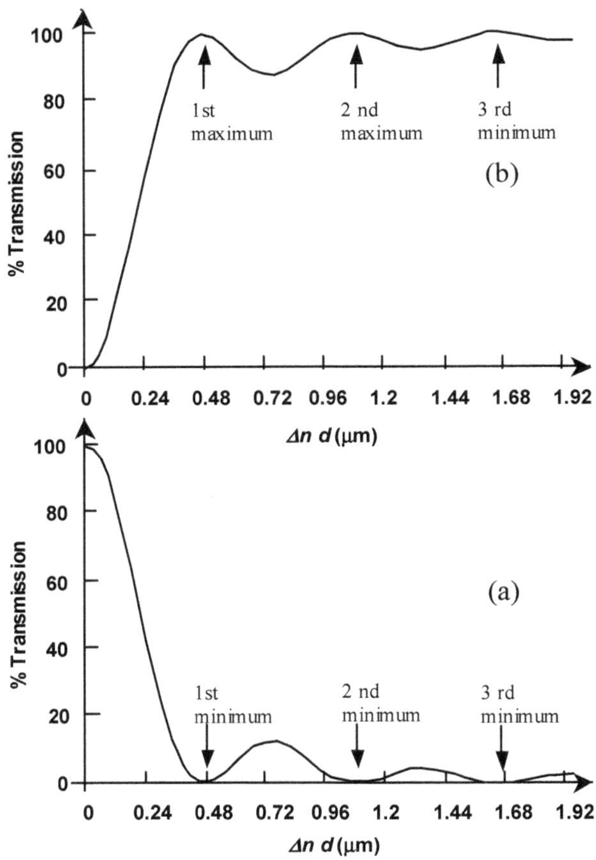

Fig. 2.10 Percentage light transmission, T, versus $\Delta n\, d$ ($\lambda = 555$ nm) of (**a**) normally white TN cell, (**b**) normally black TN cell. First, second, and third minima (maxima) are indicated

The behavior of transmission (also brightness or luminance) of the TN cell in the normally white LCD as a function of the applied voltage is shown in Fig. 2.11. No change in transmission occurs until the threshold voltage, V_{th}, is reached. An expression for the threshold voltage was already given in (1.8). Transmission then decreases as the voltage increases until saturation is reached. Threshold voltage is typically 1.5-to-2.5 V, and saturation occurs at about 4-to-5 V. A gray level scale is hence obtained, being controlled by the voltage magnitude.

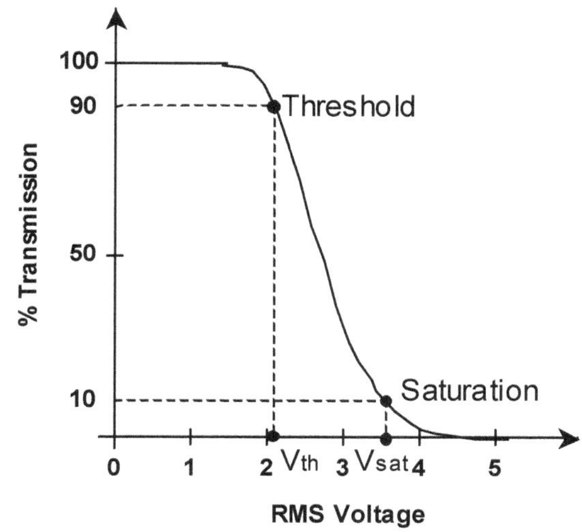

Fig. 2.11 Electro-optic transfer function (EOTF) for a typical normally white TN cell showing Transmission as a function of the RMS applied voltage (at 1 kHz)

Subsequent research has been carried out both for lowering the threshold voltage and increasing the slope of the transfer curve (important parameter for passive matrix addressing, see Section 2.7.1). Referring again to Fig. 2.11, we emphasize the fact that the applied voltage is an RMS[10] value since, as explained in 1.3.6, an AC driving voltage must be applied to the LC layer. If the frequency of the driving voltage is sufficiently high, the LC molecules are too slow to follow the waveform of the voltage, therefore they react to the energy. Note that the LC layer provides the same light transmission irrespective of the voltage polarity. Typical frequency values which are between 50 and 100 Hz do not effectively influence the director profile.

---

[10] Recall that the RMS value of a periodic waveform, $v(t)$, with period T is

$$V_{RMS} = \sqrt{\frac{1}{T}\int_0^T v^2(t)dt}$$

If $v(t)$ is a square waveform (as it is the case for actual driving voltages) V_{RMS} coincides with the maximum voltage.

It should be noted that the TN T/V characteristic suffers from a negative temperature coefficient. Indeed both V_{sat} and V_{th} decrease with increasing temperature. This leads to a translation of the T/V characteristic along the horizontal axis and may cause a reduction of contrast.

2.2.3 LC Capacitance

An effect related to the reorientation experienced by the LC molecules is the change of the capacitance offered by a single LC pixel, C_{LC}. This capacitance depends on the applied voltage because of the dielectric anisotropy of the LC mixture, as qualitatively shown in Fig. 2.12, for a *p*-type material, where $C_{LC,max}$ typically ranges form $2C_{LC,min}$ to $3C_{LC,min}$, with values of hundreds of femtofarads.

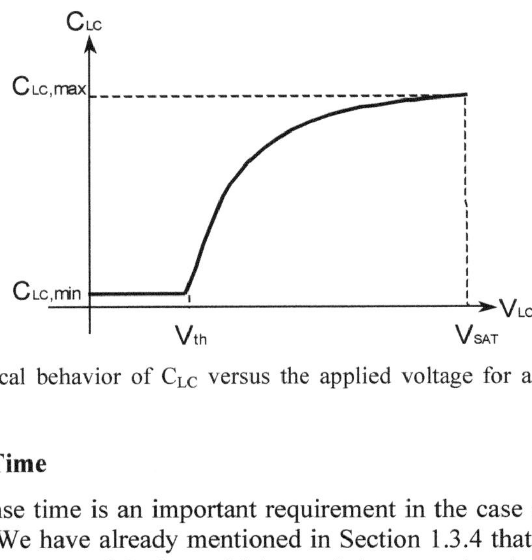

Fig. 2.12 Typical behavior of C_{LC} versus the applied voltage for a normally white display

2.2.4 Response Time

A fast LC response time is an important requirement in the case of reproduction of moving images. We have already mentioned in Section 1.3.4 that the response time increases with the square of the layer thickness (it is proportional to γd^2).

To be more precise, when the applied AC voltage has an RMS value much greater than the LC threshold, and its frequency is sufficiently low, the response time is characterized by three parameters, the delay time (t_{dr} and t_{df}), the rise time (t_r) and the fall time (t_f), as illustrated in Fig. 2.13. The delay time in the OFF to ON transition is the time occurring from the application of the voltage to the transmission of light reaching the 10% of the final value. Similarly, the fall delay time is the time to reach 90% light transmission after removing the applied voltage. The rise (fall) time is defined as the time to increase (decrease) the transmission from 10% to 90% (90% to 10%).

The rise and fall times for a normally white TN cell are given by

$$t_r = \frac{\gamma d^2}{\varepsilon_0 \Delta\varepsilon V^2 - \pi^2 k_{EQ}} \tag{2.5a}$$

$$t_f = \frac{\gamma d^2}{\pi^2 k_{EQ}} \tag{2.5b}$$

where, with the notation defined in Chapter 1, parameters γ, $\Delta\varepsilon$ and k_{EQ} are respectively the rotational viscosity, anisotropy of the dielectric constant and average elastic constant of the LC material, and V is the RMS voltage. In the operation of a TN cell, all the three splay twist and bend constants play a role in the reorientation mechanism and therefore k_{EQ} is a combination of k_{11}, k_{22} and k_{33}.

Typical values for t_r and t_f are in the range of 10–50 ms. However, (2.5a) shows that t_r increases when the applied voltage decreases. This means that the rise time between two adjacent gray levels is much greater than that required for a complete switch from white to black.

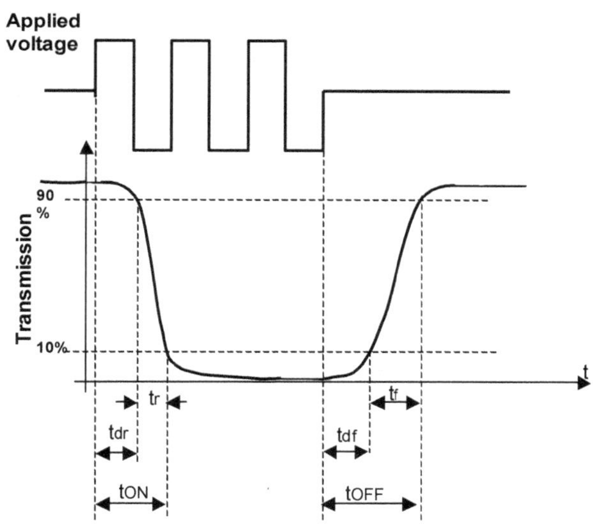

Fig. 2.13 Typical response of a normally white TN LC cell

2.2.5 Viewing Angle

We conclude this section by discussing the behaviour of a TN LCD under different **viewing angles**. The T/V curve that we have considered up to now is obtained by looking at the display with the optimum viewing angle of 90° (i.e., the observer is parallel to the LCD front panel). In reality, luminance can vary with the viewing angle both in the horizontal and in the vertical direction. This effect was common in the first generation of notebook LCDs, where different grey levels and even color inversion could be perceived by varying the viewing angle. For this particular application, the horizontal gray consistency is however of greater importance than the vertical one. At this purpose, the LCD glass substrates are rubbed at 45° (and consequently, also the polarizer and analyzer filters are rotated by 45°) so that the mid-layer LC molecule is approximately in a vertical position with respect to the observer, as illustrated in Fig. 2.14a.

In this manner, when a voltage is applied across the LC cell (see Fig. 2.14b), the mid-layer molecule reorients under the electric field in a symmetrical way with respect to the observer in the positions (2) and (3), i.e., at the left and at the right of the ideal 90° position (1). This reduces horizontal gray level inconsistencies especially for medium gray levels that are the most critical. In contrast, if the viewing angle is varied vertically, different birefringence is caused by the LC molecule with respect to an observer in positions *d* and *e*, who therefore perceives different gray levels. Alternatively, one can describe the effect by considering that at an oblique viewing angle, the transmission axes of the crossed polarizers are no longer orthogonal to each other and, as a result, light leakage would occur [W1995, KJ1998].

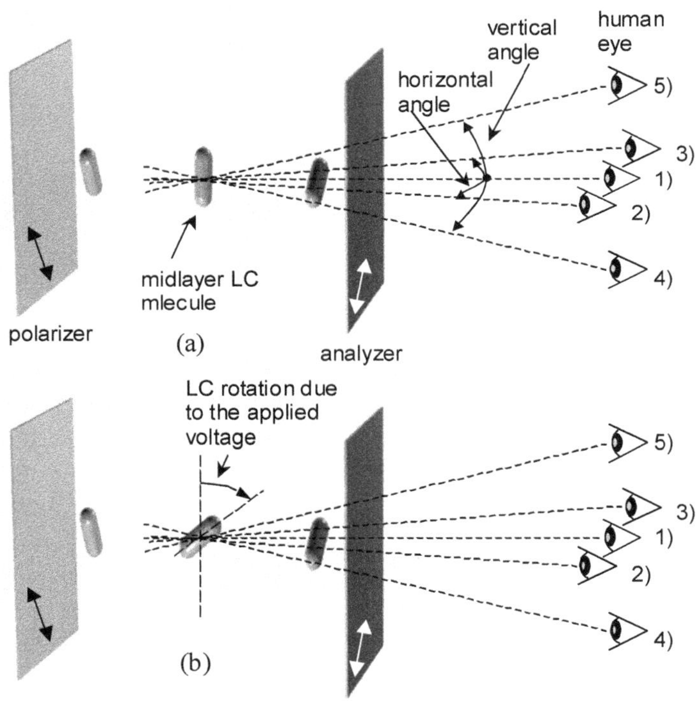

Fig. 2.14 Illustration of horizontal and vertical viewing angles: (**a**) The mid-layer LC molecule is vertical with respect to the observer at any position and therefore the same birefringence is experienced (same gray level). (**b**) Rotation of the mid-layer LC molecule causes a considerable difference in birefringence but only for vertical viewing angles

We show in Fig. 2.15 the luminance of a normally white TN LCD versus vertical viewing angle for several applied voltages.

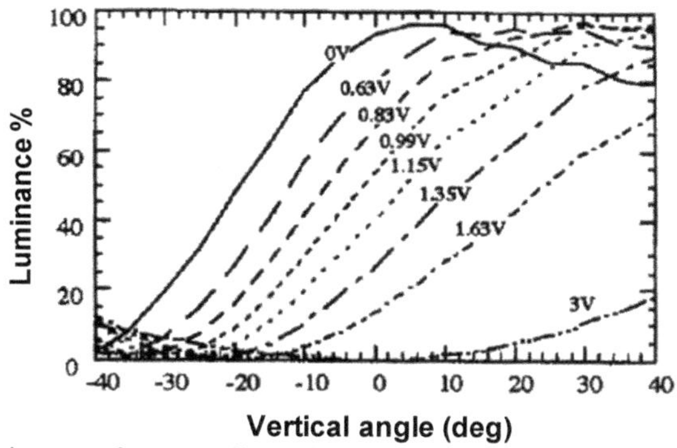

Fig. 2.15 Luminance of a normally white TN LCD versus vertical viewing angle for different applied voltages

2.2.6 Application of the TN Mode

The nematic mixture used in TN LCDs must exhibit a new set of specifications that include: a high $\Delta\varepsilon$ (for a low threshold voltage), a low viscosity (for short response times), low k_{33}/k_{11} and $\Delta\varepsilon/\varepsilon_{\perp}$ (for a steep slope of the T/V curve, [B1973] [SG1982]), and high k_{33}/k_{22}. Besides, since the transmission curve depends on temperature, a low rate of change is required (e.g., <<1% in threshold voltage per degree centigrade).[11]

The TN displays adopted in watches and calculators used liquid crystals with relatively large n_a values of 0.2 and cell thickness of about 8 μm, this yield the product $\Delta n\, d$ equal to 1.6 μm (and $\Delta n\, d\,/\lambda$ about equal to 3). Such a value falls approximately in the third minimum. Compared to the first two minima, the third one is broader, thus the display contrast is more insensitive to mismatches of $\Delta n\, d$ and hence to small thickness variation in the cell gap.

For TN active-matrix driving schemes, fast response times and, therefore, thin cell gaps are important. In this case, the first minimum condition is often chosen. In the 1980s, a distributed spacer technology was developed to accurately control cell gap uniformity from 1 to 100 μm, depending on the LC mixture and application. Figure 2.16 shows the cross section of a typical LC cell. Spacers are microscopic spheres made of polymer or glass. After the spacers are scattered on one of the glass substrate, it is attached and sealed to the other glass substrate. A fill hole is left to introduce the LC material in the gap and is immediately sealed by adhesive epoxy resins that have no chemical interaction with the LC material to avoid contamination. Finally, polarizers are attached to the liquid crystal cell.

[11]In practice, the variation of operating voltages with temperature is often compensated electronically or by using an optically active dopant whose pitch decreases with increasing temperature.

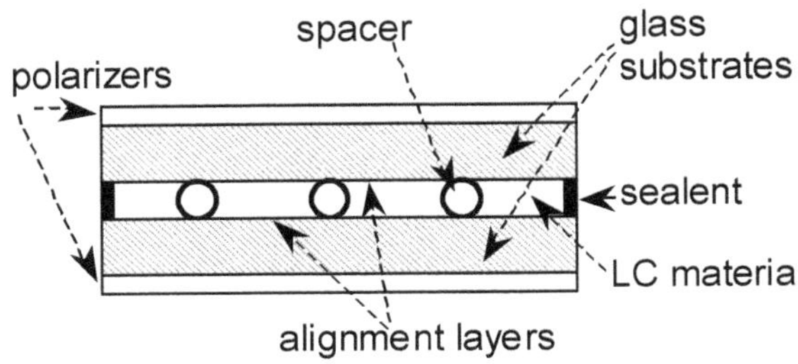

Fig. 2.16 Simplified representation of a cross section of a LC cell

The main initial drawbacks of the TN structure are the requirement of two polarizers[12] and the surface alignment of the LCs. However, the TN configuration does not require any ionic current to generate an optical signal thus requiring an extremely low electrical power (a few mW/cm^2). The contrast between OFF and ON state was higher than that of DSM and the response time was in the order of 60 ms. Thanks to these favorable properties, the article [SH1971] had a great impact in the community of liquid-crystal scientists. For instance, it induced T. Wada to switch from DSM to TN mode. Also **Suwa Seiko** (the well-known Japanese watch producer) which had been working on the DSM in developing digital watches, switched to the TN mode (see Section 2.2.1).

Moreover, when the U.K. launched research on synthesizing room-temperature liquid crystals, the TN mode was chosen (see Section 2.3). Currently, the TN mode and its derivatives are implemented in virtually all LCDs manufactured throughout the world. The TN has since been cited in hundreds of liquid-crystal patents and became the foundation of the multibillion-dollar LCD industry.

As mentioned at the beginning of the section, J. Fergason had independently discovered the TN effect in almost the same period. In 1970, Fergason had gathered some fellow scientists in a small firm called ILIXCO (International Liquid Xtal Company, Kent, Ohio, now LXD inc.) to create a marketable display. In 1971, ILIXCO produced the first LCDs based on the TN effect. The first TN digital clock was a Fergason's prototype, an eight-digit display that was unveiled at an electronics show in Cleveland in 1971.

TN field effect was filed for patent by Schadt and Helfrich in December 1970. Fergason filed an identical patent in the USA in February 1971. This caused a legal battle over ownership of the invention. However, ILIXCO employed only 100 people, and Fergason had not sufficient resources for legal battles. Eventually, Fergason sold his patent rights to Hoffmann-La Roche in 1972. The company promised him a million dollars in cash and half of all U.S.A. royalties, plus a smaller amount for foreign sales. Hoffmann-La Roche licensed the technology mainly to the Pacific area and European firms.

Many years later Fergason said [H2002]:

[12]The use of polarizers reduces the potential brightness since they allow less than half of the backlight to pass through. Besides, polarizers are optimized to work with light propagating perpendicular to the display and this is also a cause of a reduction of the viewing angle.

"There are still some arguments about who invented the liquid-crystal display phenomenon and all that sort of thing, but there are no arguments about who built the first ones and who invented the display itself. I invented the display and thus the phenomenon."

2.2.7 The LC Digital Watch

In the early 1970s, balance wheels and mechanisms of wrist watches began to be replaced by electronic components and quartz crystals. In 1970 Hamilton Watch Company (Lancaster, Pennsylvania) announced the world's first electronic digital watch, the Pulsar, with a bright-red light-emitting-diode (LED) display. It was commercially available in 1972 and very expensive (as much as a small car). It couldn't display the time continuously because of the extremely high power dissipation. The user had to push a button to read the time (and this was sometimes embarrassing, as pushing the button when people had visitors could give the erroneous impression that he was impatient for the visitor to leave). As a consequence, LED watches were soon abandoned.

The development of digital LCD watches is detailed in **Yoshio Yamazaki**'s article included in [S1979] and summarized in [K2002].

Yamazaki had joined Suwa Seiko in 1965 as the unique chemical engineer among precision-mechanics engineers. His original activity was to develop chemicals (such as lubricants) and plating techniques for the watch package. In 1968, he read in a Japanese newspaper of the RCA announcement. He soon started working, though without any background, in order to exploit LCs in low power watches and stimulated Suwa Seiko to establish a group to develop LC watches. Yamazaki became responsible for the development of liquid-crystal materials and display panels. If the development failed, he had been told that the responsibility would be all his.

The candidate for the modes of operation at the time were DSM, guest-host mode, and TN mode. In DSM, it was soon understood that the Schiff's base hydrolysis problem could be avoided if the device was correctly sealed. However, unlike a pocket calculator, a watch had to operate continuously and since the DSM required an ionic current, this made the battery drain rather quickly. Other problems were a somewhat high driving voltage and a slow response time. The viewing angle was narrow, which meant that one had to adjust the angle of view to see the display.

By adopting the guest-host mode, consisting of nematic liquid crystals and pleochroic dyes, the display would be colored. Operating voltages were relatively low and the viewing angle was enlarged. However, the contrast was not large enough and life expectancy was unknown.

Yamasaki chose the TN mode for its extremely low power consumption and high contrast. Indeed, the power dissipation of the TN cell plus its driving circuits could be reduced to less than one-fifth the power dissipation of a DSM cell plus its driving circuits.

In September 1973, the "06LC" digital LC watch (see Fig. 2.17) was announced. It was one of the first products to exploit the TN mode and the first with a 6-digit display (the first TN 4-digit LCD watch was marketed in 1972 by Gruen whose

LCD was supplied by Fergason's ILIXCO). The watch was well accepted and became a commercial success mostly because of its long battery life (using a 1.5-V 190-mAh battery, the watch would operate for two years, assuming the user turned on a viewing lamp ten times a day each for one second).

Fig. 2.17 The "06LC" from Seiko, the first 6-digit watch. Courtesy of Seiko Epson

2.3 New High-Performance LC Materials

At the end of 1960s, U.K. was paying annual royalties to RCA, on the shadow-mask color TV tube, which amounted to "*more than the development costs of the Concorde*" [K2002]. In 1970, the Ministry of Defence decided to start a program to invent a solid-state alternative to the shadow mask tube. At this scope, the Royal Radar Establishment (RRE, the search division of the British Armed Forces) and a team at Hull University composed of **George W. Gray, Ken J. Harrison** and a post-doctoral fellow **John A. Nash** were involved. The interest in TN cells had increased the need for *p*-type LCs. By the spring of 1972, Gray had tried most of the LC materials he knew, but they were without exception deficient in some aspect. The Hull group extrapolated the feature common to the unsuccessful experiments and gradually new structures were proposed. **Pentyl-cyanobiphenyl** had an operating temperature range from 22°C to 35°C and pentyloxy-cyanobiphenyl worked from 48°C to 69°C [GHN1973].

The materials were used with excellent results in TN cells, with a threshold voltage as low as 1.1 V_{RMS}. Hull and RRE were conscious that they had something of commercial value. Code words were used to describe the materials: 5CB (pentyl-cyanobiphenyl), 6CB, 7CB, 8CB, 70CB, 80CB, etc. A binary mixture of 5CB with 80CB worked from 4°C to 50°C and a quaternary mixture of 7CB, 8CB, 50CB, and 70CB from –3°C to 52°C. There was an immediate demand of these mixtures and a chemical company, the British Drug House Ltd. (BDH), was approached for the mass production.

The RRE team leaded by **Cyril Hilsum** (a well-known semiconductor physicist) recruited **Peter Raynes** in 1971, who made a fundamental contribution developing an analytical method (based on an extension of the Schröder-van Laar equation) to find optimum mixtures from an available selection. It allowed the prediction of the properties of a mixture starting from the thermodynamic data of the individual component materials [HRH1974]. Stable mixtures with operating ranges as wide as 60°C, with a melting point near or a little below zero were obtained using this method. A great improvement, but that was still insufficient, as many display manufactures required reaching –10°C. Gray understood that the **terphenyl** system would give a much higher clearing point and when mixed with biphenyls would give a wider range mixture. In 1973, Hull had synthesized **pentyl-cyanoterphenyl** (5CT, see last row in Table 2.1). In August 1974, Raynes made a eutectic of 5CB, 7CB, and 80CB and 5CT that was named E7 (see Table 2.1). It exhibited a nematic range between –9°C to 59°C.

Biphenyls and terphenyls had a great success and have been used in TN-based displays to the present day. However, they were not the unique LC substances used in the industry.

In 1970, **Dietrich Demus** and his group of Halle University developed negative esters [DWD1971]. Unlike Schiff's bases, the esters did not exhibit the hydrolysis problem and were adopted in a DSM display for use in watches. In 1973, they also synthesized cyclohexane esters [DKS1973], which have been widely used in pocket calculators to adjust the V_{th} of their TN cells. As the area of applications of LCs expanded beyond pocket calculators and watches, the time response at low temperatures became important. From 1976, various cyclohexanoates were developed by **Rudolf Eidenschink** and coworkers at Merck [EEK1977, PEK1977]. The most important of those was **phenyl cyclohexane** (PCH), (whose structure was already shown in Table 1.1), **biphenyl cyclohexane** (BCH), see Fig. 2.18, and **cyclohexyl cyclohexane** (CCH). They had a lower viscosity thereby yielding fast time response, greater stability and better optical performance. They were widely used in TN cells and subsequently in active matrix displays and represented the basic structures from which most LCs were derived.

Table 2.1 Composition of E7

Individual compound		Nematic range (°C)	Percent amount
5CB	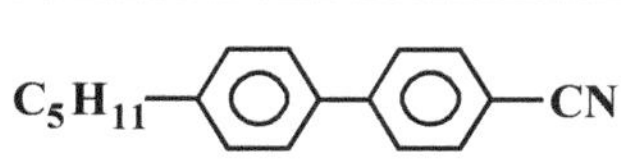	22–35	51%
7CB	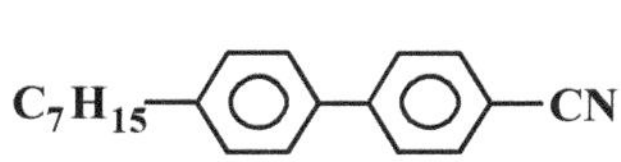	28–42	25%
80CB	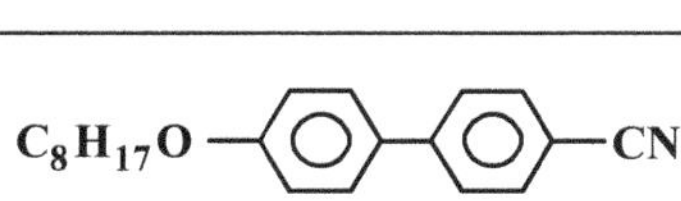	54–80	16%
5CT	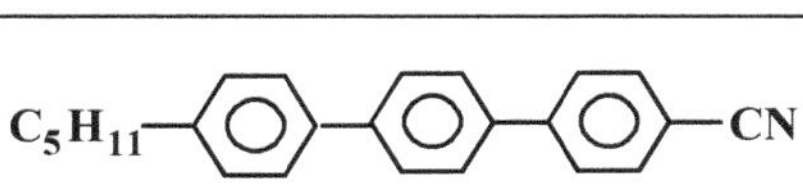	130–239	8%

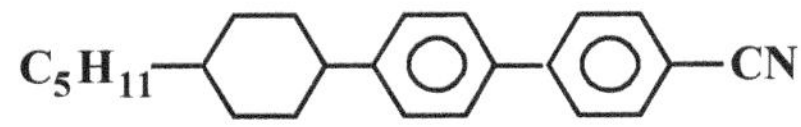

Fig. 2.18 Structure of BCH-5

2.4 Further Developments

In this section we briefly summarize the main developments that have been introduced during the last 30 years. Note that some of the following topics will be further discussed and expanded in the subsequent chapters.

2.4.1 Passive Addressing and Super Twisted Nematic Mode

After the introduction of the TN mode and cyanobiphenyl, LCs became well established in applications such as pocket calculators and digital watches. For instance, the production of pocket calculators in Japan had been 10 million units in 1973, reaching 50 million units by 1981 [K2002]. Both applications required simple displays (to represent numerals or simple characters) and used **direct addressing** schemes, in which each **segment** forming a character was directly driven. In direct-addressed LCDs, a good contrast (in excess of 100:1) could be achieved by driving the LC into saturation.

The next step was towards displaying more complex characters and even images. At this purpose engineers moved to the **(passive) matrix addressing** scheme, in which each **pixel** (i.e., the smallest addressable part of the display) was placed at the intersection of a particular row and column addressing line. The pixel is hence turned on only if both the row and column are electrically selected.

The TN mode was in principle suitable for this scope, but only for a limited number of matrix lines. The scanning limitations of multiplexed LCDs were theoretically analyzed by **Paul Alt** and **Peter Pleshko** of IBM [AP1974] and Hideaki Kawakami of Hitachi [K1976], as will be detailed in the next chapter, who found that the number of lines that can be multiplexed depends on the steepness of the transmission curve.

In 1983, **Terry Scheffer** and **Jurgen Nehring** at Brown Boveri (BBC), Baden, Switzerland, discovered and implemented the principle of the supertwisted birefringence effect (SBE) lately called the **Super Twisted Nematic (STN)** mode [S1983, ADK1983, SN1984]. As an extension of the standard TN mode, STN exploited an ideal 270° twist rotation (in practice, the rotation is comprised between 180° and 270°) achieved by doping the LC mixture with a chiral nematic LC. The STN cell exhibited a slope of the T/V curve approaching infinity, as illustrated in Fig. 2.19a.

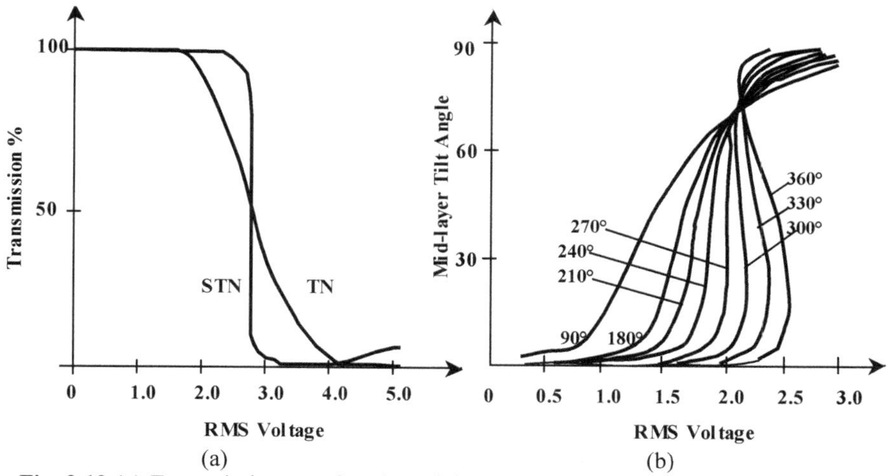

Fig. 2.19 (**a**) Transmission as a function of the RMS applied voltage (at 1 kHz) for a typical normal white TN (twist angle= 90°) cell and STN (twist angle = 270°) cell. (**b**) Tilt angle of the LC mid-layer molecules as a function of the RMS applied voltage for different twist angles

The BBC group constructed 121.5 mm × 243.0 mm screen displays with a 540 × 270 matrix [SNK1985]. STN provided a better contrast ratio and viewing angle (due to the unconventional orientation of the polarizers) and was soon adopted by several companies like Sharp, Hitachi and Seiko. However, STN required specific materials to be synthesized. To this end, Hoffmann-La Roche succeeded in developing alkenyl mixtures [SPG1985]. STN was the first technology that could be fitted to a portable computer. Its variants are now used in virtually all types of hand-held devices (the second half of 1990s witnessed the success of notebooks and mobile telephones equipped with low-power low-cost STN displays, and subsequently other best-selling products emerged like video cameras, games, portable CD/MP3 players, GPS navigation equipments, and electronic books).

Unfortunately, STN was not suitable for television applications. The steep T/V characteristic could not provide good gray scale. Because of the high twist angle, the response time was in the range of 200 ms (as opposed to 50 ms of TN) and hence it was not adequate for moving pictures. The leakage current among adjacent pixels also caused ghosting or crosstalk. In addition, early STN LCDs produced yellow and blue images, rather than black and white ones. A true black and white display (that with the addition of the RGB filters becomes a color display, see Fig. 2.20) was then realized by adding a second STN layer with the STN chains twisted in the opposite direction (**Double Super Twisted Nematic, DSTN**). These displays were much thicker, heavier and expensive and were soon abandoned after the development of the **Film-compensated STN (FSTN)**, in which a black and white display was realized with a single STN layer and a compensation film. This approach was not effective for large size panels. To overcome this drawback, Seiko and Sharp invented a structure called **Triple Super Twisted Nematic** mode (**TSTN**) in which the STN panel is sandwiched between two sheets of film [S1987, ONW1987, WOW1988, KSY1988, KSO1988].

Nowadays, STN technologies are all of these last two types, FSTN and TSTN.

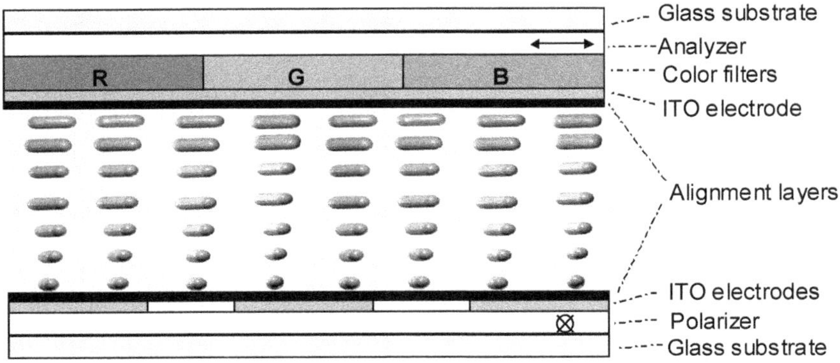

Fig. 2.20 The use of RGB filters to realize a color TN LCD

Lately, research on the passive matrix displays gave also a new technique called the **dual scanning STN** in which the display is divided into two (upper and lower) half portions and driven by independent column drivers. Of course, this cuts the number of rows in each portion by half, reducing multiplexing and thereby improving contrast (by increasing the drive margin, see Chapter 3).

2.4.2 Active Matrix Addressing Schemes

In the early 1990s the market scenario for different technologies became clearer. The simple matrix (passive matrix) addressing evolved into an **active matrix** addressing (**AM**) with a **thin film transistor (TFT)** array realized on the glass for large-size, high-quality displays. Each pixel has its own dedicated TFT allowing each column line to access one pixel. The TFT was fabricated in the corner of each LC pixel on the rear backplane of the two glass substrates. The opaque TFT (which reduces the pixel transparent area) requires a brighter backlight, but the increased isolation between adjacent pixels reduces the current leakage between the top and bottom substrates. Moreover, thanks to the increased controllability allowed by the TFT, conventional TN alignment could be used with higher time responses and to display motion pictures for the first time. As a result, nearly all LCDs for notebooks, personal computers and TV screens up to 19 in. exploit TN technology at the time of writing.

Nonetheless, the commercial exploitation of AM schemes was not easy. The idea of an AM drive for use in LCDs has already been conceived by **Bernard Lechner** and his colleagues at RCA in 1971. They proposed to use an array of TFTs to control cells operating in DSM [LMN1971]. In 1973, **T. Peter Brody** (who coined the term active matrix) and his colleagues of the Westinghouse Electron Tube Division achieved the first operating AMLCD (still picture) 6 in.× 6 in. having

14 000 dots [BAD1973]. They used TFT implemented with cadmium selenide (CdSe), but given the difficulties with reproducibility and controlling film formation, the high off-current of the TFTs, and reliability issues, no CdSe-based TFT display was ever commercialized. A great part of the Westinghouse Electron Tube Division was closed in late 1976 and, with the change of top management, any LCD development at Westinghouse ceased.

Brody some time later said: [13]

Westinghouse was very short-sighted, we were too far ahead.

It should be noted, in passing, that despite their poor performance (to nowadays standards), the Westinghouse prototypes contained all the building blocks that are still used today. Their main limitations were due principally to the poor TFT performance. Again, management in American and European companies quickly rejected this technology assumed too complex and expensive. Some Western critics of that time referred to the Asian investment in the TFT technology even as a *race to lose money*. Actually, because of their complex technology (exposed to faults and with a consequent raise in costs), TFT LCDs were introduced in the market only in the late 1980s but they truly revolutionized display technology and paved the way for ubiquitous computing.

Masataka Matsuura and others at Sharp in 1983 implemented a 2 in. × 2 in. 50 × 50 dot matrix display formed on a glass substrate [MYN1983]. Instead of CdSe used at Westinghouse for the realization of the TFT array (that was toxic and hence not suitable for mass production), they used tellurium, which had also a higher mobility. Unfortunately, tellurium TFT had a high sub-threshold current that did not satisfied the required target ON-to-OFF current ratio of 10 to 1. Silicon TFTs were considered the ideal candidate in this application, since both the silicon physics and technology were well understood. However, crystalline silicon was not suitable for large displays because it cannot be grown upon a glass substrate.[14]

To overcome crystalline Si limitations, **poly Si** and **amorphous Si** (a-Si) were studied to realize TFTs. In 1983, Shinji Morozumi and his colleagues at Seiko Epson succeeded in demonstrating the first full-color (poly-Si) display, a 2.13-in. device with a 240 × 240 dot matrix [MOY1983]. It was the first commercial application of AMLCDs, however the early poly-Si TFTs required high-temperature processing and therefore expensive quartz substrates.

---

[13]Thin-film transistors were first applied to a display as a control matrix to an electroluminescent display, not to an LCD. Brody published a thorough review of the early days of the TFT in [B1984].

[14] In 1982, Daini Seikosha Co, (a former name of Seiko), used crystalline silicon and formed the transistor arrays and driver circuits on a common silicon substrate [YKMKN1982]. From this crystalline-silicon LCD technology, Seiko developed a wristwatch television [S1982]. Its monochromatic blue-and-white display had 152 × 210 pixels and was worn by "007-Her Majesty's Secret Agent" in the movie "Octopussy."

The first idea of a-Si TFTs dates back to 1979, when professor **Walter E. Spear** and **Peter G. LeComber**, at the University of Dundee in Scotland, while studying the applications of a-Si for solar cells, indicated the possibility of using a-Si in a TFT array for LCDs [LSG1979, SMS1981]. Amorphous silicon was more attractive than (high-temperature) poly Si, because it needed lower process temperatures (below 400°C) allowing its deposition on low-cost, large area glass substrates. Besides it required a reduced number of process steps. Unfortunately, Dr. LeComber could see only the beginning of the escalation of AMLCD based on a-Si TFT, since he died of heart attack in 1992. In 1982, Canon developed a 96-mm × 96-mm display with 240 × 240 dot matrix based on this a-Si technology [ONO1982] and Matsushita demonstrated a 378 × 240 3-in. full-color display in 1986 [HNM1986]. These displays were mainly used in small portable television sets. A significant breakthrough was achieved in the late 1980s when several companies including IBM, NEC, Sharp, Toshiba and Hitachi, marketed laptop computers with 10-in. AMLCDs. This killer application caused an increase in investments both in manufacturing infrastructures and research.

In the late 1990s several companies succeeded in producing **low-temperature poly-Si (LTPS)**, requiring process temperatures below 600°C and compatible with low-cost glass. The main advantage of poly-Si is the higher carrier mobility, and hence current drive capability, that allows integration of some of the drive circuits directly on glass.

As already said, to obtain fast response and a wide gray scale, the AM drive exploited the TN mode. Liquid-crystal materials for use in the AM addressing scheme must exhibit a gradual T-V curve (this turns out into a low k_{33}/k_{11} ratio). They also need small threshold, low viscosity and high resistivity. The LCD industry first went back to PCH that had been used in the TN-mode segment scheme. However, its resistivity was not as high as that required by an AM drive. Then, **fluorinated materials** exhibiting high resistivity were developed at Merck in 1989 [PWR1990],[15] see Fig. 2.21.

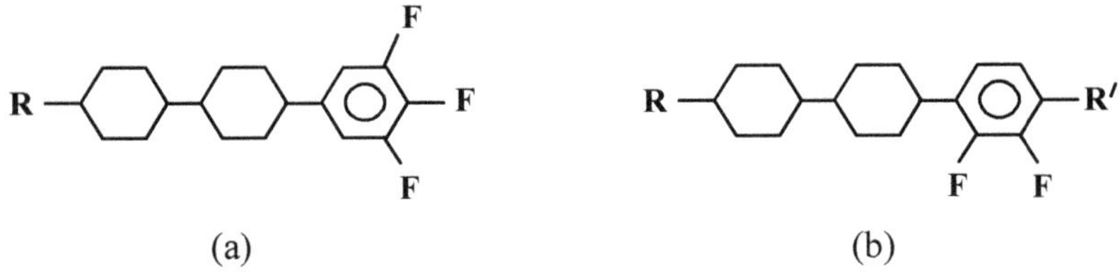

(a) (b)

Fig. 2.21 Examples of fluorinated compounds of *p* type (**a**) and *n* type (**b**)

[15]These compounds had high dielectric anisotropy due to the polarized C-F bond and contributed to make Merck (which settled laboratories in Japan, Taiwan, Korea and Hong Kong, and which purchased the patent portfolio from BBC and **Hoffmann-La Roche** in 1985 and 1996, respectively) one of the largest manufacturers of liquid crystals for displays in the world with a global market share of about 70% at 2008.

2.4.3 The First LCD for Television Set

In 1972, a group headed by **A.G. Fisher** from Westinghouse published the basic concept that a color TV would be possible by combining a red–green–blue color filter and an LC device acting as an optical shutter [FBE1972]. Later, in 1981, a group from Tohoku University, Japan, headed by **Tatsuo Uchida**, announced that a color display was possible using a technique that forms microcolor filters inside the LC cells [I2007].

In 1983, a group headed by **Shinji Morozumi** at Seiko Epson Corporation proposed a pocket color TV using microcolor filters and poly-silicon TFTs formed on a quartz substrate [MOY1983]. However, it had to be processed at high temperatures precluding the use of large and inexpensive glass substrates. As a result, following the announcement by the University of Dundee group, Japanese companies focused their attention on a-Si TFTs.

Among these, Sharp had long been working on a-Si for solar cells. With the end of the oil crisis, the engineers were shifted to the LCD area. Rather luckily in the context of our subject, Sharp was not previously involved with the production of CRTs. In 1986 Sharp introduced a 3-in. pocket TVs (240 × 384 pixel) [FTY1986], and it was planned to gradually develop 4-in., 5-in. and 6-in. displays.

In the same year, Isamu Washizuka (already cited in Section 2.1.4) became Division General Manager of the newly formed Liquid Crystal Division at Sharp. Washizuka was looking for a more impressive jump: a 14-in. display, as he knew that that was the minimum size for home-use televisions. Washizuka wanted only existing manufacturing equipments to be used. A 14-in. panel was subdivided into nine equal areas and the stepper as for a 3-in. panel was used nine times. The major problem was the length of the conduction lines that with the existing deign rules had statistically a low yield (due to breaks). Redundancy was exploited to face this problem. Ordinarily, one pixel is made up of three dots, each having a red, green, and blue filter. The Washizuka team divided one dot into four identical subdots that were driven by two source lines carrying identical signals and by two gate lines also carrying identical signals. Both the source and gate lines were driven from the two ends of the panel. As a result, the pixel had 12 subdots and the panel was made of 1284 × 960 subdots. By using four identical subdots, a defect in one subdot was not noticeable. After some refinements, including the design of a new LSI driving circuit, a 14-in. full-color full-motion display was demonstrated and the panel was shown to the president. On June 24, 1988, the news was publicly announced [S1988] and presented in an international conference [NOH1988]. The display was 27-mm thick, including a backlight, and its weight was 1.8 kg, respectively about 1/13 and 1/4 of a conventional CRT display.

At that time, there was another extremely important announcement. General Electric reported one mega-pixel AMLCD in 6.25 in. in 1988 [CP1988], which proved a-Si TFTs having a high potential for the implementation of high resolution displays.

In 1991 Sharp introduced the first wall-hanging television set, the "Liquid Crystal Museum" [S1991], in 1995 Sharp produced the first 28-in. AMLCD. In *The Wall Street Journal*, Heilmeier remarked [H1993]:

I think you need to give the credit to the people who persevered and worked on LCDs for 25 years. I don't spend too much time wringing my hands about it, but I have a lot of satisfaction knowing we had the same vision in the 1960s

In 2001 Samsung produced the first 42-in. AMLCD. In 2003 LCDs surpassed CRTs in terms of revenue for desktop monitors. In 2005 Samsung announced a 82-in. AMLCD panel [K2005] and the first 100-in. AMLCD for HDTV was demonstrated by LG Philips in 2006 [JKJ2007]. As a result of this rapid growth, the market of CRTs collapsed much more rapidly than it had been expected and many CRT TV manufacturers closed some of their plants. For instance, Sony eliminated its 17-in. and 19-in. CRT TV lines and launched its Bravia line of LCD TVs realized with screens from its joint venture with Samsung. Besides, these implementations eliminated the conventional wisdom that the Plasma Display technology was the only amenable for large TVs, whereas LCD was suitable for both small and medium TVs.

2.4.4 Improving the Viewing Angle

Displays for portable equipments are not conceived for group viewing, and therefore the limited view angle and gray inversion are not their main concerns. In contrast, image consistency over a wide angle of view is one of the most important properties required by TV applications. Considerable research was then carried out to improve the viewing angle of large size AM LCDs (e.g., for TV and CAD monitors) to 170° and beyond in all directions. At this purpose, optical compensation films for TN mode [W1995, W1996, CKJ1998], and new LC modes such as the In Plane Switching (IPS) by Hitachi [KWW1992], and the Vertical Alignment (VA) by Fujitsu and Merck [OKS1997], were developed.

The IPS mode has been found superior to the VA mode in true wide viewing angle, moving picture, response time, and color shift [K2004]. Excellent surveys on this subject can be found in [LZW2005, PT2005].

2.4.4.1 In Plane Switching

The basic concept was introduced in 1973 [S1973, S1974], then it was extended to display devices by a group headed by Günter Baur from the Fraunhofer Institute in Germany [KWW1992] and finally applied to AMLCDs by Hitachi [OK1995].

In a conventional TN or STN TFT cell, electrode pairs are placed one above the other on separate substrates. In the IPS cell, both electrodes (called interdigital electrodes) are placed parallel to each other on the bottom substrate, as illustrated in Fig. 2.22.

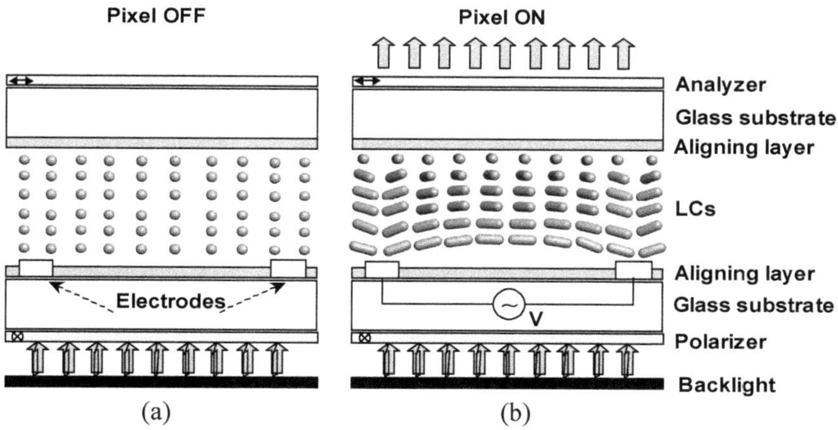

Fig. 2.22 In Plane Switching cell. (**a**) OFF state, (**b**) ON state. Cross-oriented polarizing filters are used. Electrodes are placed only in the back plane and align in the same plane the LC molecules

In the voltage-OFF state, the LC molecules are aligned homogeneously in the same plane and the incident light is blocked by the crossed analyzer, resulting in a normally black mode. The main difference with TN occurs when a voltage is applied: the LC molecules can rotate through 90° to align themselves with the field, while remaining parallel to the substrates. As a result, the birefringence offered by the LC molecules (and then the light transmission) has a reduced dependence on the viewing angle.

Though the contrast ratio offered by IPS is good, having two electrodes in each cell reduces the aperture ratio of the cell significantly (see Fig. 2.23). A brighter backlight is than needed which, causing higher power consumption, makes IPS unsuitable for battery-powered applications. However, power consumption is not the main concern for large screen displays. Consequently, the technology has been adopted, and further developed, by companies such as LG, Philips LCD and IPS Alpha (a joint venture of Hitachi,[16] Matsushita and Toshiba).

[16]The residual phase retardation difference at oblique incident angles causes also color shift. To reduce his problem in the IPS mode, Hitachi proposed the chevron or zigzag-shaped electrodes in a multidomain structure [AKO1997, MNS2000], which is often referred to as the Super-IPS mode.

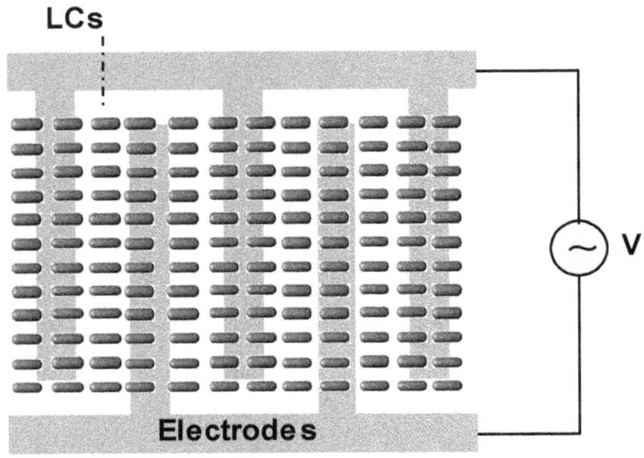

Fig. 2.23 In Plane switching electrodes scheme, top view. Typical electrode width is 4 μm and electrode gap is 8 μm

2.4.4.2 Vertical Alignment

In competition with the IPS mode of Hitachi, Fujitsu and Merck announced in 1997 the development of a new LC mode, referred to as Vertical Alignment (VA) [OKS1997]. As the name suggests, here the LC molecules are aligned homeotropically and under an electric field they rotate through 90° to lie parallel with the substrates, as illustrated in Fig. 2.24. This new mode produces a display with wide viewing angle (though intrinsically lower than IPS) and high contrast, with higher brightness and somewhat lower power consumption than IPS (but still high for battery-powered applications). Besides, this mode does not require rubbing. This is a unique advantage because the ion contamination and mechanical scratches introduced by the rubbing process are avoided. Unfortunately, the absence of pretilt makes the switching of the LC molecules into controlled directions difficult. To set a pre-orientation direction, one of the possible solutions consists in subdividing the pixel area into multi domains (at least two) that differ by having opposite pretilt angles. At this purpose, the multidomain VA mode (MVA) exploits protrusions [TKS1998] as illustrated in Fig. 2.25. It requires a relatively high precision in pixel alignment and has a reduced aperture ratio because of the opaque physical protrusions.

It should be noted that although IPS and MVA LCDs exhibit a relatively wide viewing angle, it is still inadequate for TV applications. To compensate the light leakage at oblique angles, optical phase compensation films are added [SKK1998]. For these modes, new generations of superfluorinated LCs were lately synthesized [BL2005], enabling extremely low switching times below 10 ms [B2004].

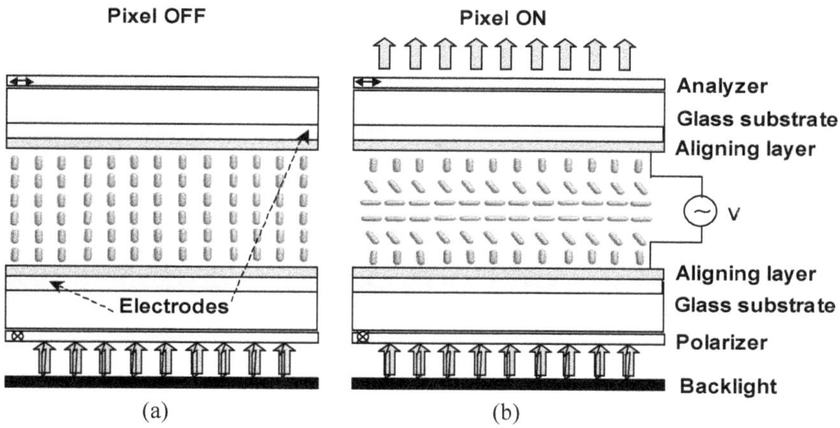

Fig. 2.24 Vertical Alignment cell. (**a**) OFF state, (**b**) ON state

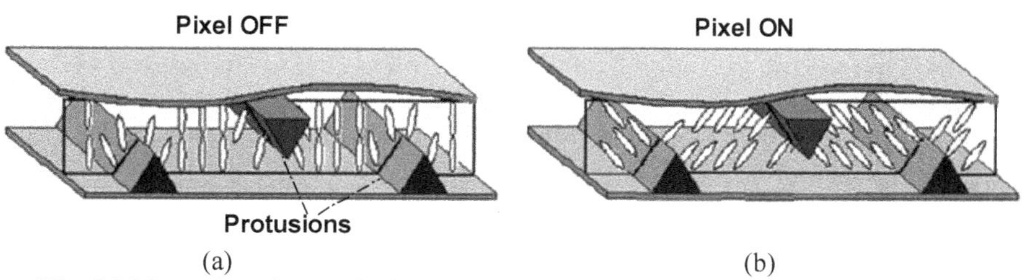

Fig. 2.25 Protusions in a Multidomain Vertical Alignment cell. (**a**) OFF state, (**b**) ON state [TKS1998]

2.5 Concluding Remarks

The advent of LCDs enabled the development of completely new portable applications thanks to the reduced thickness and power requirements. Pocket calculators, clocks and watches were the first commercial products, but other important more recent applications are digital cameras, camcorders and mobile phones. Microdisplays is another class of LCDs that have been developed. They are approximately defined as sub-1.5-in. displays, viewed indirectly either through magnifying optics near the eye (for one user) or through a projection system[17] (for one or more users). As the LCD price continues to lower and capabilities to improve, LCDs tend to invade different market segments.

[17]The first business-grade LCD front projectors appeared in the early 1990s, but now are common in many meeting rooms and even classrooms.

An introduction to the different types of flat panel displays and their main performance parameters can be found in the Appendix A and B, respectively.

In this chapter, we recapitulated the evolution of LC-based displays starting from the first attempts such as the guest-host mode and DSM, to TN, STN, and recent developments such as IPS and VA. Their first use in selected significant applications was discussed as well.

The development of the subject culminated into the most challenging effort, required to accomplish the original Sarnoff dream: the TV hanging on a wall. Compared to traditional cathode-ray tube displays, LCD ones offer several advantages: they are light weighting, non emissive, intrinsically flat, easier to read, and environmentally friendly, besides, they have a longer life, limited by the (usually exchangeable) backlights. Of course there are also several shortcomings, such as limited view angle, brightness, and contrast, and high manufacturing cost, but these points are becoming negligible as research continues. In general, the natural research evolution is from a conceived new mode of operation to the synthesis of appropriate materials (LC molecules and mixtures, compensating optical films, etc.), then prototypes, and finally commercial products.

As examples of high-performance applications, the overall requirements for cellular phone and notebook computer displays, desktop monitors, and large screen TVs are summarized in Table 2.2 [LZW2005]. Note that at the time of writing, nearly all notebooks and monitors up to 19 in. and most hand-held devices adopt Active Matrix TN mode, whereas multidomain technologies are dominant in screens larger than 20 in. Passive Matrix is adopted almost exclusively in low-cost portable equipments.

Table 2.2 Performance criteria for LCDs used in portable electronic products, computers and large-size TV sets

Performance	Portable Devices	Desktop Computers	TV
Weight	+	–	–
Thickness	+	–	–
Power	+	–	–
Resolution	=	=	–
Max. Brightness	=	=	+
Size	–	=	=
Color saturation	–	=	+
Contrast Ratio	–	+	+
View Angle	–	+	+
Response Time	–	+	+

+: highly required, =: medium required, –: less required.

To close this chapter we give an example of the principal components of an LCD *module*, with *display glass*, *driver* and often a *backlight.*[18] We show in Fig. 2.26 the pixel matrix, polarizers, backlight and metal chassis for an LCD module of a cellular phone. In this case the backlight is made up of a reflective panel lighted by six Light Emitting Diodes (LEDs) as detailed in Fig. 2.27. For larger displays, the backlight is implemented through one ore more fluorescent lamps, as summarized in note 18.

Fig. 2.26 A panel disassembled: (**a**) the pixel matrix, (**b**) and (**c**) the two polarizers detached, (**d**) the metal chassis and (**e**) the backlight

[18]In a transmissive LCD all of the light seen by the user comes from the backlight. The backlight unit is important for three main reasons: cost (around 40% of the total cost of large LCD modules), power consumption (around 75% of the total power) and picture quality (responsible for motion blur and color saturation). The dominating backlight technology exploits mercury-based Cold-Cathode Fluorescent Lamps (CCFLs). It offers a good power efficiency of 60–80 lm/W at 10 000 cd/m^2. The main drawbacks are the high power consumption and environmental side effect. The alternative LED backlight technology has a lower power efficiency and higher cost (but these drawbacks are continuously reduced by research). The intrinsic advantages are related to the longer life, lower operating voltage, color improvement and high speed, which allows to dynamically reducing the backlight according to image data (**Backlight Dimming**, see 5.6). White LED becklights have already established their dominance in cell phones and iPODs, [SKK2003, A2008].

Fig. 2.27 Backlight system made up of a reflective panel lighted by six LEDs

References

[A2008] M. Anandan, "Progress of LED backlights for LCDs," *Journal of the SID*, Vol. 16, No. 2, pp. 287–310, 2008.

[ADK1983] H. Amstutz, D. Heimgartner, M. Kaufmann, and T.J. Scheffer, "Liquid Crystal Display," *Foreign Application Priority Data July 12*, 1983; *U.S. Patent #4 634 229*, issued 1987.

[AKO1997] S. Aratani, H. Klausmann, M. Oh-e, M. Ohta, K. Ashizawa, K. Yanagawa, and K. Kondo, "Complete Suppression of Color Shift in In-Plane Switching Mode Liquid Crystal Displays with a Multidomain Structure Obtained by Unidirectional Rubbing," *Japanese Journal of Applied Physics*, Vol. 36, pp. L27–L29, 1997.

[AP1974] P.M. Alt and P. Pleshko, "Scanning Limitations of Liquid-Crystal Displays," *IEEE Transaction on Electron Devices*, Vol. 21, No. 2, pp. 146–155, 1974.

[B1973] D.W. Berreman, "Optics in Smoothly Varying Anisotropic Planar Structures: Application to Liquid-Crystal Twist Cells," *Journal of OSA*, Vol. 63, No. 11, pp. 1374–1380, 1973.

[B1984] T.P. Brody, "The Thin Film Transistor - A Late Flowering Bloom," *IEEE Transaction on Electron Devices*, Vol. 31, No. 11, pp. 1614–1628, 1984.

[B2004] W. Becker (editor), "New LC Mixture Developments," *Merck Liquid Crystal Newsletter*, No. 19, p. 6, 2004.

[BAD1973] T.P. Brody, J.A. Asars, and G.D. Dixon, "A 6×6 in 20 Lines-per-inch Liquid-Crystal Display Panel," *IEEE Transaction on Electron Devices*, Vol. 20, No. 11, pp. 995–1001, 1973.

[BK1996] A. Buka and L. Kramer (editors), *Pattern Formation in Liquid Crystals*, Springer, 1996.

[BL2005] B. Bremer and L. Lietzau, "1,1,6,7-Tetrafluoroindanes: Improved Liquid Crystal for LCD-TV Applications," *New Journal of Chemistry Articles,* Vol. 29, pp. 72–74, 2005.

[BZK1988] E. Bodenschatz, W. Zimmerrnan, and L. Kramer, "On Electrically Driven Pattern-Forming Instabilities in Planar Nematics," *Journal de Physique*, Vol. 49, p. 1875, 1988.

[C1969] E.F. Carr, "Influence of Electric Fields on the Molecular Alignment in the Liquid Crystal p-(Anisalamino)-Phenylacetate," *Molecular Crystals and Liquid Crystals*, Vol. 7, pp. 253–268, 1969.

[CKJ1998] J. Chen, K. Kim, J. Jyu, J. Souk, J. Kelly, and P. Bos, "Optimum Film Compensation Modes for TN and VA LCDs," *SID Digest*, Vol. 29, No. 1, pp. 315–318, 1998.

[CP1988] D.E. Castleberry and G.E. Possin, "A 1 Mega-Pixel Color a-Si TFT Liquid-Crystal Display," *SID Digest*, pp. 232–234, 1988.

[D1946] J.F. Dreyer, "Optical Device and Method and Manufacture Thereof," *U.S. Patent #2 400 877,* issued 1946.

[D1947] J.F. Dreyer, "Polarizer and Polarizing Lamp," *U.S. Patent #2 432 867,* issued 1947.

[D1949] J.F. Dreyer, "Optically Polarizing Dichroic Film and Method of Preparing the Same," *U.S. Patent #2 481 830*, issued 1949.

[D1950] J.F. Dreyer, "Flexible Noncrystalline Self-Contained Polarizing Films and Methods of Making and Using the Same," *U.S. Patent #2 524 286*, issued 1950.

[D1951] J.F. Dreyer, "Laminated Polarizer," *U.S. Patent #2 553 961*, issued 1951.

[D1952a] J.F. Dreyer, "Dial Indicator Having Polarizing Elements," *U.S. Patent #2 603 129*, issued 1952.

[D1952b] J.F. Dreyer, "Variable Light Transmission Device Comprising Relatively Movable Polarized Members," *U.S. Patent #2 617 329*, issued 1952.

[D1957] J.F. Dreyer, "Mirror Having a Dichroic Layer," *U.S. Patent #2 776 598*, issued 1957.

[DDG1971] E. Dubois-Violette, P.G. de Gennes, and O. Parodi, "Electrodynamique des Nématiques (Electrodynamics of Nematics)," *Journal de Physique*, Vol. 32, p. 305, 1971.

[DKS1973] H.J. Deutscher, F. Kuschel, H. Schubert, and D. Demus, "Nematische Flüssig-Kristalline Substanzen (Nematic Liquid-Crystalline Substances)," *German Patent #105 701*, issued 1974.

[DWD1971] H.J. Deutscher, W. Weissflog, D. Demus, and G. Petzl, "Kristallin-Flüssige Nematische Verbindungen (Nematic Liquid Crystal Compounds)," *German Patent #86 269*, issued 1971.

[EEK1977] V.R. Eidenschink, D. Erdman, J. Krause, and L. Pohl, "Substituted Phenylcyclohexanes – A New Class of Liquid-Crystalline Compounds," *Angewandte Chemie, International Edition*, Vol. 16, No. 2, p. 100, 1977.

[F1964] J.L. Fergason, "Liquid Crystals," *Scientific American,* Vol. 211, No. 2, pp. 76–86, 1964.

[FBE1972] A.G. Fisher, T.P. Brody, and W.S. Escott, "Design of a Liquid Crystal Color TV Panel," *Proceedings of IEEE Conference on Display Devices*, pp. 64–66, 1972.

[FTY1986] F. Funada, Y. Takafuji, K. Yano, H. Take, and M. Matsuura, "An Amorphous-Si TFT Addressed 3.2-in. Full-Color LCD," *SID Digest*, pp. 293–295, 1986.

[GC1970] J.E. Goldmacher and J.A. Castellano, "Electro-Optical Compositions and Devices," *U.S. Patent #3 540 796*, issued 1970.

[GHN1973] G.W. Gray, K.J. Harrison, and J.A. Nash, "New Family of Nematic Liquid Crystals for Displays," *Electronics Letters*, Vol. 9, No. 6, pp. 130–131, 1973.

[GT1974] C.H. Gooch and H.A. Tarry, "Optical Characteristics of Twisted Nematic Liquid-Crystal Films," *Electronics Letters*, Vol. 10, No. 1, pp. 2–4, 1974.

[GT1975] C.H. Gooch and H.A. Tarry, "The Optical Properties of Twisted Nematic Liquid Crystal Structures with Twist Angle Equal or Less than 90°," *Journal of Physics D: Applied Physics*, Vol. 8, pp. 1575–1584, 1975.

[H1969] W. Helfrich, "Conduction-Induced Alignment of Nematic Liquid Crystals: Basic Model and Stability Considerations," *Journal of Chemical Physics*, Vol. 51, No. 9, p. 4092, 1969.

[H1976] G.H. Heilmeier, "Liquid Crystal Displays: An Experiment in Interdisciplinary Research that Worked," *IEEE Transactions on Electron Devices*, Vol. 23, No. 7, pp.780–785, 1976.

[H1993] G.H. Heilmeier, "George Heilmeier. The Liquid Crystal Display," *Wall Street Journal*, 1993.

[H2002] L. Hamilton, "Liquid Crystals: How a Useless Lab Anomaly Grew Up to Become an Indispensable Part of Hundreds of Everyday Devices," *American Heritage of Invention and Technology*, Vol. 17, No.4, pp. 20–29, 2002.

[HNM1986] S. Hotta, S. Nagata, Y. Miyata, K. Yokoyama, K. Adachi, T. Chikamura, M. Yoshiyama, A. Nishkawa, and K. Kawasaki, "Full-Color Multi-Gap LC-TV Display Panel Addressed by a-Si TFTs," *SID Digest*, pp. 296–297, 1986.

[HRH1974] D. Hume, P. Raynes, and K.J. Harrison, "Eutectic Mixtures of Nematic 4′-Substituted 4-Cyanobiphenyls," *Journal of the Chemical Society, Chemical Communications*, pp. 98–99, 1974.

[HS1972] W. Helfrich and M. Schadt, "Optical Device," *Swiss Patent #532 261*, issued 1972.

[HZ1968] G.H. Heilmeier and L.A. Zanoni, "Guest-Host Interactions in Nematic Liquid Crystals. A New Electro-Optic Effect," *Applied Physics Letters*, Vol. 13, No. 3, pp. 91–92, 1968.

[HZB1968] G.H. Heilmeier, L.A. Zanoni, and L.A. Barton, "Dynamic Scattering: a New Electrooptic Effect in Certain Classes of Nematic Liquid Crystals," *Proceedings of IEEE*, Vol. 56, No. 7, pp. 1162–1171, 1968.

[I2007] Y. Ishii, "The World of Liquid-Crystal Display TVs-Past, Present, and Future," *Journal of Display Technology*, Vol. 3, No. 4, pp. 351–360, 2007.

[J1999] B. Johnstone, *We were burning: Japanese Entrepreneurs and the Forging of the Electronic Age*, Cornelia & Michel Bessie Series, Perseus Books Group, 1999.

[JKJ2007] H.C. Jin, I.B. Kang, E.S. Jang, H.M. Moon, C.H. Oh, S.H. Lee, and S.D. Yeo, "Development of 100-in. TFT-LCDs for HDTV and Public-Information-Display Applications," *Journal of the SID 15/5*, pp. 277–280, 2007.

[K1976] H. Kawakami, "Method of Driving Liquid Crystal Matrix Display Device," *U.S. Patent #3 976 362*, issued 1976.

[K2002] H. Kawamoto, "The History of Liquid-Crystal Displays," *Proceedings of IEEE*, Vol. 90, No. 4, pp. 460–500, 2002.

[K2004] W. Kim, "Technology Overview: LCDs for TV Application," *Journal of the SID*, Vol. 12, No. 4, pp. 449–453, 2004.

[K2005] S.S. Kim, "The World's Largest (82-in.) TFT-LCD," *SID Digest*, Vol. 36, No. 1, pp. 1842–1847, 2005.

[KS1969] H. Kelker and B. Scheurle, "A Liquid-Crystalline (Nematic) Phase with a Particularly Low Solidification Point," *Angewandte Chemie, International Edition*, Vol. 8, No. 11, pp. 884–885, 1969.

[KSO1988] H. Koh, K. Sawada, M. Ohgawara, T. Kuwata, H. Tsubota, M. Akatsuka, and K. Matsuhiro, "A 960×240 Pixel Multicolor Supertwisted-Nematic Display," *SID Digest*, pp. 53–56, 1988.

[KSY1988] N. Kimura, T. Shinomiya, K. Yamamoto, Y. Ichimura, K. Nakagawa, Y. Ishii, and M. Matsuura, "Multicolor Display by Double-Layered Supertwist-Nematic LCD," *SID Digest*, pp. 49–52, 1988.

[KWW1992] R. Kiefer, B. Webber, F. Windscheid, and G. Baur, "In-Plane Switching of Nematic Liquid Crystals," *Proceedings of Japan Displays '92*, pp. 547–550, 1992.

[LL1936] B. Levin and N. Levin, "Improvements in or Relating to Light Valves," *British Patent #441 274*, issued 1936.

[LMN1971] B.J. Lechner, F.J. Marlowe, E.O. Nester, and J. Tults, "Liquid Crystal Matrix Displays," *Proceedings of IEEE*, Vol. 59, No. 11, pp. 1566–1579, 1971.

[LSG1979] P.G. Le Comber, W.E. Spear, and A. Ghaith, "Amorphous-Silicon Field-Effect Device and Possible Application," *Electronics Letters*, Vol. 15, No. 6, pp. 179–181, 1979.

[LZW2005] R. Lu, X. Zhu, S.T. Wu, Q. Hong, and T.X. Wu, "Ultrawide-View Liquid Crystal Displays," *IEEE/OSA Journal of Display Technology*, Vol. 1, No. 1, pp. 3–14, 2005.

[M1911a] C. Mauguin, "Sur les Cristaux Liquides de Lehmann (On the Liquid Crystals of Lehmann)," *Bulletin de la Société Française de Minéralogie*, Vol. 34, pp. 71–117, 1911.

[M1911b] C. Mauguin, "Orientation des Cristaux Liquides par le Champ Magnétique (Orientation of Liquid Crystals by a Magnetic Field)," *Comptes rendus de l'Académie des Sciences*, Vol. 152, pp. 1680–1683, 1911.

[MNS2000] Y. Mishima, T. Nakayama, N. Suzuki, M. Ohta, S. Endoh, Y. Iwakabe, and H. Kagawa, "Development of a 19-in.-Diagonal UXGA Super TFT-LCM Applied with Super-IPS Technology," *SID Digest*, Vol. 31, No. 1, pp. 260–263, 2000.

[MOY1983] S. Morozumi, K. Oguchi, S. Yazawa, T. Kodaira, H. Ohshima, and T. Mano, "B/W and Color LC Video Displays Addressed by Poly Si TFT's," *SID Digest*, pp. 156–159, 1983.

[MYN1983] M. Matsuura, Y. Takafuji, K. Nonomura, F. Funada, and T. Wada, "A Liquid Crystal Display Panel Using a Te-TFT Array," *SID Digest*, pp. 148–149, 1983.

[NOH1988] T. Nagayasu, T. Oketani, T. Hirobe, H. Kato, S. Mizushima, H. Take, K. Yano, M. Hijikigawa, and I. Washizuka, "A 14-in-Diagonal Fullcolor a-Si TFT LCD," *Proceedings of International Display Research Conference*, pp. 56–58, 1988.

[OK1995] M. Oh-e and K. Kondo, "Electro-Optical Characteristics and Switching Behavior of the In-Plane Switching Mode," *Applied Physics Letters*, Vol. 67, No. 26, pp. 3895–3897, 1995.

[OKS1997] K. Ohmuro, S. Kataoka, T. Sasaki, and Y. Koike, "Development of Super High-Image-Quality Vertical-Aligned Mode LCD," *SID Digest*, Vol. 28, p. 845, 1997.

[ONO1982] Y. Okubo, T. Nakagiri, Y. Osada, M. Sugata, N. Kitahara, and K. Hatanaka, "Large-Scale LCDs Addressed by a-Si TFT Array," *SID Digest*, pp. 40–41, 1982.

[ONW1987] O. Okumura, M. Nagata, and H. Wada, "Neutralized STN (NTN)LCD Full Color Picture Image," *Institute of Image Information and Television Engineers*, Vol. 11, No. 27, pp. 79–84, 1987.

[PEK1977] L. Pohl, R. Eidenschink, G. Krause, and D. Erdmann, "Physical Properties of Nematic Phenylcyclohexanes, a New Class of Low Melting Liquid Crystals with Positive Dielectric Anisotropy," *Physics Letters A*, Vol. 60, No.5, pp. 421–423, 1977.

[PF1972] P.A. Penz and G.W. Ford, "Electromagnetic Hydrodynamics of Liquid Crystals," *Physical Review A*, Vol. 6, No. 1, pp. 414–425, 1972.

[PT2005] D. Pauluth and K. Tarumi, "Optimization of Liquid Crystals for Television," *Journal of SID*, Vol. 13, No. 8, pp. 693–702, 2005.

[PWR1990] H.J. Plach, G.Weber, and B. Rieger, "Liquid Crystal Mixtures for Active Matrix Displays Using New Terminally Fluorinated Compounds," *SID Digest*, pp. 91–94, 1990.

[RCA1968] "RCA Announces Breakthrough in Liquid Crystal Field; Demonstrates Thin-Screen Displays of Print, Pictures, Moving Images," *RCA Press Conference*, 1968.

[S1973] R.A. Soref, "Transverse Field Effect in Nematic Liquid Crystals," *Applied Physics Letters*, Vol. 22, pp. 165–166, 1973.

[S1974] R.A. Soref, "Field Effect in Nematic Liquid Crystals Obtained with Interdigital Electrodes," *Journal of Applied Physics*, Vol. 45, No. 12, pp. 5466–5468, 1974.

[S1979] A. Sasaki, *Liquid Crystal Bases and Applications*, Ohm Inc., 1979.

[S1982] "Succeeded in The Development of a Television Wrist Watch," *Seiko Corporation News Release*, 1982.

[S1983] T.J. Scheffer, "Liquid Crystal Display with High Multiplex Rate and Wide Viewing Angle," *Proceedings of International Display Resolution Conference*, pp. 400–403, 1983.

[S1987] "Double Super Twist Nematic LCD. High Quality Real Black-and-White LCD Unit Suitable for Word Processor and PC Applications," *Sharp Corporation News Release*, 1987.

[S1988] "Sharp has Developed a 14″ TFT Color LCD Unit," *Sharp Corporation News Release*, 1988.

[S1991] "Realizing a Dream of Television Hanging on the Wall 8.6 Inch Liquid Crystal Museum," *Sharp Corporation News Release*, 1991.

[SG1982] M. Schadt and P.R. Gerber, "Class Specific Physical Properties of Liquid Crystals and Correlations with Molecular Structure and Static Electrooptical Performance in Twisted Cells," *Z. Naturforsch*, Vol. 37a, pp. 165–178, 1982.

[SH1971] M. Schadt and W. Helfrich, "Voltage-Dependent Optical Activity of a Twisted Nematic Liquid Crystal," *Applied Physics Letters*, Vol. 18, pp. 127–128, 1971.

[SHO1973] Y. Suzuki, M. Hirasawa, and K. Odagawa, "Clocked CMOS Calculator Circuitry," *International Solid State Circuit Conference Digest*, Vol. 16, pp. 58–59, 1973.

[SKK1998] Y. Saitoh, S. Kimura, K. Kusafuka, and H. Shimizu, "Optimum Film Compensation of Viewing Angle of Contrast in In-Plane-Switching-Mode Liquid Crystal Display," *Japanese Journal of Applied Physics*, Vol. 37, No. 9a, pp. 4822–4828, 1998.

[SKK2003] H. Sugiura, H. Kaneko, S. Kagawa, M. Ozawa, K. Niki, and H. Tanizoe, "Prototype of a Wide Gamut Monitor Adopting LED-Backlighting," *SID Digest*, Vol. 34, No. 1, pp. 1266–1269, 2003.

[SMS1981] A.I. Snell, K.D. Mackenzie, W.E. Spear, P.G. Le Comber, and A.J. Hughes, "Application of Amorphous Silicon Field Effect Transistors in Addressable Liquid Crystal Display Panels," *Applied Physics*, Vol. 24, No. 4, pp. 357–362, 1981.

[SN1984] T.J. Scheffer and J. Nehring, "A New Highly Multiplexable Liquid Crystal Display," *Applied Physics Letters*, Vol. 48, No. 10, pp. 1021–1023, 1984.

[SNK1985] T.J. Scheffer, J. Nehring, M. Kaufmann, H. Amstutz, D. Heimgartner, and P. Eglin, "24×80 Character LCD Panel Using the Supertwisted Birefringence Effect," *SID Digest*, pp. 120–123, 1985.

[SPG1985] M. Schadt, M. Petrzilka, P.R. Gerber, and A. Villinger, "Polar Alkenyls: Physical Properties and Correlations with Molecular Structure of New Nematic Liquid Crystals," *Molecular Crystals and Liquid Crystals*, Vol. 122, pp. 241–260, 1985.

[TMB2000] S.J. Tavener, T. Mullin, G.I. Blake, and K.A. Cliffe, "Numerical bifurcation study of electrohydrodynamic convection in nematic liquid crystals," *Physical Review E*, Vol. 63, 011708, pp. 1–12, 2000.

[TKS1998] A. Takeda, S. Kataoka, T. Sasaki, H. Chida, H. Tsuda, K. Ohmuro, T. Sasabayashi, Y. Koike, and K. Okamoto, "A Super-High Image Quality Multi-Domain Vertical Alignment LCD by New Rubbing-Less Technology," *SID Digest*, Vol. 29, No. 1, pp. 1077–1100, 1998.

[W1963a] R. Williams, "Liquid Crystals in an Electric Field," *Nature*, Vol. 199, pp. 273–274, 1963.

[W1963b] R. Williams, "Domains in Liquid Crystals," *Journal of Chemical Physics*, Vol. 39, No. 2, pp. 384–388, 1963.

[W1995] S.T. Wu, "Phase-Matched Biaxial Compensation Film for LCD's," *SID Digest*, Vol. 26, pp. 555–558, 1995.

[W1996] S.T. Wu, "Design and Fabrication of Phase-Matched Compensation Films for LCDs," *SID Digest*, Vol. 27, pp. 21–24, 1996.

[WOW1988] H. Watanabe, O. Okumura, H. Wada, A. Ito, M. Yazaki, M. Nagata, H. Takeshita, and S. Morozumi, "Full-Color LCD with Neutralized STN (NSTN)," *SID Digest*, pp. 416–419, 1988.

[YKM1982] T. Yamazaki, Y. Kawahara, S. Motte, H. Kamamori, and J. Nakamura, "A Liquid Crystal TV Display Panel with Drivers," *SID Digest*, pp. 48–49, 1982.

Chapter 3

Passive LCDs and Their Addressing Techniques

Passive LCDs can be divided into two main groups: **segmented LCDs** and **passive-matrix LCDs (PMLCDs)**. The former are today used in monochromatic applications like simple calculators, digital watches and simple instrumentation and climate control equipments. They are not suitable for high-information-content and graphical applications because of their limited resolution. High-performance calculators, organizers, notebooks, portable computers, projectors and LCD-TVs, exploit almost invariantly active-matrix displays. Nonetheless, PMLCDs have still a market in applications that meet the following criteria: simple fabrication process, limited resolution, dimensions and price, 1 or 2 colors, and limited tolerable viewing angle. For instance, they are used in mobile telephones, Personal Digital Assistants (PDAs), word processors, automotive applications, logic analyzers, oscilloscopes, electrocardiographs, etc.

The main target of this chapter is to describe techniques that are amenable to address passive LCDs. **Addressing** implies the conversion of the information to be displayed into sequential voltage pulses in order to individually switch on (or switch off) the display elements (segments or pixels). There are three main methods by which this can be achieved: **direct addressing**, **passive-matrix addressing**, and **active-matrix addressing**. All these techniques are closely related to the type of display to be driven. In this chapter we will discuss the first two approaches, while the related driver circuit schemes and implementation issues will be treated in Chapter 4. The active-matrix addressing techniques and correlated drivers will be treated in Chapters 5 and 6, respectively.

3.1 Seven-Segment Displays and Direct Addressing

The first alphanumeric displays had seven segment electrodes and exploited direct addressing (also referred to as static drive) schemes. A seven-segment display is made up of 2 substrates of glass, each with ITO (transparent) electrodes patterned on their surface. The substrates are sandwiched together leaving a small gap between them which is filled with LC material. The electrode arrangement is illustrated in Fig. 3.1. Each segment is formed by two ITO electrodes placed over two glass plates

D.J.R. Cristaldi et al., *Liquid Crystal Display Drivers*, DOI 10.1007/978-90-481-2255-4_3,

(front and rear substrates) filled with the LC mixture. In the simplest case, the rear electrode is common to all segments to form a single back plane which is taken out as a single connection. Electrodes on the other glass plate are taken out independently in order to connect them externally. The intersection of top and bottom electrodes forms a segment. In general, a display with n segments needs $n+1$ connections and n drivers to electrically drive each segment.

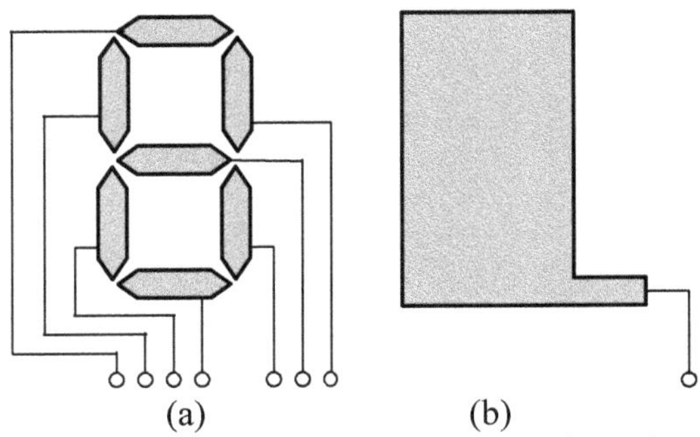

Fig. 3.1 Electrodes of a seven-segment LCD: (**a**) upper electrodes, (**b**) common electrode

To avoid ionization of the LC molecules, process that reduces the LC display life, the waveform across the segment must be dc free. Therefore, to turn on the segment a pulse driving voltage is used.

Figure 3.2 shows a possible voltage sequence applied to one of the segments. A square waveform with 50% duty cycle is applied to the backplane. The refresh frequency, $1/T$, is chosen sufficiently low to contain power consumption but high enough to avoid flickering.

A square voltage with the same characteristics is also applied to the upper electrode. Of course, the voltage across the segment (the last waveforms in Fig. 3.2) is the difference between the voltages applied to the upper and back plane electrodes. When these waveforms are out of phase a net voltage is applied to the segment. If this voltage is greater than V_{SAT}, then the segment is driven into the ON state (Fig. 3.2a). Usually, the voltage across the ON segments is chosen to be about three times the threshold voltage (V_{TH}). A segment is turned OFF only when the waveforms applied to the segment electrode and the back-plane are in phase (Fig. 3.2b).

A simple driving circuit suitable for a seven-segment LCD is depicted in Fig. 3.3. It exploits seven XOR gates (whose truth table is shown in the same figure) to generate the voltage applied to the segment, according to the input bit and clock signal.

As long as the display is limited to numerals, it is possible to achieve a reasonably high image contrast even with the gradual T/V curve offered by the TN method.

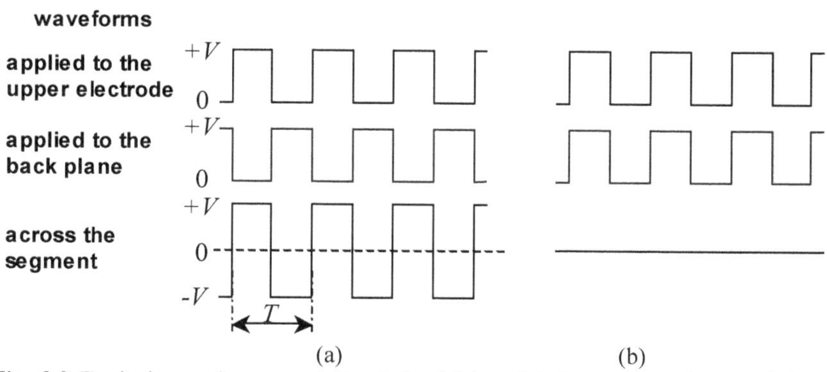

Fig. 3.2 Typical waveforms used for static driving: (**a**) the segment is turned ON, (**b**) the segment is turned OFF

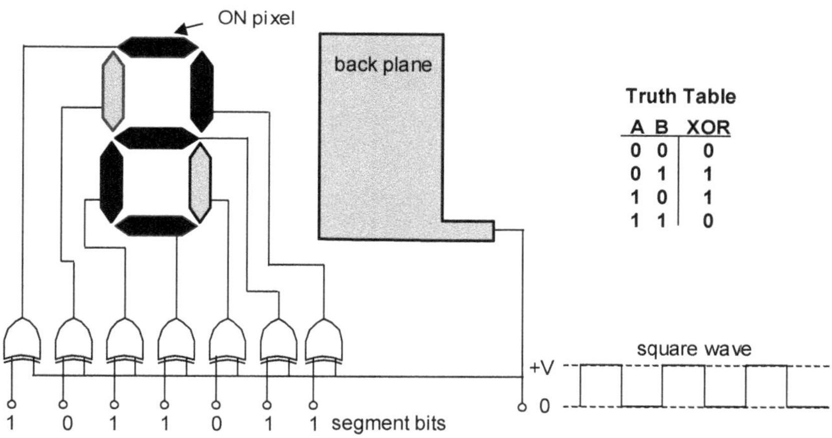

Fig. 3.3 Simple driving circuit for a seven-segment LCD

It is worth mentioning that power consumption can be reduced by modifying the waveforms applied to the ON segments. By adding a delay in the waveform that is applied to the ON segments with respect to the waveform applied to the backplane (Fig. 3.4), the instantaneous voltage across the ON segment is zero for a time duration which corresponds to the delay. The ON segment terminals are hence effectively shorted before reversing the driving polarity each half cycle and the ON segment is discharged without drawing any current from the power supply. About 40% reduction in power consumption has been achieved using this technique [M1982]. Of course, the segment voltage applied to an OFF pixel must not be delayed; otherwise the power-saving advantage will be lost.

The need for displays with higher information content requires an increased number of segments to which an increased number of leads must be connected. Therefore, more complex addressing approaches are mandatory due to the high cost of using many drivers and the absence of sufficient area devoted to the electrical contacts. To alleviate the problem, the driving method was changed to a multiplexed one.

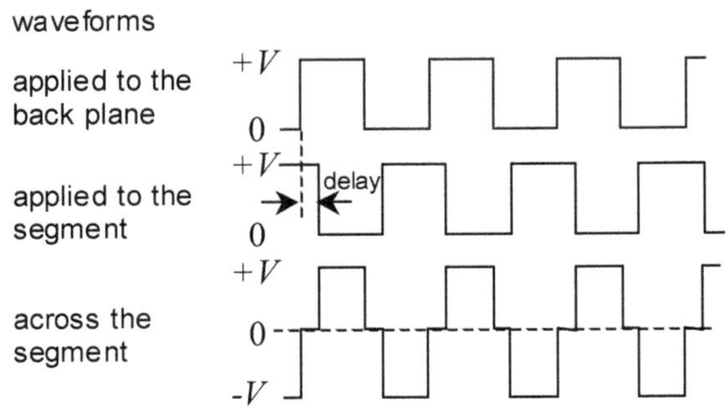

Fig. 3.4 Waveforms used to reduce power consumption in direct addressing

Figure 3.5 shows such **multiplexed (or duty-cycle) driving** approach. The bottom electrode is now divided into two parts and, thus, it is driven at a 1/2 duty cycle. On the other side, couples of upper electrodes are joined together. As a result, the number of leads has been reduced from eight for the static driving to six for the 1/2 duty-cycle drive.

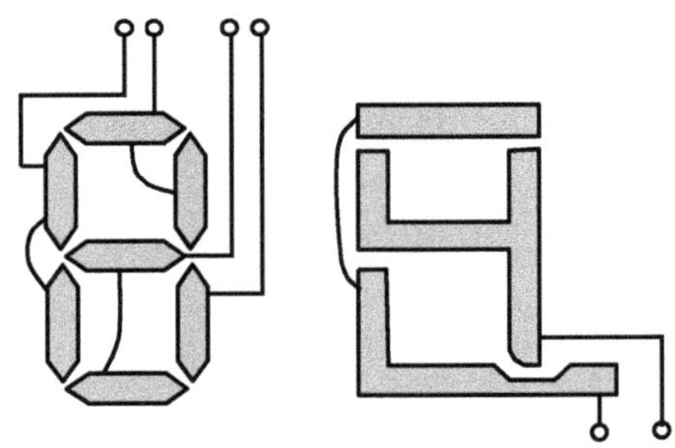

Fig. 3.5 Segment scheme for 1/2 duty-cycle drive: (**a**) upper and (**b**) lower electrodes

3.2 Addressing of Passive-Matrix LCDs

In passive-matrix displays, segments are replaced by **pixels** arranged in a matrix with **row** and **column electrodes**, as shown in Fig. 3.6. Similarly to a seven-segment display, a passive-matrix LCD is made up of 2 substrates of glass filled with LC material and with ITO electrodes patterned on their surface. One substrate's electrodes are patterned into N rows and the other substrate's electrodes are patterned into M columns, resulting in a matrix with $N \times M$ addressable intersections called pixels. This arrangement reduces the total number of connections from $N \times M + 1$ (when the pixels are addressed individually) to $N + M$. As a result, the duty-cycle segment scheme ultimately evolved into a passive-matrix addressing scheme.

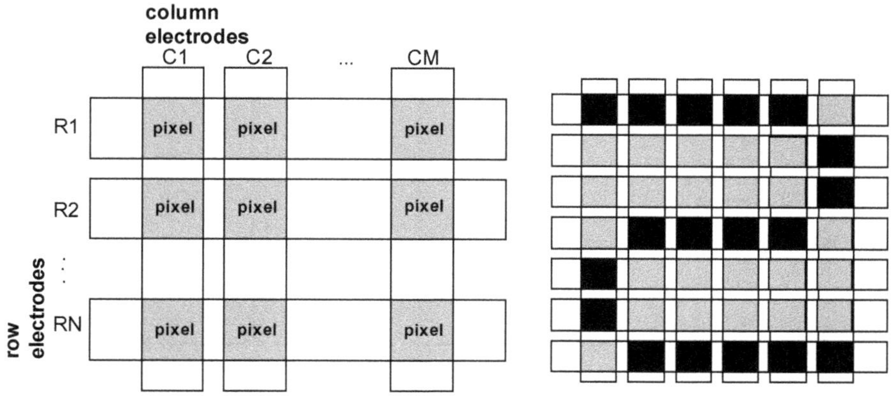

(a) (b)

Fig. 3.6 Passive matrix LCD: (**a**) schematic representation of the pixels made up of orthogonal rows and columns of electrodes, (**b**) example of a displayed numeral

Integrated circuits called **drivers** are attached to the periphery of the display to both the columns and the rows allowing for pulses of voltage to be sent to the pixels. At this purpose, a set of bonding pads is fabricated on each end of the row and column signal bus-lines to attach LCD Driver IC chips.

In its simplest version, each row (representing the common electrode) is sequentially scanned by a select pulse at a frame time period T, each pulse being of duration T/N. The data (alternatively called column, signal or video) voltages are applied in parallel for all the columns. Similarly to the direct addressing, the pixel is turned on provided that the associated row and column signals are out of phase. This is schematically shown in Fig. 3.7.

Usually, the scanning frequency $1/T$ is chosen between 60 and 90 Hz (60, 72, 75, 85 Hz are common, though rates up to 200 Hz may be supported). Some attention is paid to prevent beating with the grid frequency (50 Hz or 60 Hz, according to the standard adopted) causing in turn flicker.

It is seen that the **select period** is T/N and the non-select period is $(N\text{-}1)T/N$. As a consequence, during $1/N$ of the frame, there is an electrical field that orients the LC molecules, but during the longer remaining frame fraction, the LC molecules are free to relax to their original orientation.

The phenomenon of the response of the liquid crystal material to signals within a frame time (often called **frame response**) is illustrated in Fig. 3.8, where the light transmission versus time is depicted. The phenomenon is more evident at high temperatures. At low temperatures the LC viscosity does not allow the LC molecules to change appreciably their position. At high temperatures the LC relaxation process is faster and there is risk of flickering, as the average transmission is reduced. Observe that the light transmission perceived by the human eye is the average over the whole frame period.

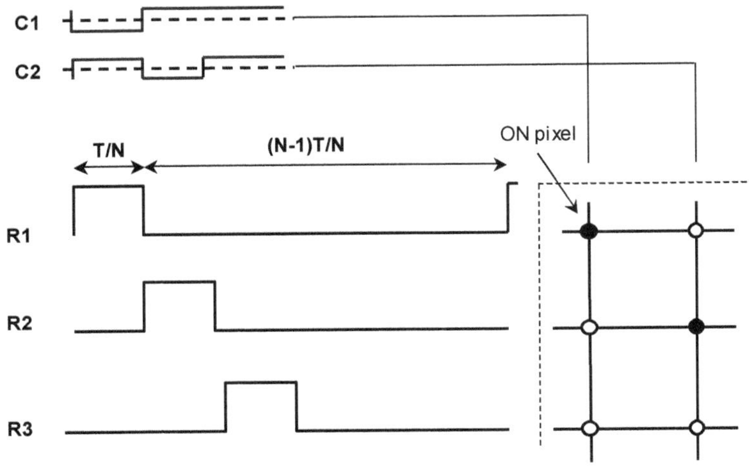

Fig. 3.7 Simple row and column waveforms for a passive matrix LCD

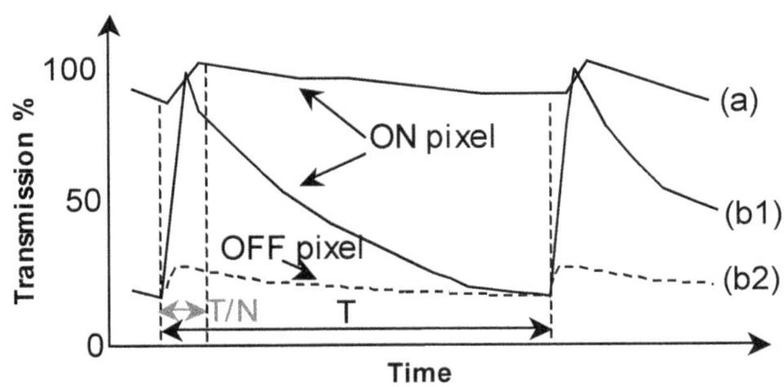

Fig. 3.8 Pixel transmission of a passive matrix under single line addressing. Curve (a) refers to a slow case (low temperature), curves (b1–b2) refer to a fast case (high temperature)

In conclusion, the state of the pixel does not change abruptly after removing (reverting) the applied voltage, due to the slow response of the LCD (in the range of several tens of milliseconds). With the adopted scanning frequencies, the period of the addressing waveforms is smaller than the response times. Thus, the state of the pixel depends on the RMS voltage across the pixel rather than the instantaneous voltage. This means also that the instantaneous voltage across an OFF pixel can exceed the threshold voltage (V_{TH}) of the LCD as long as the RMS voltage across the pixel is below V_{TH}.

Based on this consideration, addressing techniques for modern passive-matrix (STN) LCDs exploit the pixel RMS response. There are several approaches, which differ basically on the method adopted to scan the rows. Namely, they are **Single Line Addressing (SLA)**, **Multiple Line Addressing (MLA)**, **Active Addressing (AA)** and **Distributed MLA (DMLA)**. In the following subsections we will analyze these techniques.

3.2.1 Single Line Addressing

The two principal techniques based on selecting one line at a time and exploiting the pixel RMS response are called **Alt and Pleshko (A&P)** and **Improved Alt and Pleshko (IA&P)**.

The basic A&P version is illustrated in Fig. 3.9, for the case of an ON pixel. Assume as usual that the frame time period is T and the number of rows is N. Within the first frame, the voltage associated to the specified row is set to its maximum value, $+V_R$, only for a fraction T/N. In this interval a voltage with opposite phase, $-V_C$, is applied to the column (in order to activate the correlated pixel). As already said, all the pixels of the same row are driven in parallel. The row and column voltages are inverted in the next frame (**frame inversion**), to obtain an almost dc free average pixel voltage. Observe that the row voltage goes to zero when the considered row is not selected. In contrast, the column voltage is in phase (OFF pixel) or in opposite phase (ON pixel) to the row voltage, but it is never zero. As we shall shortly see, this particular choice makes the square voltage (proportional to the energy) transferred to the i-th pixel irrespective of the particular configuration of the other pixels of the same column. Indeed, we remember that the column voltage is applied to all the pixels of the column and not only to the selected pixel; for this reason pixels receive a small amount of energy even during the unselected time frame.

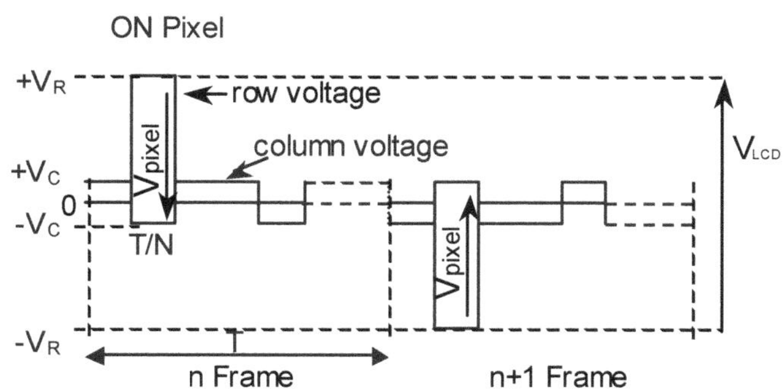

Fig. 3.9 Row and column voltages used to switch ON a pixel in the case of A&P technique

The **maximum driving voltage**, usually called V_{LCD} is equal to $2V_R$, provided that $V_R \geq V_C$, as is for hypothesis the case

$$V_{LCD,A\&P} = 2V_R \tag{3.1}$$

This value can be reduced (almost halved) by adopting the IA&P technique [KNK1976], whose waveforms are illustrated in Fig. 3.10.

The voltages in the first frame (i.e, when the polarity of the row voltage in A&P is positive) are the same as in the A&P approach. In the subsequent frame (i.e, when the polarity of the row voltage in A&P is negative) both row and column IA&P voltages are obtained through a shift of a quantity $V_R - V_C$ on both row and column voltages. To avoid the use of negative voltages that otherwise would require a triple-well IC technology, the voltage levels in Fig. 3.10 must be shifted by $+V_C$. Alternatively, starting from the standard A&P one can shift all the row and column voltages by $+V_C$ in the first frame and by $+V_R$ in the second frame. These shifts in the addressing waveforms do not alter the RMS voltage across the pixels. Again, we have no DC component. However, in this case V_{LCD} decreases to

$$V_{LCD,IA\&P} = V_R + V_C \tag{3.2}$$

Remembering the typical T/V curve, redrawn in Fig. 3.11 for convenience, we observe that to obtain an acceptable contrast ratio, the ON voltage applied to the pixel must be greater than the saturation voltage. And conversely, the OFF voltage must be lower than the threshold voltage. In summary, we get $V_{OFF} \leq V_{TH}$ and $V_{ON} \geq V_{SAT}$.

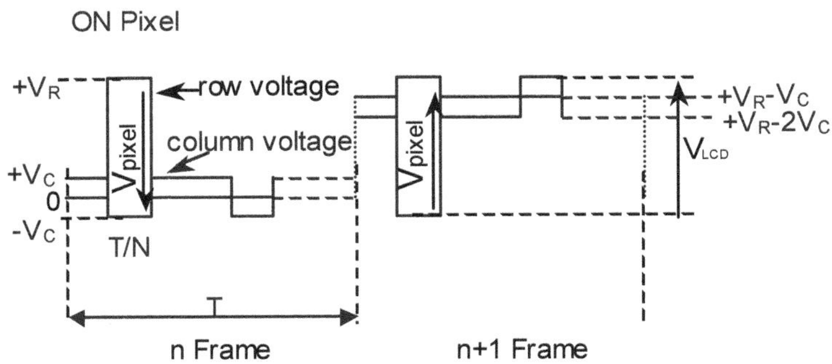

Fig. 3.10 Row and column voltages used to switch ON a pixel in the case of IA&P technique

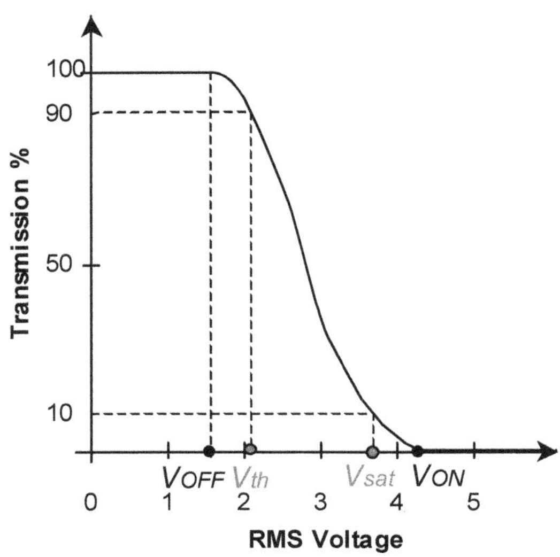

Fig. 3.11 Transmission as a function of the RMS applied voltage with superimposed V_{ON} and V_{OFF} values

When the LC responds to an RMS voltage, we must evaluate V_{pixel}, V_{ON} and V_{OFF} as RMS values within a frame. If N is the number of rows, and $V_{pixel,i}$ is the instantaneous pixel voltage during the addressing subperiod of the i-th row, we can write

$$\overline{V}_{pixel} = \sqrt{\frac{1}{N}\sum_{i=1}^{N} V_{pixel,i}^2} \tag{3.3}$$

We observe that in the case of ON pixel, the applied voltage is $|V_R+V_C|$ for the selection period whereas it is V_C for the remaining frame time. Similarly, in the case of OFF pixel, the applied voltage is $|V_R-V_C|$ for the selection period and it is V_C for the remaining frame time. Therefore, the expressions of the ON and OFF RMS voltages are, respectively

$$\overline{V}_{ON} = \sqrt{\frac{1}{N}\left[(V_R+V_C)^2+(N-1)V_C^2\right]} = \sqrt{\frac{1}{N}\left(V_R^2+2V_RV_C+NV_C^2\right)} \tag{3.4a}$$

$$\overline{V}_{OFF} = \sqrt{\frac{1}{N}\left[(V_R-V_C)^2+(N-1)V_C^2\right]} = \sqrt{\frac{1}{N}\left(V_R^2-2V_RV_C+NV_C^2\right)} \tag{3.4b}$$

By comparing the last expressions of (3.4a) and (3.4b), it is seen that the contribution of the row voltage is V_R^2, whereas that of the column voltages including the unselected time periods is NV_C^2. These contributes are present in both ON and OFF pixels. The only difference that distinguishes the ON from the OFF pixel is due to the term $\pm 2V_RV_C$.

3.2.2 Multiple Line Addressing

The MLA technique, or MLS (Multiple Line Selection), was first proposed by **Temkar N. Ruckmongathan** [R1988]. In the generic MLA-p approach, the N rows are divided into N/p subgroups with p rows in each subgroup, as conceptually illustrated in Fig. 3.12. A set of p orthogonal waveforms is used to address rows in a subgroup together, within a time interval equal to pT/N, while the remaining $N-p$ unselected rows are grounded (i.e., they are set to the mid range voltage 0V).

We recall that a set of functions, $F_i(t)$ with $i=1\ldots p$, are orthogonal to each other in a certain interval of time T if the integral of the product of any two of the functions is 0 if the functions are different, and a constant if they are the same function.

$$\frac{1}{T}\int_0^T F_i(t)F_j(t)dt = 0 \quad \text{for } i \neq j \tag{3.5a}$$

$$\frac{1}{T}\int_0^T F_i(t)F_j(t)dt = \text{const} \quad \text{for } i = j \tag{3.5b}$$

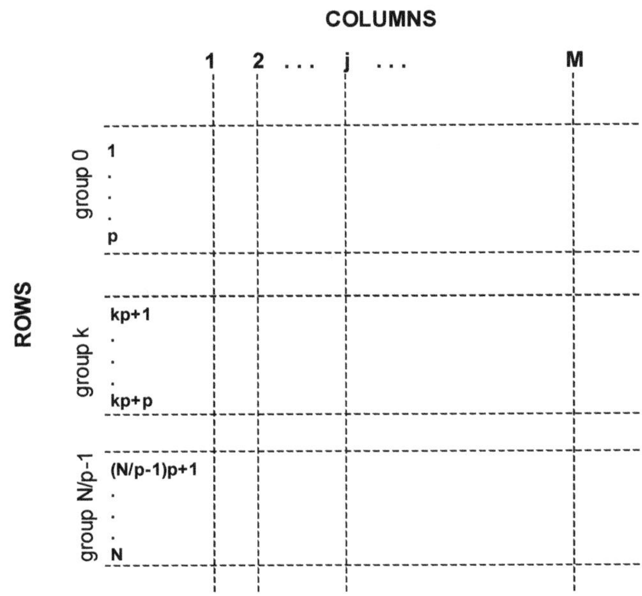

Fig. 3.12 The N rows of the LCD matrix are grouped into N/p subgroups each made up of p rows

From the signal processing theory, we can represent a given signal as a linear combination (a weighted sum) of these orthogonal functions. If we suppose the signal we wish to represent is the sequence of pixel values in a particular column, we apply the orthogonal functions to the rows and, in principle, the required weights to the columns. In reality, the column voltages are not weights, but they add or subtract a certain amount from the voltage that is being applied to the row line.

Compared to those orthogonal functions based on sinusoidal waveforms, the **Rademacher**, **Hadamard** and **Walsh** functions contain only two levels, which are simply described by +1 or –1 elements.[1] In particular, the last two are that of choice as orthogonal basis set[2] because they can be represented by square matrices that can be easily manipulated in the digital domain [B1984]. Specifically, the matrices have orthogonal columns (and rows), in which the select voltages, $+V_R$ or $-V_R$, are proportional to the elements, +1 or –1, that are the columns of the matrices. Each matrix column is hence called a **row select pattern** (or **select vector**).

As far as the column waveforms is concerned, we define as the **data pattern** in the k-th row subgroup (see again Fig. 3.12) the vector $\mathbf{d}_{\mathbf{kp+i}} = (d_{kp+1}, d_{kp+2}, \ldots\, d_{kp+p})$, in which each element is +1 (–1) if the pixel is OFF (ON). It can be shown that the required column voltage levels equals the nonzero entries in a column and are hence

[1] Examples of Rademacher, Hadamard and Walsh matrices are given in the Appendix C.

[2] Sparse matrices were also suggested to reduce hardware complexity and number of column voltages [RS2005], see Appendix C.

p+1 using Hadamard and Walsh functions. For instance, they are 3 for MLA-2, 5 for MLA-4, and so on.

The main advantage of the matrix representation is that the required column signals, starting form the digital data to be displayed, are computed with a simple algorithm.

Let us denote as **M**(i,j) the matrix (usually a Hadamard or Walsh matrix) whose dimension is $n \times n$. The column signal is the *dot product* between the row select pattern and the data pattern in the selected subgroup. For each column j the voltage to be applied is given by

$$C_k(j) = \frac{V_C}{p} \sum_{i=1}^{n} M(i,j) \cdot d_{kp+i} \tag{3.6}$$

To illustrate the above concepts we shall examine in detail two examples.

Example 1 MLA-2 technique

We depict in Fig. 3.13 the set of waveforms adopted in a MLA-2 technique (in the k-th row subgroup). For simplicity, we draw only the frame interval $2T/N$. In the remaining frame interval, $T(N\text{-}2)/N$, the row waveforms are all zero. The row waveforms are obviously orthogonal, and represent the only possible combination (unless for other three combinations obtained multiplying by −1 one or both waveforms). Like in the A&P technique, we must invert all the waveforms for every subsequent frame to obtain a zero mean applied voltage to the LC.

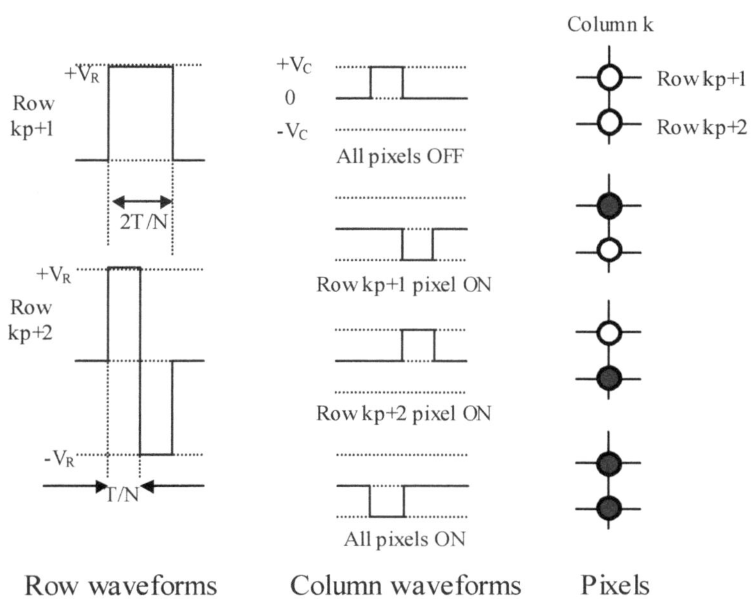

Fig. 3.13 Row and column voltages for all possible configurations of two pixels using the MLA-2 technique

The associated matrix is

$$\mathbf{M}(i,j)=\begin{bmatrix}+1 & +1\\ +1 & -1\end{bmatrix}$$

The data pattern is summarized in the first two columns of Table 3.1, for all possible pixel combinations.

Table 3.1 Data pattern and column voltages for the MLA-2 configuration considered in Fig. 3.13

	d_{kp+1}	d_{kp+2}	$C_k(1)$	$C_k(2)$
All pixels OFF	+1	+1	$+V_C$	0
Row *kp*+1 pixel ON	−1	+1	0	$-V_C$
Row *kp*+2 pixel ON	+1	−1	0	$+V_C$
All pixel ON	−1	−1	$-V_C$	0

Using (3.6) we get

$$C_k(1)=\frac{V_C}{2}\left[M(1,1)d_{kp+1}+M(2,1)d_{kp+2}\right]$$

$$C_k(2)=\frac{V_C}{2}\left[M(1,2)d_{kp+1}+M(2,2)d_{kp+2}\right]$$

From the above equations, the given orthogonal matrix and the data pattern, we can compute the values of $C_k(1)$ and $C_k(2)$ reported in the last two columns of Table 3.1, and to which correspond the waveforms in Fig. 3.13.

Example 2 MLA-4 technique

An example of MLA-4 driving waveforms is illustrated in Fig. 3.14.

Again, for the sake of simplicity we represent the row waveforms of the *k*-th subgroup in a reduced frame interval, $4T/N$, in the remaining frame interval $(N–4)T/N$ they are zero. Like the A&P technique, we must invert the waveforms at each subsequent frame.

Besides, for the sake of simplicity we label with 1, 2, 3 and 4 the rows of the considered subgroup.

The associated matrix is the Hadamard of type 2, as it is called in the Appendix C.

$$\mathbf{M}(i,j)=\begin{bmatrix}+1 & -1 & -1 & -1\\ -1 & +1 & -1 & -1\\ -1 & -1 & +1 & -1\\ -1 & -1 & -1 & +1\end{bmatrix}$$

The data pattern is exemplified in the first four columns of Table 3.2, for a set of possible pixel combinations.

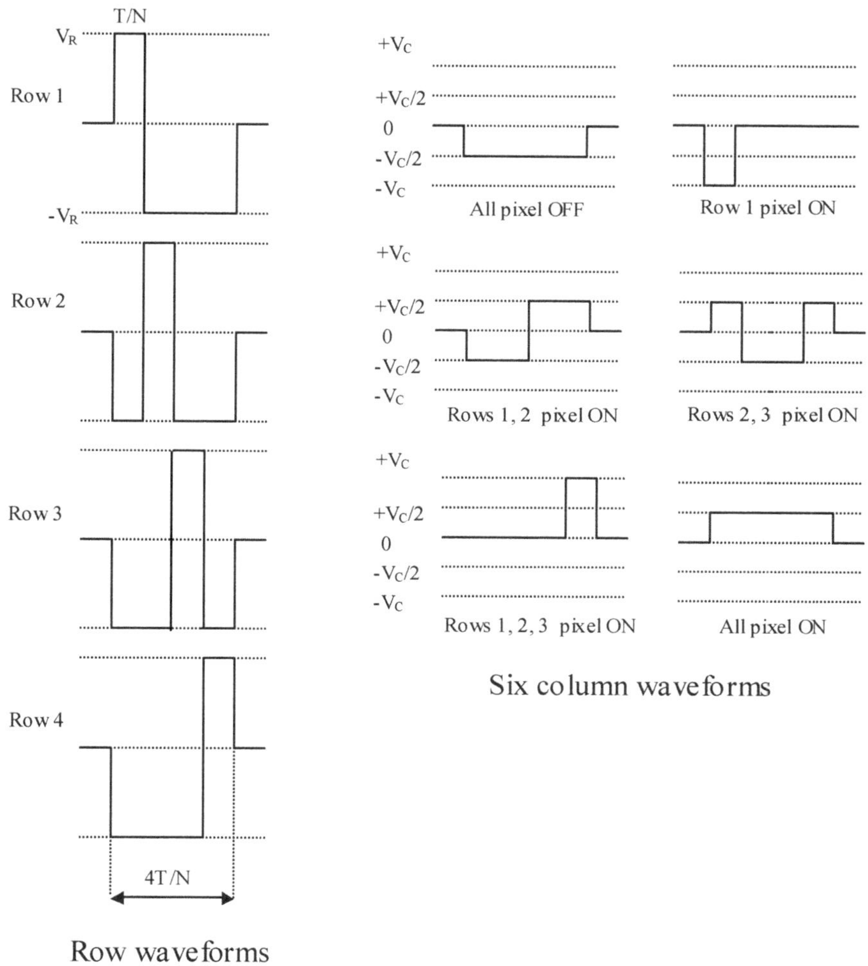

Fig. 3.14 Row and six selected column waveforms adopted in a MLA-4 driving technique

Table 3.2 Data pattern and column voltages for the MLA-4 configuration considered in Fig. 3.14

	d_1	d_2	d_3	d_4	$C(1)$	$C(2)$	$C(3)$	$C(4)$
All Px OFF	+1	+1	+1	+1	$-V_C/2$	$-V_C/2$	$-V_C/2$	$-V_C/2$
Row 1 Px ON	−1	+1	+1	+1	$-V_C$	0	0	0
Rows 1, 2 Px ON	−1	−1	+1	+1	$-V_C/2$	$-V_C/2$	$+V_C/2$	$+V_C/2$
Rows 2, 3 Px ON	+1	−1	−1	+1	$+V_C/2$	$-V_C/2$	$-V_C/2$	$+V_C/2$
Rows 1, 2, 3 Px ON	−1	−1	−1	+1	0	0	0	$+V_C$
All Px ON	−1	−1	−1	−1	$+V_C/2$	$+V_C/2$	$+V_C/2$	$+V_C/2$

Using (3.6) we get

$$C(1) = \frac{V_C}{4}\left[M(1,1)d_1 + M(2,1)d_2 + M(3,1)d_3 + M(4,1)d_4\right]$$

$$C(2) = \frac{V_C}{4}\left[M(1,2)d_1 + M(2,2)d_2 + M(3,2)d_3 + M(4,2)d_4\right]$$

$$C(3) = \frac{V_C}{4}\left[M(1,3)d_1 + M(2,3)d_2 + M(3,3)d_3 + M(4,3)d_4\right]$$

$$C(4) = \frac{V_C}{4}\left[M(1,4)d_1 + M(2,4)d_2 + M(3,4)d_3 + M(4,4)d_4\right]$$

From the above equations, given the row matrix and the selected values of $d_{1,2,3,4}$ summarized in Table 3.2, we compute the values of $C(1)$, $C(2)$, $C(3)$, $C(4)$ reported in the last four columns of Table 3.2, to which correspond the column waveforms in Fig. 3.14.

3.2.3 Further Considerations on MLA

Efforts have been done to reduce the number of column voltages required by the MLA approach [SSK2003]. A simple theoretical way to minimize them is by the introduction of one zero in each matrix column [RS2005], so as to obtain a so called sparse matrix (see Appendix C). In this case, only 4 different column voltages are required ($+V_C$, $+V_C/2$, $-V_C/2$ and $-V_C$) instead of 5 ($+V_C$, $+V_C/2$, 0, $-V_C/2$ and $-V_C$).

A special case that should be mentioned is the **MLA-3** technique. It is actually a pseudo MLA-4 technique, in which the fourth row is a ghost row used to reduce the maximum column voltage as described below. We have seen that the MLA-*p* technique requires p+1 different column voltage levels. Therefore, as can be seen in Fig. 3.14, the MLA-4 technique requires 5 column voltage levels, namely $+V_C$, $+V_C/2$, 0, $-V_C/2$ and $-V_C$. Also form the same figure, it can be inferred that $|V_C|$ is needed when the number of ON pixels is odd (e.g., pixels of Row a or of Rows a-b-c are ON), whereas $|V_C/2|$ is required for an even number of ON pixels (e.g., Rows a-b, Rows b-c, or Rows a-b-c-d pixels are ON). In the MLA-3 technique, the (ghost) pixel related to the ghost row can be controlled in such a way that the number of ON pixels is *always* even, so that we need only 3 column voltage levels instead of 5. The drawback of this approach is that if N is the actual number of display rows, we are effectively driving a number of $4N/3$ rows (including the ghost rows). Therefore, from (3.4a) to (3.4b) the ON and OFF RMS voltages are reduced. As a result, the MLA-3 technique is typically used for $N<128$.

To conclude this section, in a similar manner as done for the A&P technique, we compute the RMS voltages $\overline{V}_{ON}$ e $\overline{V}_{OFF}$ applied to an ON and OFF pixel, respectively. In general, whatever orthogonal row waveforms and associated column voltages are used, the expressions of $\overline{V}_{ON}$ and $\overline{V}_{OFF}$ are

$$\overline{V}_{ON} = \sqrt{\frac{1}{N}\left[(p-1)V_R^2 + (V_R + V_C)^2 + \left(\frac{N}{p} - 1\right)V_C^2\right]}$$
$$= \sqrt{\frac{1}{pN}\left[p^2 V_R^2 + 2pV_R V_C + NV_C^2\right]} \tag{3.7}$$

$$\overline{V}_{OFF} = \sqrt{\frac{1}{N}\left[(p-1)V_R^2 + (V_R - V_C)^2 + \left(\frac{N}{p} - 1\right)V_C^2\right]}$$
$$= \sqrt{\frac{1}{pN}\left[p^2 V_R^2 - 2pV_R V_C + NV_C^2\right]} \tag{3.8}$$

It is worth noting that for p=1, (3.7) and (3.8) are equal respectively to (3.1) and (3.2) obtained with the A&P technique. This indicates that the A&P technique can be seen as a particular case of the MLA with addressed rows equal to 1. For this reason, in the followings we will carry out our computations starting from (3.7) and (3.8), but the obtained results will be valid both for MLA and A&P techniques.

As we will see in Section 3.3, the MLA technique provides as main advantage the reduction of V_{LCD}, with respect to the A&P and IA&P techniques. This allows low-voltage (and cheaper) CMOS processes to be employed for implementing the driving circuits and power consumption to be reduced. Moreover, reducing V_{LCD} reduces crosstalk (see Section 3.3.3).

The main drawback of the MLA techniques relies in the increased complexity to manage the row subgroups and the increased number of column voltage levels. Although design and implementation of an MLA controller is rather simple, provided that p is an integer power of two, the additional digital circuitry (logic circuits, latches, etc.) increases the silicon area. Besides, and perhaps more importantly, the hardware complexity and power dissipation of the row and column driver circuits increase too.

Consequently, the approach has proven to be rather unpractical for p>6.

3.2.4 Active Addressing

In the Active Addressing (AA) all the lines are selected simultaneously [SC1992, GM1994] and can be seen as an extension of MLA technique with p=N. The main advantage is related to both high brightness uniformity and contrast it allows, even with fast responding LCDs. However, the number of voltages in the column waveforms is $N+1$, this increases greatly the hardware complexity of the column driver, augmenting in turn costs. As a result, the active addressing technique is seldom used in commercial applications.

3.2.5 Distributed MLA

The above MLA techniques are often referred to as Concentrated MLA. In practical implementations, a variant of the MLA approach, called Distributed MLA (DMLA), is often exploited. In a distributed MLA-p technique the frame time period T is divided into p subperiods. Also the row waveforms are divided into p subwaveforms. The i-th subwaverform is distributed into the i-th subperiod.

Figure 3.15 illustrates the Row-a waveform, already shown in Fig. 3.14, for the Concentrated and the Distributed cases. It is seen that the waveform is decomposed into four portions applied to Row a during four subsequent subperiods.

The advantages of such distributed schemes are substantial. The overall image quality is greatly improved with respect to the single-line (and also to concentrated MLA) addressing techniques because now the refresh rate is p times greater and hence the frame response effect is greatly reduced (see Fig. 3.16). Consequently, flickering is much less noticeable and very stable images are displayed. In addition, most of the spectrum of the drive signals becomes concentrated around p times the frame frequency. This is an advantage, because the frequency dependence of the electro-optical transmission curve of STN-LCDs requires narrow-band drive signals to prevent visual artefacts. Lastly, this feature allows passive-matrix LCDs to be made up with fast responding LC mixtures, enabling in turn the possibility of managing also moving pictures without the use of more expensive (but also with better performance) TFT technologies. Compared to the Concentrated MLA, the drawbacks are the increased power consumption of about 1.5–2 times, depending on the set of waveforms chosen.

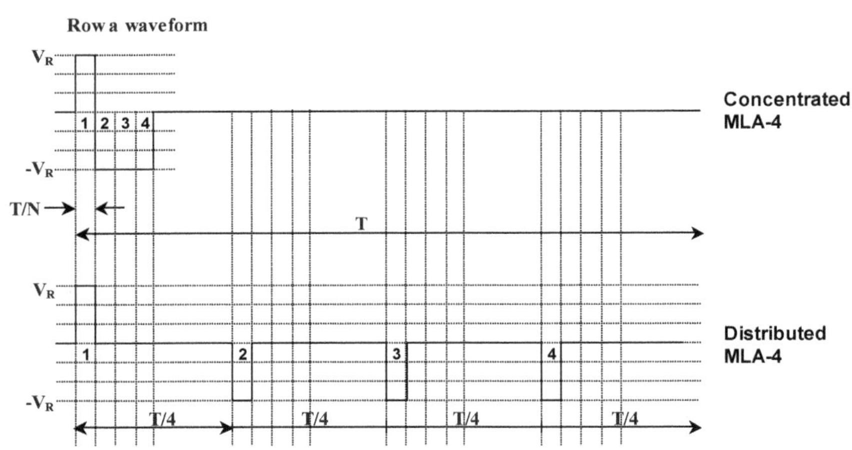

Fig. 3.15 Row-a waveform of Fig. 3.14 for a Concentrated MLA-4 approach (upper trace) and Distributed MLA-4 approach (lower trace)

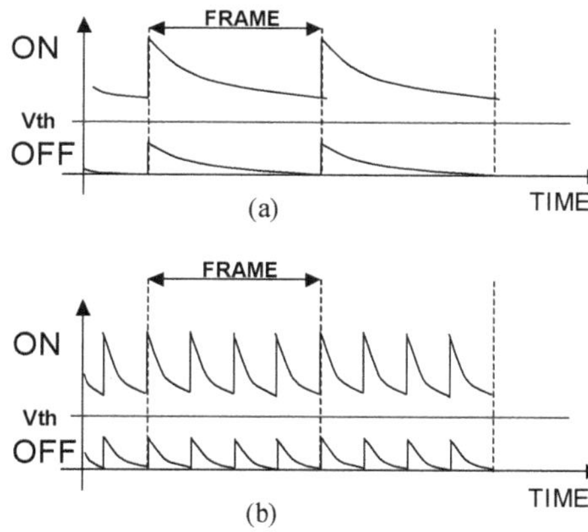

Fig. 3.16 Time distribution of pixel voltages in: (**a**) the A&P technique, and (**b**) MLA-4 technique

3.3 Limitations of Passive-Matrix Addressing

In this section we shall derive some fundamental limitations of the driving techniques already discussed [AP1974, NK1979].

Before going further in our developments it is useful to define two electrical parameters, namely the drive margin, *DM*, and the bias ratio, *r*.

The **drive margin** (also called the **selection ratio**) is defined as the ratio of the RMS values of the ON pixel voltage to the OFF pixel voltage.

$$DM = \overline{V}_{ON} / \overline{V}_{OFF} \tag{3.9}$$

Given the display physical properties it can be easily understood that the contrast increases with the drive margin.

The **bias ratio** is defined as the ratio of the maximum row voltage to the maximum column voltage.

$$r = V_R / V_C \tag{3.10}$$

Assuming conventionally $V_R \geq V_C$, a bias ratio approaching 1 means that similar peak voltages are used for both row and columns.

Substituting (3.7) and (3.8) in (3.9) we obtain the expression of *DM* as a function of p, V_R, V_C and N.

$$DM = \sqrt{\frac{p^2 V_R^2 + 2pV_R V_C + NV_C^2}{p^2 V_R^2 - 2pV_R V_C + NV_C^2}} \tag{3.11}$$

Using (3.10) in (3.11) we also get the alternative expression below

$$DM = \sqrt{\frac{p^2r^2 + 2pr + N}{p^2r^2 - 2pr + N}} \tag{3.12}$$

3.3.1 Maximum Drive Margin

In this section we shall search the value of r that maximizes the drive margin (i.e., the contrast). It is found by equating to zero the derivative of DM with respect to r:

$$\frac{\partial}{\partial r}DM = 0 \Rightarrow \ (2pr+2)\left(pr^2 - 2r + \frac{N}{p}\right) - (2pr-2)\left(pr^2 + 2r + \frac{N}{p}\right) = 0$$

and after simple algebra we get the expression of r maximizing DM

$$r = \frac{\sqrt{N}}{p} \tag{3.13}$$

Substituting (3.13) into (3.12) we obtain the maximum drive margin

$$DM_{MAX} = \sqrt{\frac{\sqrt{N}+1}{\sqrt{N}-1}} \tag{3.14}$$

The above equation shows that the maximum of DM is independent of the parameter p, (i.e., it is an intrinsic property not depending on the addressing technique) but it depends only on the row number N. As an important consequence, it is seen that DM decreases with N. For large N values, DM (and its maximum) tends to the unity. In other words, $\overline{V}_{ON}$ and $\overline{V}_{OFF}$ tend to coincide and contrast nullifies. Therefore, there must be a limit for the maximum number of addressable rows ensuring an acceptable contrast.

For example, if N = 100, the maximum ratio between $\overline{V}_{ON}$ and $\overline{V}_{OFF}$ is only $\sqrt{11/9}$ =1.106. That is, $\overline{V}_{ON}$ and $\overline{V}_{OFF}$ differ for less than 11%. To achieve a reasonable contrast ratio, a very steep electro-optic transfer characteristic is hence required. Due to the gradual T/V characteristic, the limit for TN LCDs is found to be about 64 rows. Supertwisted nematic LCDs have a much steeper characteristic and can be used with a number of multiplexed rows up to 240 (STN displays with 480 rows, make use of the dual scan technique and are hence conceived as two 240-row displays).

Given N, and once chosen the driving technique, maximizing parameter DM causes also the maximization of the drive voltage V_{LCD}. To obtain the expression of V_{LCD} we must find the expression of voltages V_R and V_C. Voltage V_R can be found from (3.8) by setting a specific value for V_{OFF} (we can set the minimum value, i.e, $\bar{V}_{OFF} = V_{TH}$).

$$V_{TH} = \frac{1}{\sqrt{N}}\sqrt{pV_R^2 - 2V_R V_C + \frac{N}{p}V_C^2} = \frac{V_C}{\sqrt{N}}\sqrt{pr^2 - 2r + \frac{N}{p}} \tag{3.15}$$

and using (3.13) we get

$$V_C = \frac{\sqrt{N}}{\sqrt{2\frac{N}{p} - 2\frac{\sqrt{N}}{p}}}V_{TH} = \frac{1}{\sqrt{2}}\frac{\sqrt{p}}{\sqrt{1 - \frac{1}{\sqrt{N}}}}V_{TH} \tag{3.16}$$

From (3.10) we can also compute the expression of V_R

$$V_R = rV_C = \frac{1}{\sqrt{2}}\frac{\sqrt{N/p}}{\sqrt{1 - \frac{1}{\sqrt{N}}}}V_{TH} \tag{3.17}$$

It is seen that for high N values, such that $\sqrt{N} \gg 1$, V_C remains almost independent of N whereas V_R increases with the square root of N. This is intuitively explained by considering that increasing N, we reduce the time interval in which the energy (proportional to the square of the voltage) can be transferred to the pixel.

To decrease the amplitude of the row voltage, V_R, we can exploit parameter p, by driving together more than one row. Indeed, increasing p, V_R decreases while V_C increases. Note that there is a value of p, that is $p=\sqrt{N}$, for which both V_C and V_R are theoretically equal.

We plot in the Fig. 3.17a V_R and V_C, normalized to V_{TH}, as a function of N for p=1, 2 and 4. Figure 3.17b shows V_R and V_C versus p, for N=100. As expected, V_R equals V_C for p=10.

Finally, the expression of voltage V_{LCD} for the basic A&P and MLA-p techniques is

$$V_{LCD} = 2V_R = \frac{\sqrt{2N/p}}{\sqrt{1 - \frac{1}{\sqrt{N}}}}V_{TH} \tag{3.18}$$

while for the IA&P technique we get

$$V_{LCD} = V_R + V_C = \frac{1}{\sqrt{2}}\frac{\sqrt{N} + 1}{\sqrt{1 - \frac{1}{\sqrt{N}}}}V_{TH} \tag{3.19}$$

Observe that as the number of rows increases, V_{LCD} increases too. However, V_{LCD} in the IA&P technique is lower than in the A&P one, this explains why IA&P is the most widely used single line addressing technique, (even if the number of voltages in row and column waveforms are respectively three and two, in the case of the A&P technique, whereas the number of voltages in both row and column waveforms are four in the IA&P technique).

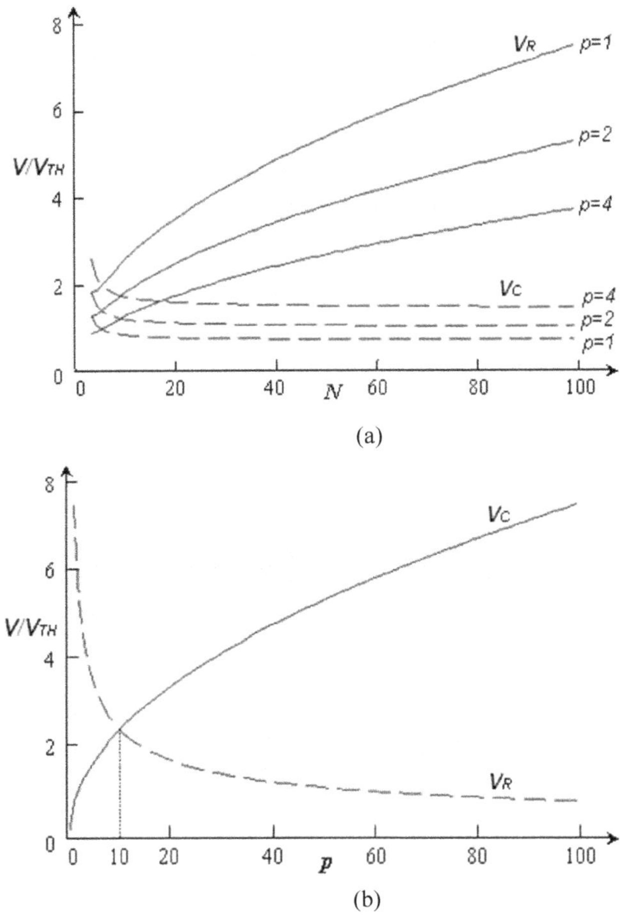

Fig. 3.17 Maximum row voltage, V_R, and maximum column voltage, V_C, normalized to V_{TH}: (**a**) plotted versus N for p=1, 2 and 4, (**b**) plotted versus p for N=100

Figure 3.18 illustrates the behaviour of V_{LCD} normalized to V_{TH} versus N for the A&P, IA&P and MLA-2,4,6 techniques. It is seen that IA&P requires lower V_{LCD} than A&P and even than MLA-2 (for N>6).

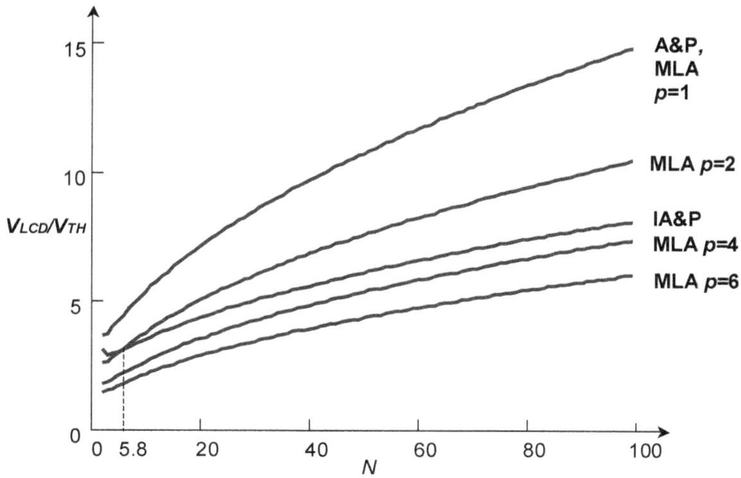

Fig. 3.18 Maximum driving voltage, V_{LCD}, normalized to V_{TH} versus N, for the cases of A&P, IA&P and MLA (p=2, 4 and 6)

3.3.2 Minimum Driving Voltage

We discussed in Section 3.3.1 the methods to obtain the maximum drive margin and hence contrast. In this subsection we will be interested in evaluating the minimum driving voltage [NTN1999]. This issue is of utmost importance in portable applications. Indeed, it allows to reduce power dissipation and to exploit standard, low-cost CMOS process for the implementation of the driving circuits.

To minimize the driving voltage, we must set equal the maximum row and column voltages, i.e., $V_R=V_C$, and hence $r=1$. From (3.12) the drive is given by

$$DM = \sqrt{\frac{p^2+2p+N}{p^2-2p+N}} \tag{3.20}$$

which reaches its maximum for $p=\sqrt{N}$, [R1993].

Besides, from (3.8) we can derive the expression of V_C, by setting $\overline{V}_{OFF} = V_{TH}$, hence we get

$$V_R = V_C = \sqrt{\frac{pN}{p^2-2p+N}}V_{TH} \tag{3.21}$$

Finally, the expression of V_{LCD} (we have a unique value for all techniques, since $V_C=V_R$) is

$$V_{LCD} = 2\sqrt{\frac{pN}{p^2-2p+N}}V_{TH} \tag{3.22}$$

We summarize in Tables 3.3 and 3.4 the main results found in this section.

Table 3.3 Summary of the main electrical parameters of passive-matrix addressing techniques

<table>
<tr><th>Param.</th><th>Maximum drive margin</th><th>Minimum driving voltage</th></tr>
<tr><td>r</td><td>$\sqrt{N}/p$</td><td>1</td></tr>
<tr><td>DM</td><td>$\sqrt{\frac{\sqrt{N}+1}{\sqrt{N}-1}}$</td><td>$\sqrt{\frac{p^2+2p+N}{p^2-2p+N}}$</td></tr>
<tr><td>V_R</td><td>$\frac{V_{TH}}{\sqrt{2}}\frac{\sqrt{N/p}}{\sqrt{1-1/\sqrt{N}}}$</td><td rowspan="2">$V_{TH}\sqrt{\frac{pN}{p^2-2p+N}}$</td></tr>
<tr><td>V_C</td><td>$\frac{V_{TH}}{\sqrt{2}}\frac{\sqrt{p}}{\sqrt{1-1/\sqrt{N}}}$</td></tr>
<tr><td rowspan="2">V_{LCD}</td><td>$\sqrt{2}\,V_{TH}\frac{\sqrt{N/p}}{\sqrt{1-1/\sqrt{N}}}$ MLA-p, A&P</td><td rowspan="2">$2V_{TH}\sqrt{\frac{pN}{p^2-2p+N}}$</td></tr>
<tr><td>$\frac{V_{TH}}{\sqrt{2}}\frac{\sqrt{N}+1}{\sqrt{1-1/\sqrt{N}}}$ IA&P</td></tr>
</table>

Table 3.4 Comparison of the electrical parameters for different addressing techniques for V_{TH}= 1.3V and N=64

	A&P MLA-1	IA&P	MLA-2	MLA-4	MLA-8	Unit
DM_{MAX}	1.134	1.134	1.134	1.134	1.134	-
V_R	7.9	7.9	5.6	3.9	2.8	V
V_C	1.0	1.0	1.4	2.0	2.8	V
V_{LCD}	15.7	8.8	11.1	7.9	5.6	V
DM	1.031	1.031	1.061	1.106	1.134	-
V_R=V_C	1.3	1.3	1.8	2.5	2.8	V
$V_{LCD,\mathrm{min}}$	2.6	2.6	3.7	4.9	5.6	V

3.3.3 Crosstalk

Crosstalk is a pattern-dependent visible defect that occurs in the displayed picture due to the mutual interference between adjacent pixels as well as between row and column tracks. It appears as differences in the brightness of theoretically equal gray-scale pixels and often produces a translucent tail added to the original picture. The effect usually worsens with increasing display size, higher resolution and faster responding LC materials.

At least two types of **static crosstalk** (i.e., affecting still pictures) can be distinguished, namely vertical crosstalk and horizontal crosstalk.

Vertical crosstalk (qualitatively illustrated in Fig. 3.19) is due to the panel electrical characteristics (parasitic resistances and capacitances) and to the IC driver performance (output resistance of the row and column drivers).

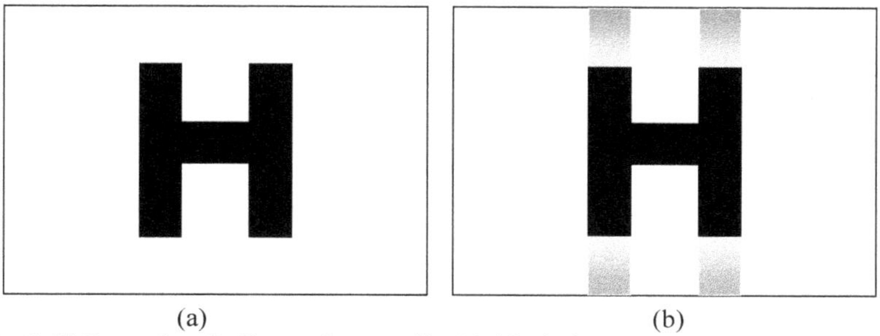

Fig. 3.19 Example of effects of crosstalk: (**a**) ideal picture, (**b**) picture with vertical crosstalk

To understand the phenomenon, we show in Fig. 3.20 the electrical model of a PMLCD panel. At every intersection of a row and a column, forming a pixel, we get a capacitor that takes into account the capacitance of the LC material. Resistors at each row, R_R, and column, R_C, are also added. They take into account the ITO resistance plus the output resistances of the row and column drivers (that are connected in series). Usually, the drivers' resistances are greater than the ITO ones.

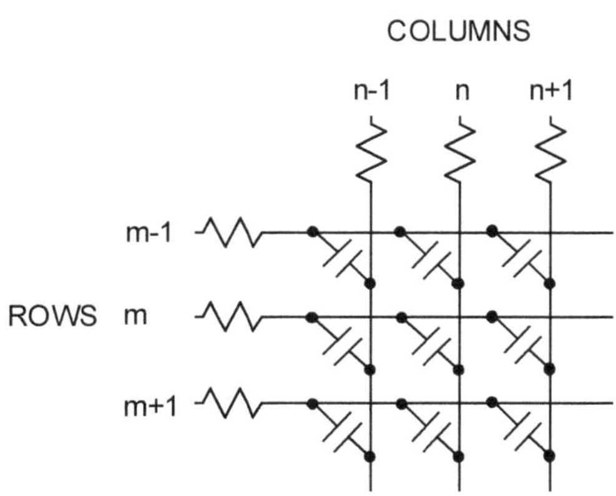

Fig. 3.20 Electrical model of a PM-LCD panel

The resulting RC time constant causes a *delay* into the pulse driving voltages which causes the applied waveform to be distorted at every column transition. In contrast, the pixel voltage remains substantially undistorted in absence of column transitions, since only the row RC delays (equal for all pixels) are introduced.

Figure 3.21 exemplifies the behaviour of row and column voltages applied to five consecutive pixels of the same column, assuming that the first and the last two pixels must be OFF (white) while the remaining second and third pixels must be ON (black). Ideal and actual column waveforms are depicted. For the sake of simplicity we do not represent actual row waveforms. As already said, they affect at the same manner all pixels and therefore do not introduce brightness discrepancies.

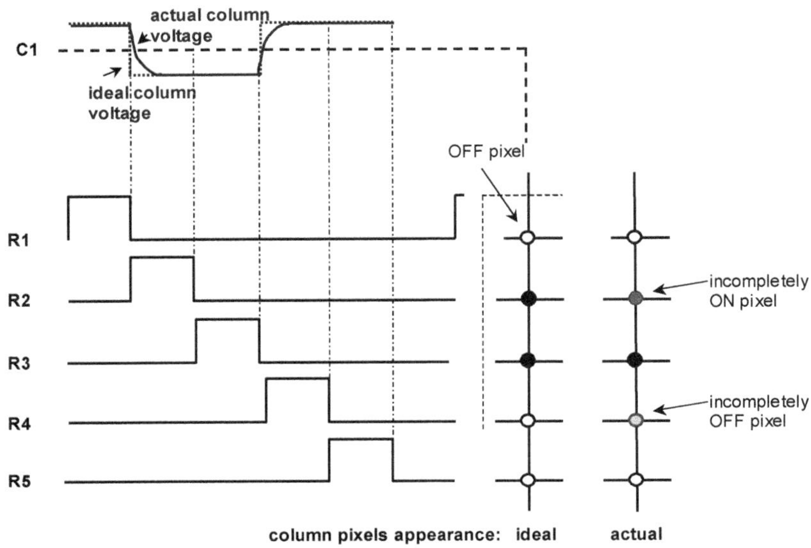

Fig. 3.21 Effect of the RC delay onto the pixel driving voltages. The ideal pixel appearance and actual one are compared at the right. Pixels driven in presence of a column transition can receive more (row 4) ore less (row 2) energy

Because of the time delays, the first consequence of a column transition is the reduction (increase) of the brightness of ON (OFF) pixels. For example, the pixel of row 2 receives less energy from the driving voltage with respect to the ideal case. The second consequence is related to the fact that the pixels in the same column are driven by one signal line. Hence, the voltage across one pixel will affect all the other pixels (and hence their RMS voltage) at the same column. Brightness non-uniformity in the entire column is therefore introduced (vertical crosstalk). Of course, vertical crosstalk becomes more apparent with an increasing number of transitions in the column voltage. In other words, it depends on the high-frequency contents of the column driving waveforms. These contents contribute in turn to vary the LC cell response because of the frequency dependence of the LC parameters like, for instance, the threshold voltage [W1997, K2000]. Vertical crosstalk also worsens with the increasing steepness of the LC T-V curve.

Horizontal crosstalk arises when differences in the LC dielectric constant for black and white pixels induce spatially-asymmetrical capacitive coupling between rows and columns. This originates perceptible visual artefacts depending on the width of dark/bright horizontal blocks along a row.

In addition, **dynamic crosstalk** (*splicing*) [PCS1995], occurring on switching from one picture to another, may affect LCD modules supporting video streaming.

To reduce crosstalk, lower RC time constants have been obtained through technological advances in display manufacturing and IC driver design. A system-level solution is accomplished by driving techniques that perform **Line Inversion,** in which the polarity of the addressing waveforms is reversed periodically [M1984, M1987, KHM1990]. Such schemes will be discussed in Section 5.4. Other more complex solutions were also developed to quite completely eliminate crosstalk [HKM1992, PCS1995, W1997, WIO1997, WS1997, AH2000, SOM2006]. In summary, the pixel voltage distortion is compensated in these approaches by superimposing a correction voltage on the column waveform whenever a voltage transition occurs. Other special addressing schemes (i.e. improvements of the column/row waveform structure) have been also devised [NMK1993, KI1997].

To get an idea of the limits imposed by crosstalk, assume the gray levels are equally distributed within the ON and OFF region. For a 4k-color STN display, each of the RGB pixels needs to display 16 levels of gray (4 bit). If the display has 80 rows and the threshold V_{TH} is 1.4 V, the maximum DM is about 1.12 from (3.14). Let us assume that V_{OFF} equals V_{TH}, then using the definition of DM given in (3.9), V_{ON} results to be about 1.57 V. Therefore, the RMS voltage difference between two gray levels is $(V_{ON}\text{-}V_{OFF})/16 \approx 10$ mV. In order to ensure that adjacent gray levels can be identified, the crosstalk effect must be less than this voltage. As a result, the greater the gray levels and rows number are, the more sensitive to crosstalk the display is.

3.4 Displaying Gray Levels

Up to now we have considered that pixels can be exclusively turned ON or OFF. In this manner we can display *bilevel* images, that is, images made up of only either black or white pixels. However, in order to either produce black and white photographs or color images from the three primary RGB colors, we need to display different gray levels.

There are three approaches that starting from ON and OFF pixels allow to generate gray levels. They are namely, the **Spatial Dithering (SD)**, **Pulse Width Modulation (PWM)** and **Frame Rate Control (FRC)** techniques; they will be described in Sections 3.4.1, 3.4.2 and 3.4.3. As we shall see, the above approaches are simple but amenable only for a limited number of gray levels. To display more gray levels they can be combined together (an example is given in Section 3.4.4) or we need to exploit the LC T/V characteristic. Referring again to Fig. 3.8, we see that the light transmittance through a pixel varies with continuity from the OFF to the ON state, giving all the intermediate gray tones. In other words, by modulating the applied RMS voltage to the LC we can vary the gray level displayed from black to white. We know that the pixel voltage is dictated by the superposition of the column and row voltage, therefore we can modulate the RMS voltage through two different techniques, namely the **Column Pulse Height Modulation** and **Row Pulse Height Modulation**, described in Sections 3.4.5 and 3.4.6.

It must be observed however, that when such schemes are applied on top of MLA, frequency multiplication may cause artefacts. At this purpose, polarity inversion (see Section 5.4) is an effective technique for reducing the driving signal bandwidth by increasing the lowest frequency of the spectrum [SOM2006].

3.4.1 Spatial Dithering

This technique works in the same way as the dithered reproduction of photographs in print. A group of adjacent sub-pixels (that can be either black or white and whose dimensions must be below the eye resolution) is selected in any combination to give different gray levels. For example, using m sub-pixels of the same area, we get $m + 1$ gray levels, as perceived by the human eye. Figure 3.22 illustrates an example of all possible combinations with 4 pixels. With this technique we trade pixel resolution for gray levels.

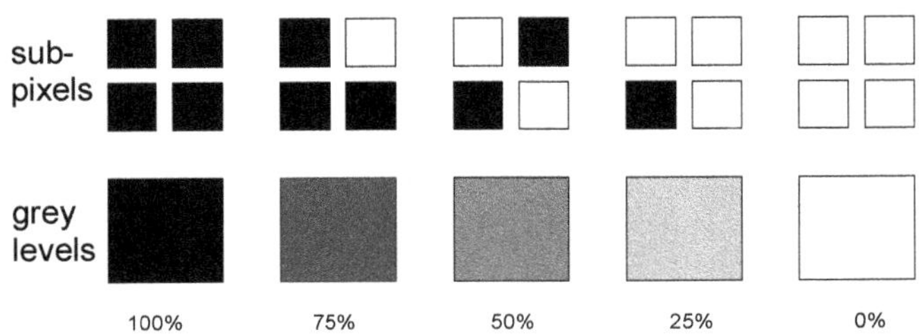

Fig. 3.22 Example of the spatial dithering technique with 4 sub-pixels

3.4.2 Pulse Width Modulation

With this technique, each *row select time* is subdivided into m equal time intervals to display m+1 gray levels [KHK1980].

For example, the row select time may be divided into four subintervals as shown in Fig. 3.23, which illustrates the five equivalent gray levels corresponding to the possible set of column waveforms.

Note that the width of the pulses decreases as the number of gray levels increases. This will result in brightness non-uniformity (due to the presence of high-frequency components).

Observe also that for a column in which all pixels are black (or, equivalently, are all white), the column voltage does not experience any commutation, i.e., it remains at $-V_C$ (or at $+V_C$). In all the other cases, in which pixels are characterized by intermediate gray levels, the column voltage must have two commutations at each row select interval. The number of commutations and hence the current consumption can be reduced by adopting the modified voltage sequence illustrated in Fig. 3.24. Here, the pixel waveforms related to even rows (pixel i, in the figure) are horizontally mirrored, so that PWM is not affected but two commutations are saved. In the complete column, a maximum reduction of 50% in the power consumption due to commutations can be achieved.

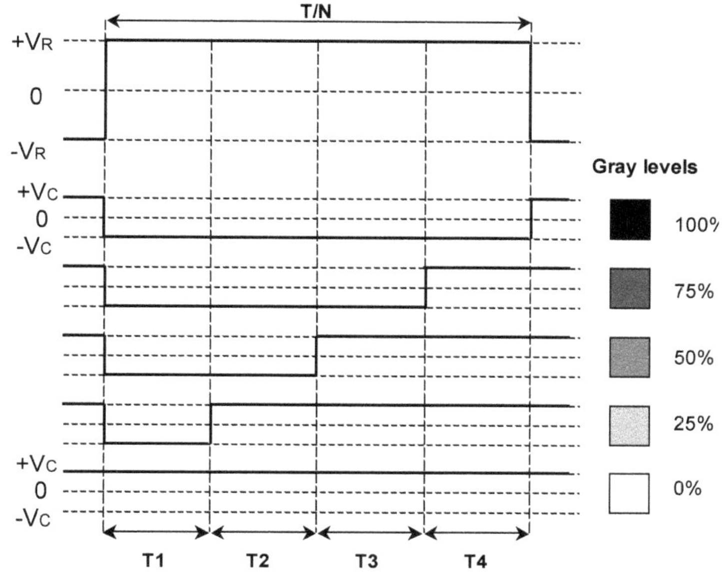

Fig. 3.23 Example of Pulse Width Modulation with the row select time divided into 4 subintervals. We can therefore obtain 5 gray levels

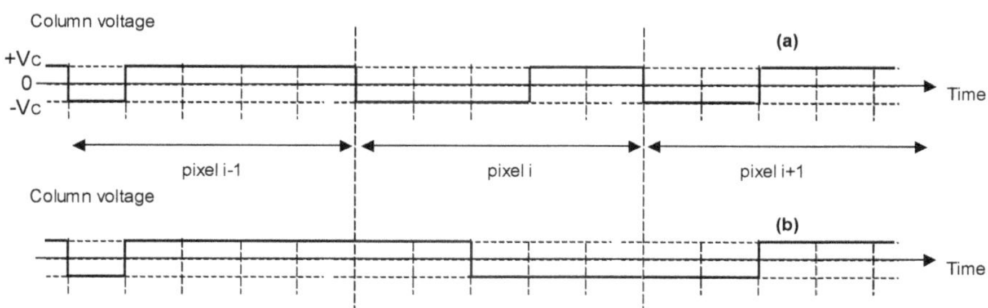

Fig. 3.24 Waveforms of a generic column: (**a**) conventional PWM voltages, (**b**) improved solution. By mirroring the column voltage of pixel *i*, we save two commutations

3.4.3 Frame Rate Control

With this technique, the gray levels are obtained by turning the pixels ON and OFF during different frames forming a superframe [SSA1983]. Specifically, m frames are used to display $(m + 1)$ gray levels. Figure 3.25 illustrates the case of 5 gray levels. This approach is widely exploited in STN LCDs, but leads to an increase in power consumption, because we need a higher scan frequency. For instance, assuming 5 gray levels, the frame will be reconstructed in 4 subsequent frames. Therefore, to avoid flickering, the scan frequency must be increased of four times with respect to the bilevel case.

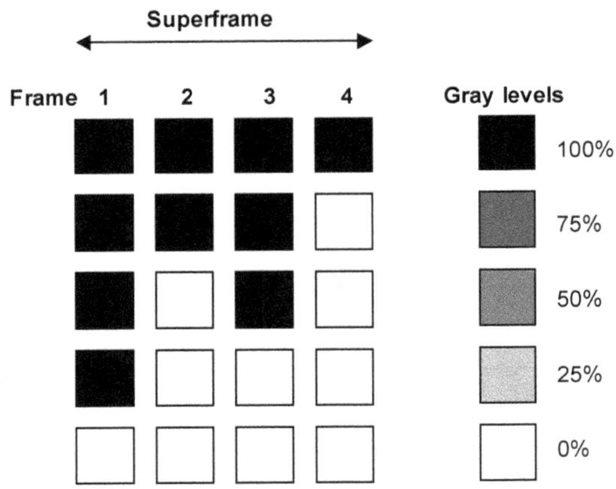

Fig. 3.25 Example of 4-FRC technique to display 5 gray levels

3.4.4 Mixed Techniques

To increase the number of gray levels, the above techniques can be used jointly. A typical situation where two of the above techniques are adopted together is given by the combination of the FRC and PWM methods. If the FRC technique exploits m_{FRC} frames and the PWM row select time is subdivided into m_{PWL} equal time intervals, then the combined effect would give rise to $m_{FRC} \cdot m_{PWL}+1$ gray levels.

An example of 3-FRC and 16-PWM mixed technique is illustrated in Fig. 3.26.

Another approach called **Phase Mixing** is the combination of spatial dithering and FRC. This helps in alleviating flicker induced by uneven RMS contributions over a superframe. For best optical performance, phase-mixing is typically applied on a RGB-subpixel basis [SOM2006]. An example with 4-FRC is illustrated in Fig. 3.27.

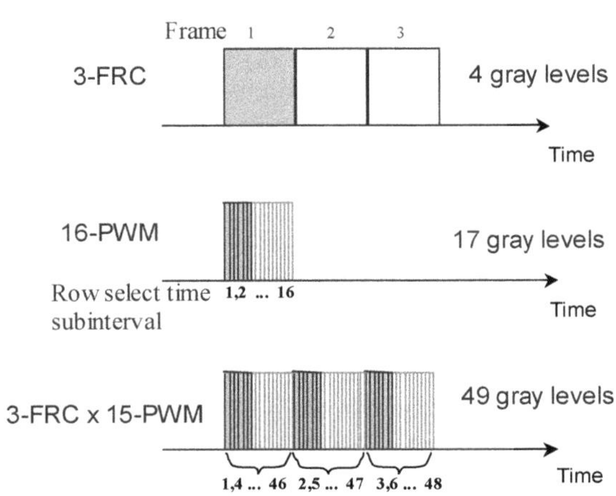

Fig. 3.26 Example of 3-FRC and 16-PWM mixed technique to display 49 gray levels

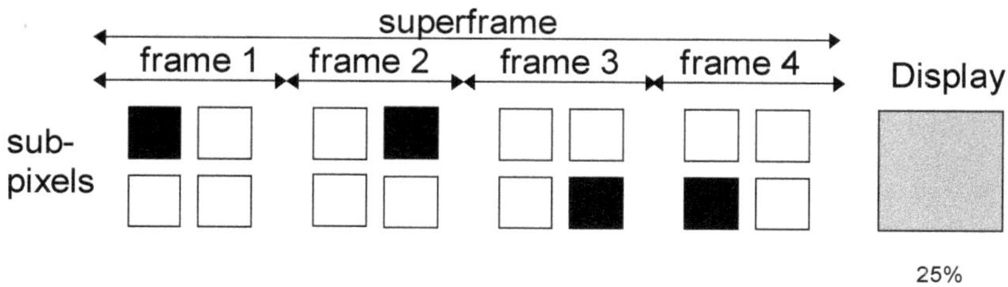

Fig. 3.27 Example of Phase Mixing with 4-FRC

3.4.5 Column Pulse Height Modulation

With this technique, the amplitude of the column voltage is modulated to vary the RMS voltage across the LC [CS1992].

The N LCD rows are scanned one row at a time. A row voltage V_R is applied to the selected row, while the remaining (N–1) rows are grounded (similarly to the A&P technique). Unlike the previous techniques, however, now each column voltage value is not restricted to $\pm V_C$, rather, it can span with continuity from $-V_C$ to V_C. Considering the k-th row, we define the parameter a_k ranging from -1 (ON pixel) to +1 (OFF pixel) that encodes the gray levels of the columns in the selected row. The RMS voltage across the k-th pixel in a column when it is selected is given by

$$\overline{V}_k = \sqrt{\frac{1}{N}\left(V_R^2 - 2a_k V_R V_C + \sum_{i=1}^{N} a_i V_C^2\right)} \qquad (3.23)$$

Comparing (3.23) to (3.4a and b), we see that the term $-2a_k V_R V_C$ represents the gray level of the k-th pixel. Unfortunately, term $\sum_{i=1}^{N} a_i V_C^2$ is not a constant and hence voltage $\overline{V}_k$ depends also on the gray levels of the other pixels in that column (it depends not only on a_k, but also on $a_1 \ldots a_N$). As a result, we need a correction term that makes the column voltage contribution constant.

One possible method is based on dividing the row select time interval into two equal subperiods in which two different column voltages are applied

$$V_{C,k1} = \left(a_k + \sqrt{1 - a_k^2}\right) V_C \qquad (3.24)$$

$$V_{C,k2} = \left(a_k + \sqrt{1 + a_k^2}\right) V_C \qquad (3.25)$$

The resulting RMS voltage across the k-th pixel in a column when it is selected is given by

$$\bar{V}_k = \sqrt{\frac{1}{2N}\left[\left(V_R - V_{C,k1}\right)^2 + \left(V_R - V_{C,k2}\right)^2 + \sum_{\substack{i=1 \\ i \neq k}}^{N}\left(V_{C,i1}^2 + V_{C,i2}^2\right)\right]}$$

$$= \sqrt{\frac{1}{2N}\left[2V_R^2 - 2V_R V_{C,k1} - 2V_R V_{C,k2} + \sum_{i=1}^{N}\left(V_{C,i1}^2 + V_{C,i2}^2\right)\right]} \tag{3.26}$$

And using (3.24) and (3.25) we get

$$\bar{V}_k = \sqrt{\frac{1}{N}\left(V_R^2 - 2a_k V_R V_C + N V_C^2\right)} \tag{3.27}$$

which is the wanted result, showing that $\bar{V}_k$ depends only on a_k.

The ON and OFF RMS pixel voltages are obtained respectively by setting a_k equal to −1 and +1. They are equal to (3.4a and b) valid for the A&P technique without gray levels. Therefore, regarding the drive margin and maximum supply voltage, the same considerations apply.

Column Pulse Height Modulation can be combined with Frame Modulation to reduce the required number of column voltage levels [HNK1995].

3.4.6 Row Pulse Height Modulation

With this technique, the amplitude of the row voltage waveforms is modulated to vary the RMS voltage across the LC. In brief, the amplitude is reduced by a factor of 2 in the successive time intervals from the most significant bit (MSB) to the least significant bit (LSB). The column signal is generated accordingly [MNK1993].

By dividing the *row select time* into m equal time intervals, 2^m gray levels are obtained. This approach can be adopted in conjunction with MLA techniques.

3.5 Concluding Remarks

In this chapter we have mainly dealt with passive-matrix displays and their driving techniques. Passive-matrix displays were the first marketed and are still used in low-cost applications due to the simple fabrication process (compared to active-matrix displays) and thanks to the fact that all the electronics can be implemented in a standard CMOS technology (avoiding processing the thin film transistors directly on glass).

We have seen that there are two principal addressing techniques that use a single line or a multiple line approach. The first requires simple signal generation, but needs high voltage levels (therefore some driver architectures and/or standard IC processes cannot be used), it is also exposed to crosstalk, and mandates for slow LC mixtures (otherwise frame response may become visible). It is hence not suitable for dynamic contents (i.e., rapidly changing frames).

In order to overcome these limitations, multiple line techniques, especially in the *distributed* variant, have been introduced. Thanks to the simultaneous selection of a group of rows, the driving voltage is reduced, each pixel receives many small pulses throughout the frame interval instead of a single large pulse and a long dead time, and transmission is increased. Besides, fast responding LCDs (for full motion applications) can be utilized. The main drawback of these techniques is related to the increased number of required column voltages, increased complexity of the signal generation and additional memory required. This limits to a maximum number of 6 the rows collectively selected. Actually, the MLA-4 technique is commonly used in the cellular phone market.

It was also shown that contrast in a passive matrix display depends on the drive margin (DM) as well as the steepness of the electro-optic characteristics. The DM is defined as the ratio of RMS voltage across ON pixels to that across OFF pixels in the display. A high DM ensures good contrast even when the electro-optic characteristic is not steep as in the case of TN LCDs. In contrast, STN LCDs have steeper electro-optic characteristics but response times of about a few hundred milliseconds, larger than those exhibited by TN LCDs that are in the range of tens of milliseconds (MLA STN LCDs have response times in the range of 20 ms).

The most relevant techniques to reproduce different levels of gray, which is an essential feature for the generation of color images, were lastly discussed.

We conclude this chapter by observing that some electronic equipment like oscilloscopes, logic analyzers, as well as electrocardiographs must basically display waveforms. In a normally white display, to show one waveform, just one pixel in each column is OFF, as it is a point on the waveform, and all other pixels in the display are driven to the ON state, as they are background pixels. The correlation properties of pseudo random binary sequences (PRBS) were exploited in an addressing technique that has an infinite drive margin, but that was not suitable for displaying multiple waveforms [SH1973]. Techniques based on single line addressing [R1986] and multiple line addressing [PR2000, PR2005] were proposed to display multiple waveforms (restricted patterns). Since the number of OFF pixels in each column is equal to the number of waveforms, W, this fact was exploited to achieve a drive margin that is independent of the matrix size.

$$DM = \sqrt{\frac{\sqrt{W}}{\sqrt{W}-1}} \tag{3.28}$$

Equation (3.28) shows that restricted pattern addressing gives a much higher drive margin than that of general pattern addressing given in (3.14). Therefore, in these applications, TN LCDs can be used even with a large display matrix size.

References

[AH2000] O. Akimoto and S. Hashimoto, "A 0.9-in. UXGA/HDTV FLC Microdisplay," *SID Digest*, Vol. 31, No. 1, pp. 194–197, 2000.

[AP1974] P.M. Alt and P. Pleshko, "Scanning Limitations of Liquid-Crystal Displays," *IEEE Transaction on Electron Devices*, Vol. 21, No. 2, pp. 146–155, 1974.

[B1984] K.G. Beauchamp, *Applications of Walsh and related functions with an introduction to sequency theory*, Academic Press, 1984.

[CS1992] A.R. Conner and T.J. Scheffer, "Pulse-Height Modulation (PHM) Gray Shading Methods for Passive Matrix LCDs," *Proceedings of Japan Display '92*, pp. 69–72, 1992.

[GM1994] P. Gulick and T. Mills, "Active Addressing(TM) of Passive Matrix Displays," *Information Display*, Vol. 10, pp. 14–17, 1994.

[HKM1992] H. Hirai, S. Kondo, A. Murayama, and H. Hatoh, "A New Driving Method for Crosstalk Compensation in Simple-Matrix LCDS," *Proceeding of Japan Display '92*, pp. 499–502, 1992.

[HNK1995] Y. Hirai, A. Nakazawa, K. Kawaguchi, H. Motegi, H. Koh, T. Kuwata, Y. Nakagawa, T. Ohbiki, M. Noguchi, and H. Araki, "A Color STN-LCD for Video Display Using Amplitude-Modulation MLS Technology," *SID Digest*, pp. 339–342, 1995.

[KHK1980] H. Kawakami, H. Hanmura, and E. Kaneko, "Brightness Uniformity In Liquid Crystal Displays," *SID Digest*, Vol. 11, pp. 28–29, 1980.

[KHM1990] Y Kaneko, M Haraguchi, H Murakami, and H Yamguchi, "Crosstalk-Free Driving Methods for STN-LCDs," *SID Digest*, Vol. 31, No. 4, pp. 333–336, 1990.

[KI1997] T. Kurumisawa and A. Ito, "Eliminating Moving Picture Cross-Talk by the Self-Calibrating Driving Method (SCDM)," *SID Digest*, pp. 33–36, 1997.

[K2000] K.E. Kuijk, "Minimum-Voltage Driving of STN LCDs by Optimized Multiple-row Addressing," *Journal of the SID*, Vol. 8, No.2, pp. 147–153, 2000.

[KNK1976] H. Kawakami, Y. Nagae, and E. Kaneko, "Matrix Addressing Technology Of Twisted Nematic Liquid Crystal Display," *Proceedings of Biennial Display Conference*, pp. 50–53, 1976.

[M1982] B.W. Marks, "Power Reduction in Liquid-Crystal Display Modules," *IEEE Transactions on Electron Devices*, Vol. 29, No. 12, pp. 1884–1886, 1982.

[M1984] P. Maltese, "Cross-Modulation and Disuniformity Reduction in the Addressing of Passive Matrix Display," *Proceedings of Euro Display '84*, pp. 15–20, 1984.

[M1987] P. Maltese, "Complex Polarity Sequences For Directly Driven Liquid Crystal Matrices," *Proceedings of Euro Display '87*, pp. 139–143, 1987.

[MNK1993] H. Mano, S. Nishitani, K. Kondo, J. Taguchi, and H. Kawakami, "An Eight-Gray-Level Drive Method for Fast-Responding STN-LCDs," *SID Digest*, pp. 93–96, 1993.

[NK1979] J. Nehring and A.K. Kmetz, "Ultimate Limits for Matrix Addressing of RMS Responding Liquid Crystal Displays," *IEEE Transactions on Electron Devices*, Vol. 26, No. 5, pp.795–802, 1979.

[NMK1993] S. Nishitani, H. Mano, and Y. Kudo, "New Drive Method to Eliminate Crosstalk in STN-LCDs," *SID Digest*, pp. 97–100, 1993.

[NTN1999] A. Nakazawa, K.Tamai, M. Nagai, M. Kitamura, Y. Hirai, S. Matsuoka, and Y. Soda, "Ultra-Low-Power STN-LCDs Using Multiple-Line Addressing for Mobile Telecommunication Applications," *Journal of the SID*, Vol. 7, No. 4, pp. 277–280, 1999.

[PCS1995] D. Prince, B. Clifton, T.J. Scheffer, and B. Leybold, "Identifying and Eliminating 'Splicing' in Multiple Line Addressed STN Displays," *Proceeding of Japan Display '95*, pp. 241–244, 1995.

[PR2000] K.G. Panikumar and T.N. Ruckmongathan, "An Addressing Technique for Displaying Restricted Pattern in RMS-Responding LCD's by Selecting a Few Rows at a Time," *Journal of the SID*, Vol. 8, No. 2, pp. 155–162, 2000.

[PR2005] K.G. Panikumar and T.N. Ruckmongathan, "A Controller for Liquid Crystal Displays in Logic Analyzers and Oscilloscopes," *IEEE/OSA Journal of Display Technology*, Vol. 1, No. 1, 2005.

[R1986] T.N. Ruckmongathan, "An LCD for Multitrace Oscilloscope," *SID Digest*, pp. 128–131, 1986.

[R1988] T.N. Ruckmongathan, "A Generalized Addressing Technique for RMS Responding Matrix LCDs," *Proceedings of International Display Research Conference*, pp. 80–85, 1988.

[R1993] T.N. Ruckmongathan, "Novel Addressing Methods For Fast Responding LCDs," *Reports of the Research Laboratory, Asahi Glass Company Ltd.*, Vol. 43, pp. 65–87, 1993.

[RS2005] T.N. Ruckmongathan and A.R. Shashidhara "Sparse Orthogonal Matrices for Scanning Liquid Crystal Displays," *Journal of Display Technology*, Vol. 1, No. 2, 2005.

[SC1992] T.J. Scheffer and B. Clifton, "Active Addressing Method for High-Contrast Video-Rate STN Displays," *SID Digest*, pp. 228–231, 1992.

[SH1973] I.A. Shanks and P.A. Holland, "Addressing Methods for Nonmultiplexed Liquid Crystal Oscilloscope Displays," *SID Digest*, pp. 112–113, 1979.

[SOM2006] S. Smorfa, M. Olivieri, R. Mancuso, and M. Lienhard, "A Physical-Level LCD Driver Model and Simulator with Application to Pixel Crosstalk Suppression," *IEEE Transactions on Consumer Electronics*, Vol. 52, No. 3, pp. 1027–1034, 2006.

[SSA1983] Y. Suzuki, M. Sekiya, K. Arai, and A. Ohkoshi, "A Liquid-Crystal Image Display," *SID Digest*, pp. 32–33, 1983.

[SSK2003] N. Sako, H. Sasama, H. Kitayama, and H. Fukasawa, "Single-chip Driver for 65k Color STN-LCD with Half Column Voltages in a MLA Drive System," *SID Digest*, Vol. 34, No. 1, pp. 324–327, 2003.

[W1997] M. Watanabe, "High Resolution, Large Diagonal Color STN for Desktop Monitor Application," *SID Digest*, Vol. 34, pp. M81–87, 1997.

[WIO1997] M. Watanabe, Y. Inoue, T. Ohaba, and K. Taniguchi, "High Resolution, Large Diagonal Color STN For The Desktop Monitor Applications," *International Display Research Conference*, M81–M87, 1997.

[WS1997] I. Washizuka and A. Sakamoto, "Driving Method to Reduce Shadowing for STN-LCD Module," *Proceedings of International Display Workshop*, pp. 297–300, 1997.

Chapter 4

Drivers for Passive-Matrix LCDs

This chapter is devoted to the description of reliable architectures and circuit solutions suitable to efficiently implement *drivers* for passive-matrix LCDs. Since passive-matrix LCDs have a limited number of rows and columns (the present maximum resolution for hand-held devices adopting passive-matrix LCDs is the Quarter-VGA having 320 rows×240 columns), the driver is commonly a single-chip[1] mixed-signal integrated circuit (**driver IC**). It includes as main functionalities: microprocessor interface, control logic, display memory, temperature compensation, voltage-level generation/regulation, and drive (row and column) outputs.

It is worth noting that research has continuously produced new materials suitable for display applications, even substantially different from LCs (see Appendix A for alternative flat panel technologies). These materials need driving circuits with somewhat different requirements than those treated here [L2001]. Nevertheless, the architecture discussed here is conceptually of broad application and many of the building blocks presented are of general significance.

In this chapter the typical driver IC architecture is introduced in section 4.1. A discussion of the constituting blocks is performed in Sections 4.2–4.6. The features required to the IC fabrication technology suitable for the application are described in Section 4.7. Finally, a survey on the chip assembling methods is given in Section 4.8.

4.1 Driver Architecture

It was described in Section 3.2 that single line or multiple line addressing approaches are possible in PMLCDs. CSTN LCDs often adopt the MLA technique which offers the best performance. For this reason, unless otherwise stated, we will consider this driving approach in this chapter.

---

[1]Because of the greater number of rows and columns, large-area (active-matrix) LCDs cannot be driven by a single chip which would require an unfeasible number of pins. As we will see in Chapter 6, they adopt a different architecture made up of a controller board and several row and column driver ICs.

D.J.R. Cristaldi et al., *Liquid Crystal Display Drivers*, DOI 10.1007/978-90-481-2255-4_4,

The generic block diagram of a passive-matrix LCD driver[2] is illustrated in Fig. 4.1. Other than the LCD panel, that is represented for reference as it does not belong to the driver, we have indicated as main blocks the **Host Interface**, **LUT** (Look-up Table)/**Dither**, Display Data **RAM**, **Color Processing**, **MLA and Grey Generation**, **Control Logic**, **Timing Controller**, **OTP** (One Time Programmable) **Memory**, **Oscillator**, **Temperature Sensor**, **Temperature Compensation**, **DC/DC Converter**, and finally **Column and Row Drivers**. It is worth mentioning that some block functions have distributed allocation when actually implemented, but are here described as concentrated for the sake of clarity. Also, the diagram does not include the power supply, but two external supplies are usually required: one for the host interface and the other for the DC/DC converter and analog blocks. The value of the former is related to the voltage levels adopted by the standardized signalling approach (usually, it is equal to 1.8 V), the latter depends on the silicon process utilized and is somewhat higher (e.g., 2.8 V). Today there is a certain tendency to adopt only a single supply, but this trend has not produced standardization yet.

Some of the blocks included in Fig. 4.1 perform functions already explained in the previous chapter and should be familiar to the reader, other blocks and their related functions will be explained in the following.

Data coming from a micro controller unit (MCU host) can represent either commands or video information. They are initially interfaced and prepared to be respectively interpreted (by the Control Logic section) or stored (into the RAM). Video data are also optimized through Color Processing techniques (e.g., Gamma Correction). On-chip oscillators and stabilized DC voltage generators are also necessary for synchronization and generation of accurate reference voltages. Proper biasing with high-voltage boosting circuits is also required to generate pixel voltage levels greater than the supply one. Column and row voltages are finally provided as outputs through the Column and Row Drivers.

The implementation of all these functions should be optimized to contain chip size and cost,[3] as well as the required number of external components, such as capacitors used by the voltage multipliers (pumping capacitors) or needed to absorb large current spikes (bypass capacitors). Reduction of power consumption[4] and crosstalk with enhanced protection to electrostatic damage (ESD) are also important issues. Lastly, easily programmable display features are often desirable from the end user. As a result, both the utilized silicon process with its minimum lithographic resolution (0.18–0.25 μm are common) and the adopted circuit design techniques play an important role in the driver economy considering that, owing to the great number of output channels, the aspect ratio of the die is usually high (e.g., 20 mm×1 mm). However, unlike one could be induced to think by the large number of output pads, an IC driver for small-area LCDs is not PAD limited for two main reasons:

---

[2]Another electrical section, the backlight inverter (a DC/AC converter) is present in an LCD, but it is implemented as a separate circuit.

[3]Given the strong competition in the flat-panel display market, the cost is perhaps the most crucial parameter. Thanks to the scaling-up and technological simplification the price of mobile devices has been steadily decreasing by approximately 30% a year.

[4]Power consumption of PMLCD drivers for mobile applications is usually around 2 mW.

(1) With the evolution of the market and technologies, the display manufacturers have reduced the assembly pitch to follow the lithographic scaling of silicon driver technologies.

(2) High-performance AMLCDs are potentially PAD-limited, but they may adopt LTPS technologies that allow the PAD number to be reduced thanks to the implementation of multiplexers directly on the glass. At this purpose, a single column line instead of three for the RGB subpixels is typically adopted; and some manufacturers have succeeded in implementing a 12-to-1 MUX.

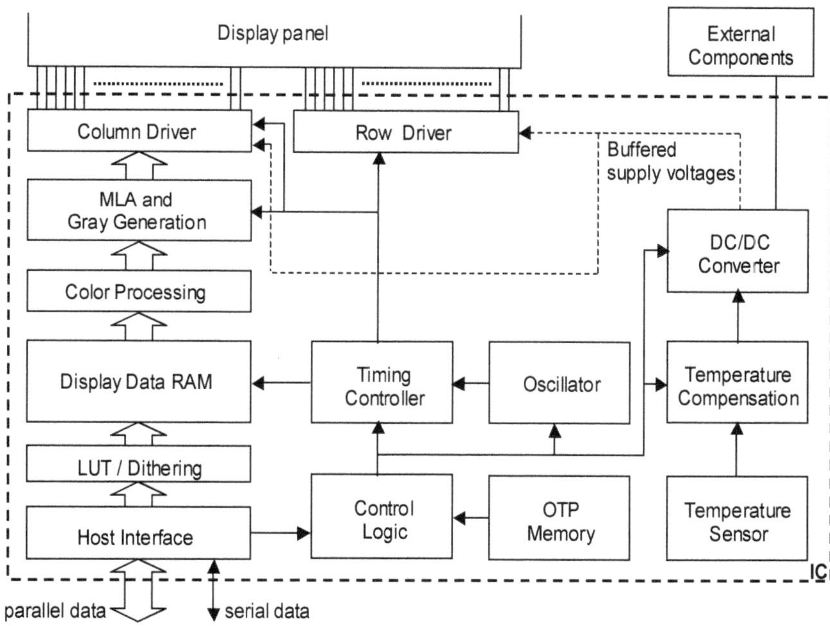

Fig. 4.1 Block diagram of a PM-LCD driver IC (*within dashed box*)

In the following sections we will explain in detail the behavior of a PMLCD driver by grouping its functions into 5 main categories, namely: (1) Host Interface, (2) Power Manager, (3) Driver Manager, (4) Image Processing and (5) Output Drivers.

4.2 Host Interface

4.2.1 Interfaces

Image data and commands are transferred from the MCU host to the driver, whereas diagnostic data and state information (e.g., driver ON/OFF condition) are transferred from the driver to the host. A bidirectional data link between the MCU and the driver is hence required. At this purpose, either parallel type or serial type communication interfaces have been exploited. In presence of multiple standards, to maximize flexibility and easy of use, a general driver IC is equipped with the most popular industry standard host bus interfaces such as the (parallel type) **68 000** and **8080** and (serial type) **3 lines 9 bit** and **4 lines SPI** (Serial Peripheral Interface).

The two well-known parallel-type interfaces (introduced by Motorola and Intel, respectively) were the first exploited. A detailed description of these interfaces is beyond the scopes of this book. Here it is sufficient to say that they adopt CMOS standard signalling levels (that have progressively reduced from 3.3 V, 1.8 V, and in the next generation products to 1.2 V) and provide maximum speed rates up to 20 MHz for each channel.

Note that from the physical point of view, speed is ultimately limited by the RC time constant of the connection lines. However, the transfer rate between the host and the driver has significantly increased over the last years, to support the progressive migration of displayed data from text, to images, video, and full color compatibility. At this purpose, parallel-type interfaces were the first adopted for their higher transfer rates obtained by transmitting more bits in parallel. The main drawbacks of parallel connections are related to their cost and high exposure to mechanical breaks (of bus lines and connector contacts), which obviously may harm the whole system. The aforementioned problems are considerably reduced by adopting serial type interfaces, as they require a limited number of physical lines and are thus cheaper and more reliable.

We have mentioned two serial-type interfaces. These are briefly described below:

– 3 lines 9 bit from Nokia
the three lines are used for *enable*, *clock*, and *data*, the latter being in the 8-bit format plus 1 additional bit for datum/command discrimination.

– 4 lines Serial Peripheral Interface (SPI)
the function of the first three lines is the same as above, the fourth line is used for datum/command discrimination. Data are in the 8-bit format.

It should be observed that these are just two examples, since new interfaces are continuously developed. For example, Motorola adopts another interface type that uses 16 bit. The first 8 bit are used to code the command and the remaining 8 bit represent the data. It simplifies the subsequent elaboration task but can be less efficient for long data series transfer.

In conclusion, it is worth noting that these "low-speed" serial interfaces are typical of PMLCDs, owing to the limited data-transmission rate required. High-speed serial interfaces, required by AMLCDs, will be discussed in Section 6.2.

4.2.2 LUT/Dithering

Before being stored into the RAM, data at the output of the host interface may be remapped to fit the RAM word length. Two techniques may be needed: *insertion of additional bits* and *rounding*.

The former approach is required if the data length is smaller than the RAM word length. In this case, we need extending a reduced data color depth into the maximum one dictated by the RAM through a **Look-Up Table** (LUT) that is an associative array giving an output value for each of a range of index values. The LUT is also called *color map* or *palette*.

For example, consider a 18-bit RAM and 8-bit video data (as in the first case illustrated in Fig. 4.2). Assume that a number of 3 bit are used for Red and Green colors and 2 bit for Blue (3R 3G 2B). Data can be hence remapped to employ all 18 bit (6R 6G 6B). Other examples are included in Fig. 4.2.

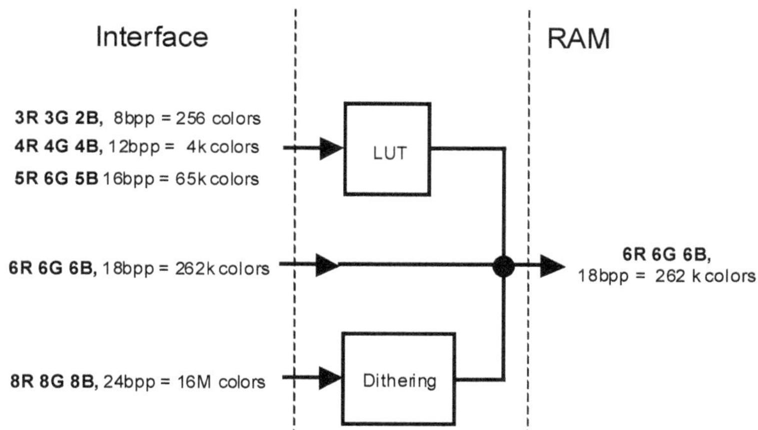

Fig. 4.2 Example of LUT/Dithering operation required in the case of a 18-bit video RAM

On the contrary, when the incoming data word length is greater than that allowed by the RAM we need to reduce the color depth, as illustrated in the last case of Fig. 4.2. At this purpose, the easiest solution is to truncate some bits. However, truncation may cause a significant error in the displayed image, like the generation of false contours. To limit this problem, several rounding (**dithering**) techniques have been developed such as conventional rounding, error feedback rounding, and dynamic rounding [J2001]. For example, error feedback rounding is performed by storing the residue of a truncation and adding it to the next addressed pixel. It substitutes less visible noise-like quantizing errors instead of contouring artefacts of simple truncation. An implementation scheme is shown in Fig. 4.3 where data word length is reduced from 16 to 8 bit. Note that the register included in the feedback path produces 16-bit output word in which the eight most significant bits are "0" and the remaining 8 bit are the 8 LSB of the previous output word.

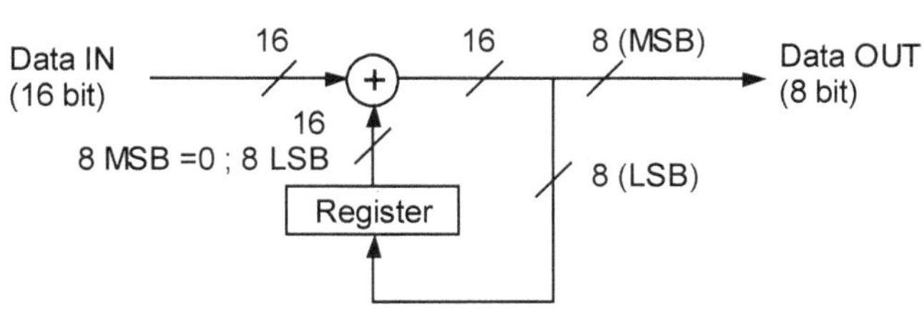

Fig. 4.3 Example of dithering algorithm using error feedback

4.3 Power Manager

The Power Manager section is devoted to the generation of all the bias and reference voltages (and currents) required by the driver circuits. Especially for hand-held devices, that can be subjected to dramatic temperature excursions during normal operation, it is essential that the reference voltages are temperature stabilized. The use of color graphics also worsens this problem since the color quality heavily depends on the absolute value of the LC threshold voltage and on the accuracy of the driving voltage level.

4.3.1 Temperature-Compensated Voltage Generator

All the internal reference voltages are generated from a reference bias circuit that is temperature compensated. A DC voltage, independent of temperature, is obtained through well-established circuits called *bandgap references*.[5]

A **bandgap** reference voltage has nominal zero temperature dependence as it is obtained by summing two voltages with proper weighting that have respectively negative and positive temperature coefficients. These voltages rely on the exponential voltage-to-current law of bipolar devices, for both positive (thermal voltage, V_T) and negative (base-emitter voltage, V_{BE}) temperature coefficients [W1971, K1973, B1974, BSU1999, MMF2001, GPC2003].

A bandgap reference circuit can be implemented in CMOS technologies by exploiting the inherent parasitic bipolar devices. Two examples of bandgap voltage generators are illustrated below.

The voltage reference in Fig. 4.4 is based on a differential amplifier. The bias currents, I_{C1} and I_{C2}, are set equal by imposing equal voltage drops across both the resistors R_3 and R_4 exploiting the virtual ground of the OpAmp.

Design equations are

$$I_{C1,2} = \frac{V_{BE1} - V_{BE2}}{R_2} = \frac{V_T}{R_2} \ln\left(\frac{A_{E2}}{A_{E1}}\right) \tag{4.1}$$

$$V_{REF} = V_{BE1} + 2R_1 I_{C1,2} = V_{BE1} + 2V_T \frac{R_1}{R_2} \ln\left(\frac{A_{E2}}{A_{E1}}\right) \tag{4.2}$$

This circuit requires that npn BJTs are available in the adopted CMOS technology.

[5]As known, this terminology is due to the fact that limit for T->0 K of the obtained reference voltage is the *bandgap* voltage of silicon.

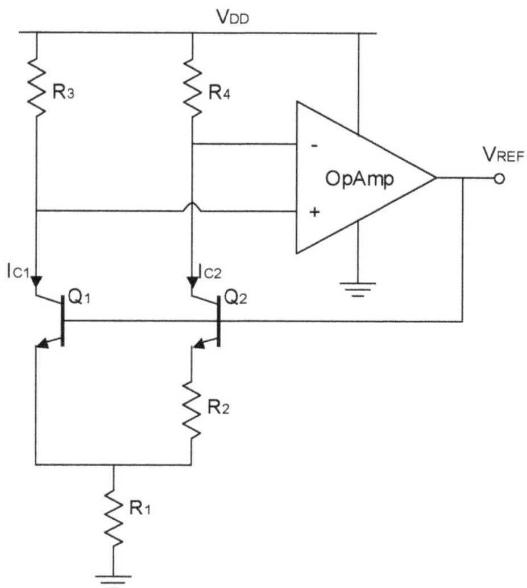

Fig. 4.4 Bandgap voltage reference with bipolar transistors

The circuit topology shown in Fig. 4.5 is fully compatible with a CMOS technology, because the npn transistors can be replaced by pnp or even simple pn junctions. Assuming the circuit properly biased, voltage V_{REF} is set to $V_{BE1}+(R_1+R_3)I_{C1}$. Considering the virtual ground of the OpAmp, current I_{C1} can be found by inspection and results in

$$I_{C1} = \frac{V_T}{R_3}\ln\left(\frac{A_{E1}}{A_{E2}}\frac{I_{C2}}{I_{C1}}\right) \tag{4.3}$$

The virtual ground also forces the voltage drops across R_1 and R_2 to be equal, that is $I_{C1}/I_{C2}=R_2/R_1$. Substituting this expression in (4.3), V_{REF} results to be

$$V_{REF} = V_{BE1} + V_T\left(1+\frac{R_1}{R_3}\right)\ln\left(\frac{A_{E1}}{A_{E2}}\frac{R_1}{R_2}\right) \tag{4.4}$$

Generally, transistors Q_1 and Q_2 are equal so that their emitter ratio is unitary and the output voltage mainly depends on resistive ratios.

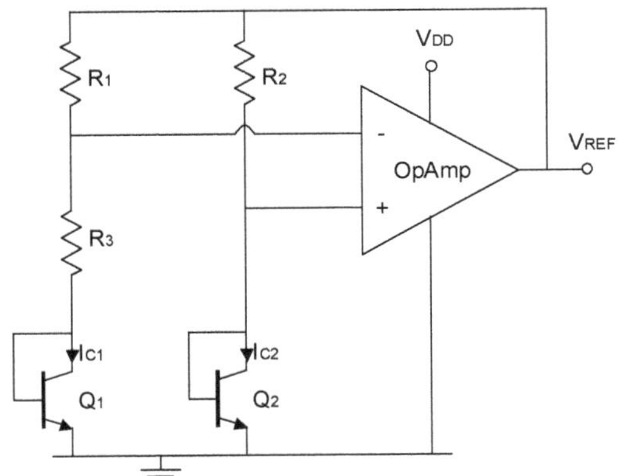

Fig. 4.5 Bandgap voltage reference circuit compatible with CMOS technologies

Figure 4.6 shows a typical behavior of the voltage versus temperature obtained by such circuits. They provide a temperature drift performance in the range of about 20–100 ppm/°C.

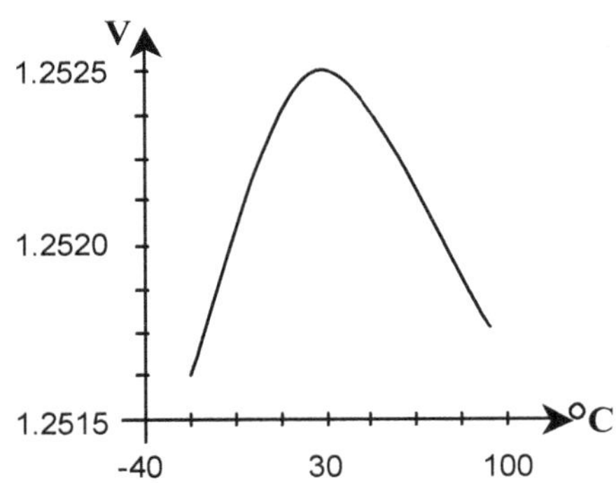

Fig. 4.6 Typical behavior of bandgap reference voltage versus temperature

4.3.2 Temperature Sensor

It was explained in Section 2.2.2 that the electro-optic (*T*/*V*) characteristic is dependent on the temperature. This problem is illustrated in Fig. 4.7a, where three curves are plotted for different temperatures $T_1 > T_A > T_2$, where T_A is the ambient temperature (be careful that letter *T* is used for both Transmittance and Temperature). From the figure it is seen that V_{TH} (and also V_{SAT}) decreases with the temperature, this is because the viscosity of the LC mixture decreases with the temperature and hence a reduced energy is required for the realignment. More specifically, the typical curve of V_{TH} versus temperature for STN LCDs is illustrated with solid curve in Fig. 4.7b.

As described in (3.18) and (3.19), V_{LCD} is proportional to V_{TH}, therefore in order to avoid that display contrast is affected by temperature variations, V_{LCD} must follow the same V_{TH} temperature behaviour. At the system level, V_{LCD} can be reconstructed by a piece-wise curve. This allows a simplification of the design and the minimization of data stored in the OTP. For example, the dashed curve in Fig. 4.7b represents one of the possible implementations of the curve into three segments having two intersections at temperatures T_L and T_H. T_L and T_H as well as the slopes of the three segments can be fine-tuned by the module maker according to the adopted LCD mixture and then stored in the OTP.

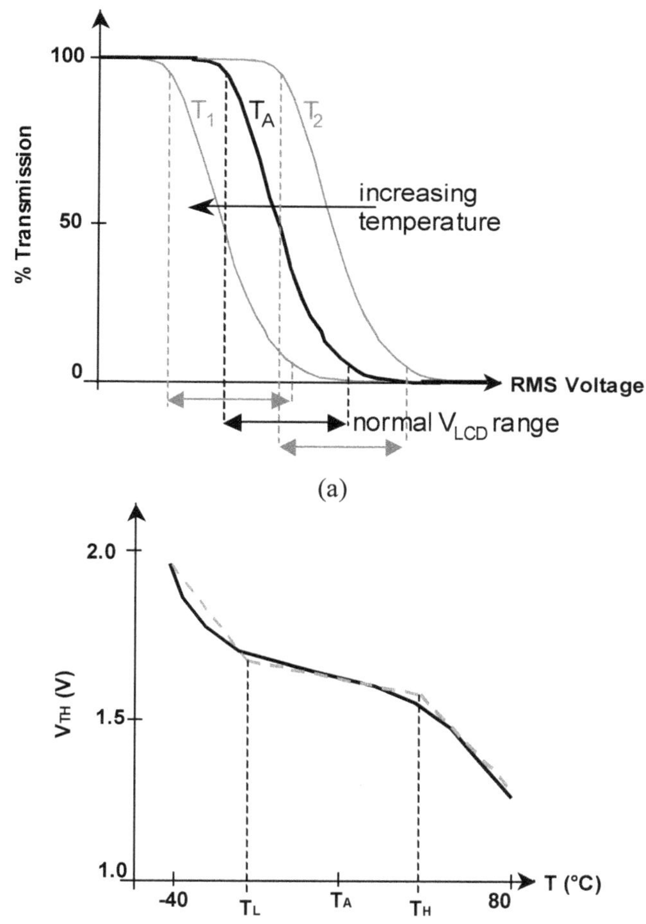

Fig. 4.7 (**a**) Electro-optic transfer function for three different temperatures, $T_1 > T_A > T_2$, highlighting the associated V_{LCD} ranges. (**b**) Threshold voltage versus temperature (*solid line*), and a possible segmentation (*dashed line*)

A temperature sensor is used to monitor the junction temperature of the driver, which is supposed to be very close to the display temperature. This information is used not only in the block that generates the temperature segmentation of V_{LCD}, but also to change the frame frequency adapting it to the operating conditions of the LCD mixture. A possible block diagram of a temperature sensor is shown in Fig. 4.8.

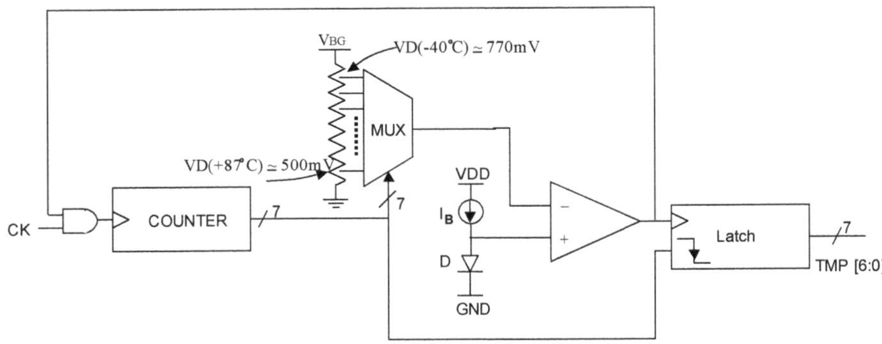

Fig. 4.8 Block scheme of the Temperature Sensor

CK is a synchronization signal coming from the Local Oscillator and feeding the 7-bit Counter. If the input feedback voltage at the AND gate is high, then the Counter output is increased by 1 bit at each clock pulse. The Counter output is then converted to an analog voltage through the resistor-string DAC and compared with the junction voltage of diode, D. The diode is forward biased through a constant current I_B and its useful voltage range is comprised between $V_D(-40°C) = 770$ mV and $V_D(+87°C) = 500$ mV with a slope of –2 mV/°C. As far as the DAC output voltage is higher than the diode voltage, the counter will increase its output. The count up is stopped as soon as the output comparator switches down and the counter result is transferred to the Latch Register.

Observe that the DAC output corresponds to 770 mV for the lowest input word, 00HEX (i.e., the output voltage decreases for increasing input word).

A digital low pass filter can be inserted after the Latch Register to average the temperature value over a few (e.g., 4) acquisitions.

With a 7-bit variable TMP[6:0], the temperature ranging from the minimum value, –40°C, to the maximum 87°C can be detected with a resolution of 1°C.

Finally, the expression of the segmented V_{LCD} as a function of the temperature is given by

$$V_{LCD} = \begin{cases} V_{LCD,TA} + (T_L - T_A)S_{TA} + (T - T_L)S_{TL} & \text{for} \quad T < T_L \\ V_{LCD,TA} + (T - T_A)S_{TA} & \text{for} \quad T_L < T < T_H \\ V_{LCD,TA} + (T_H - T_A)S_{TA} + (T - T_H)S_{TH} & \text{for} \quad T > T_H \end{cases} \tag{4.5}$$

in which $V_{LCD,TA}$ is the nominal V_{LCD} voltage at ambient temperature, whereas S_{TA}, S_{TL}, and S_{TH} are the slopes of the segments defined by the module maker and stored in the OTP. Figure 4.9 illustrates an example of the segmented V_{LCD}. In the figure are also included V_{LCD,TA_MAX} and V_{LCD,TA_MIN}, that are the boundaries of the range in which V_{LCD} may vary as a function of the threshold voltage and the number of display rows. In general, for a color STN LCD with 128 rows the maximum V_{LCD} variation at ambient temperature is comprised between 5 V and 15 V.

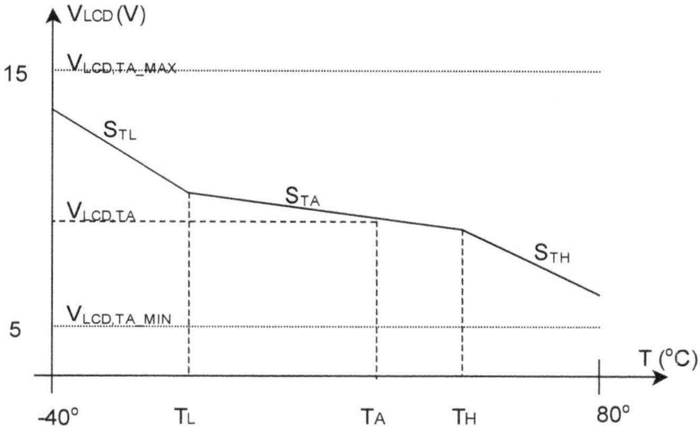

Fig. 4.9 Example of a segmented temperature programmable V_{LCD} voltage. The temperature range is split into three segments. Parameters $V_{LCD,TA}$, S_{TL}, S_{TA} and S_{TH} are the only stored into the OTP

4.3.3 DC/DC Converter

For battery-operated portable devices, the external supply is kept low to contain power consumption, therefore integer multiplication factors (×2, ×3, ×4 and even more) in both positive and negative amplitudes are required to generate the row and column voltages. A DC/DC converter that uses only switches (diodes originally) and capacitors is referred to as charge pump [D1976]. This inductorless architecture is amenable for integration and causes relatively low electromagnetic interferences (EMI). It is popular in low noise applications such as EEPROM, implantable pacemakers and, of course, small-size (up to around 8 in.) LCDs [GP1996, M1997, WW1998, YKC2003, SK2008].

The simplified scheme of a voltage multiplier ideally providing an output voltage, V_{out}, equal to $3V_{DD}$ is depicted in Fig. 4.10. The circuit includes three stages, each made up of a diode and a capacitor. C_{IN} models the input generator capacitance, while C_{P1} and C_{P2} are the **pumping capacitors** and C_B is the **bulk capacitor** with function of storage and filtering, acting as a voltage generator for the load. A square-wave oscillator provides two phases that drive one terminal of the pumping capacitors. After a certain transient, dictated by the charging time of the generator-diode-capacitor system and the frequency of the oscillator, C_{P1} is charged to V_{DD}-V_D, and C_{P2} to $2V_{DD}$-$2V_D$ and C_{OUT} to $3V_{DD}$-$3V_D$, in which V_D is the diode voltage drop. The typical behavior of V_{OUT} is depicted in Fig. 4.11. More details on charge pumps can be found in Chapter 7.

Fully integrated multipliers (i.e., without requiring external components) are of course preferred. But this depends on the current drive capability required. Indeed, to obtain large output currents we need large out-of-chip capacitors to avoid fast discharging.

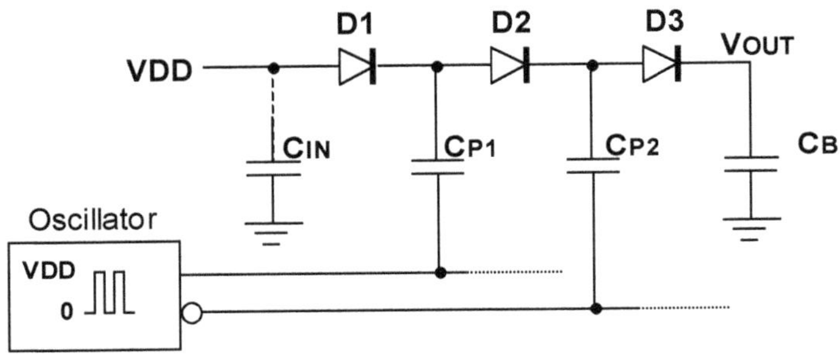

Fig. 4.10 Simplified scheme of a ×3 charge pump

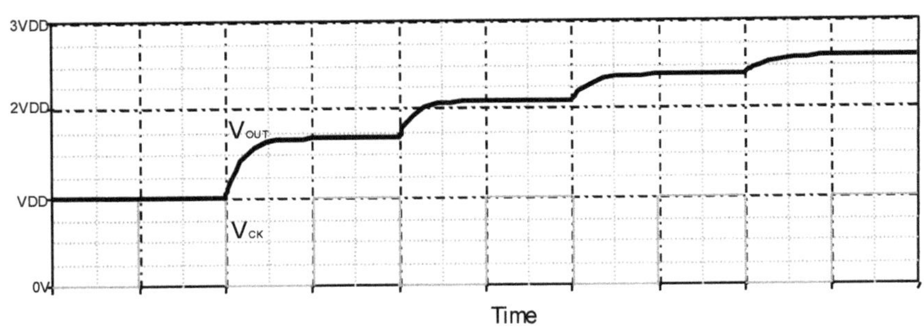

Fig. 4.11 Typical behavior of charge pump output voltage, V_{OUT}, and square wave, V_{CK}

Observe that owing to the high-valued bulk capacitor, a buffer amplifier is not needed to drive the row/column loads. This is instead required for voltages that are obtained by resistive partition. At this purpose, see Fig. 4.12 that illustrates a possible way to generate the voltages for a MLA-4 LCD.

The scheme generates the six required supply voltages $\pm V_R$, $\pm V_C$, and $\pm V_C/2$ (V_{COM} is the common voltage externally connected to ground). Four of them, $\pm V_R$ and $\pm V_C$, are obtained from charge pumps configured in closed loop, CP1, CP2, CP3, CP4, respectively. Each loop being constituted by an error amplifier, a voltage-controlled oscillator (VCO), and a charge pump. The loop works as follows. Consider for instance the loop generating V_R, and assume that it tends to decrease because of the load current demand. Then, also V_R' decreases, the output voltage of the error amplifier EA1 increases as well as the VCO1 frequency and, as a result, CP1 pumps more charges to its bulk capacitor, increasing (regulating) V_R.

The other two voltages, $\pm V_C/2$, are obtained from simple resistive partitioning (resistors R_1–R_2) followed by a Linear Drop-Out Regulator (LDO) buffer. Buffer B1 is supplied between $+V_C$ and V_{COM}, whereas buffer B2 is supplied between V_{COM} and

$-V_C$. Two de-multiplexers and resistive dividers are also embedded into the above mentioned feedback loops. Their role is to set proper row, $+V_R$, and Bias Ratio, defined in (3.10), through coded words V_{LCD}[…] and BR[…]. The last resistor divider (R_5–R_5) is used to generate $-V_R$ from $+V_R$.

It is important to observe, in conclusion, that this architecture adopts a *centralized* driving voltages buffering, as opposite to the *distributed* buffering approach that is usually followed in AMLCDs (see Section 6.3.1).

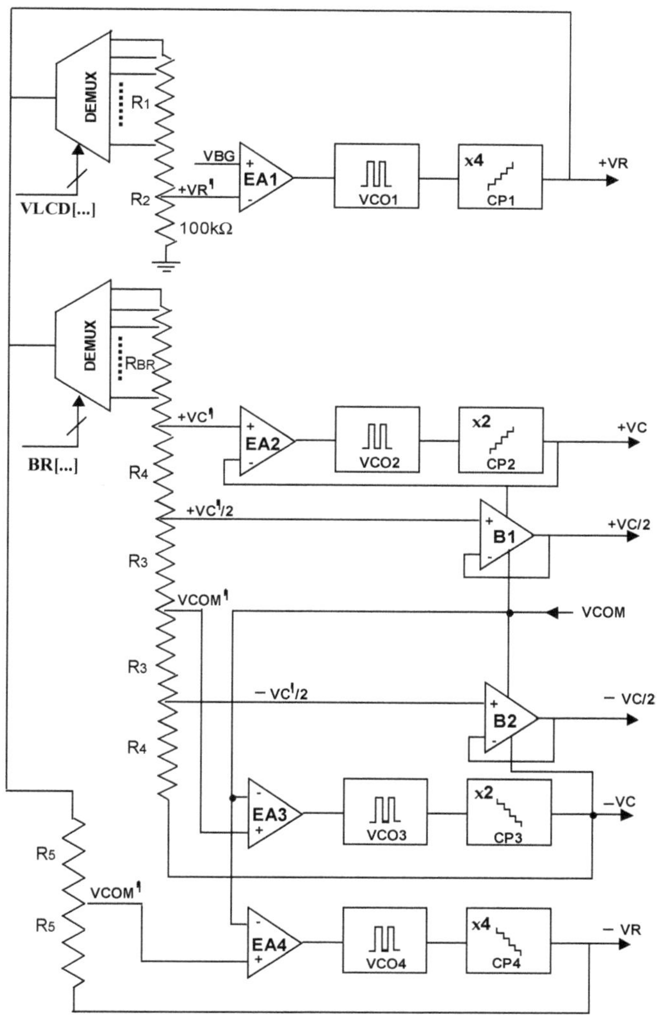

Fig. 4.12 Power Supply for LCD driver using MLA-4 technique

4.4 Driver Manager

4.4.1 One Time Programmable Memory

We have seen that the effectiveness of the LCD matrix in color reproduction depends mainly on the absolute values of parameters such as the LC threshold voltage and on the accuracy of the driving voltage levels. Any tolerance of these values must be hence compensated for. This task is performed by recording permanently in a dedicated non-volatile memory block the value of critical parameters, such as characteristics of the LC mixture, gamma curve data (see Section 4.5.2), the values of the reference voltages, etc., as well as configuration parameters. At this purpose, it is used a Memory called OTP (One Time Programmable) memory which can be implemented in a poly-fuse or anti-fuse technology.

The values of the aforementioned parameters are first measured by the module makers, for each production lot of LC mixture, as these may change from one lot to another one, and then written in the OTP. For instance an OTP word can be 22 bit long: 16 are used to code effective data and 6 are used to implement an Error Correction Code (ECC).

After each system reset or power on, the OTP is automatically initialized. Parameters stored in the non-volatile memory are read and latched in a volatile register situated in the Control Logic section (discussed in Section 4.4.3) and then the OTP is switched off to avoid current consumption. In the end-user application there is the possibility to upgrade the parameters stored in the OTP. This depends on the status of a "Protection Mode" bit present in the OTP. If it is low, said parameters can be upgraded, if it is high, the parameters cannot be upgraded.

4.4.2 Oscillator (Internal Clock Generator)

An internal oscillator, preferably fully integrated (without requiring external components) is included in the driver IC. It generates the master clock signal (at the maximum frequency), from which all the other synchronization signals are derived by frequency division.

A possible scheme of a relaxation oscillator is shown in Fig. 4.13 [P2002]. It is made up of two capacitors C_1 and C_2 alternatively fed by two constant current sources, I_{OSC}, or discharged by MOS transistors, M_1 and M_2, acting as switches. Assuming one of the switches (e.g., M1) be OFF, then the voltage across the associated capacitor increases linearly with the law $(I_{OSC}/C_1)t$ and when it reaches the threshold, V_{TH}, of M1A the drain voltage of M1A goes to "0" as well as the outputs of NOT_3 and NOT_5 (output). Hence, M1 is switched ON and C_1 is discharged, while M2 is switched OFF and capacitor C_2 starts to charge with the same law as C_1. Assuming a duty cycle of 50%, C_1=C_2=C, and the oscillation frequency is given by

$$f_{OSC} = \frac{I_{OSC}}{2CV_{TH}} \tag{4.6}$$

and is electrically controlled by I_{OSC}. For instance, assuming V_{TH}=0.8 V, a 5-MHz oscillation frequency is obtained with I_{OSC}=8 μA and C=1 pF.

Observe that M1A and M2A perform logically as NOT gates, but they are in common source configuration in order to provide a threshold voltage independent of V_{DD} and to limit the cross-conduction current to I_B.

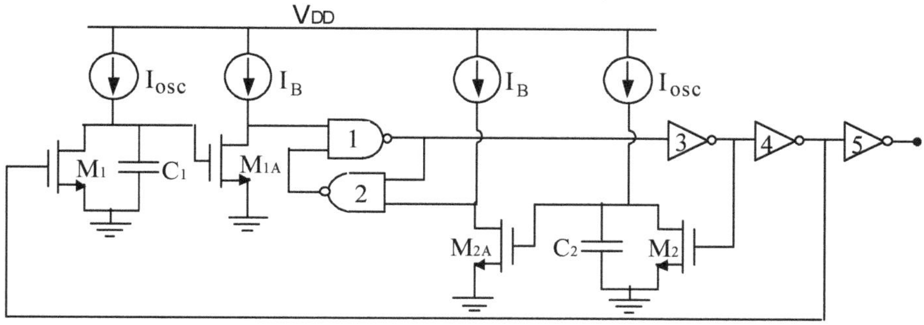

Fig. 4.13 Schematic of a relaxation oscillator [P2002]

Current I_{OSC} must be accurately set starting from the bandgap voltage, V_{BG}, an adjustable resistor and current mirrors, as illustrated in Fig. 4.14. Voltage V_{REF} is derived from V_{BG} by resistive partition and replicated across R_{ADJ}, where I_{REF} is generated ($I_{REF}=V_{REF}/R_{ADJ}$). R_{ADJ} is a resistor tuned during testing procedure to compensate for process variations in mass production.

The multiplexer, MUX, is usually implemented through a binary-tree selector (a three-bit version is depicted in Fig. 4.15).

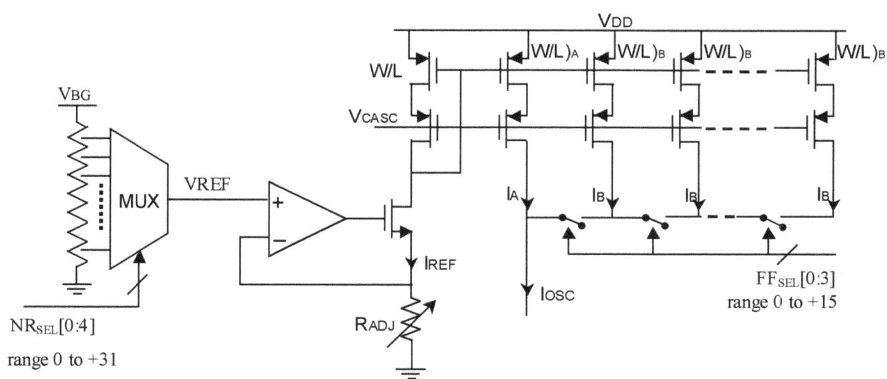

Fig. 4.14 Scheme for the generation of I_{OSC}

Referring to Fig. 4.14, if $I_A=k_1I_{REF}$ and $I_B=k_2I_{REF}$ we write the expression of I_{OSC}

$$I_{OSC} = I_A + FF_{SEL} \cdot I_B = I_{REF}\left(k_1 + k_2 \cdot FF_{SEL}\right)$$
$$= \frac{NR_{SEL}}{N}\frac{V_{BG}}{R_{ADJ}}\left(k_1 + k_2 \cdot FF_{SEL}\right) \tag{4.7}$$

where N is the number of display rows, NR_{SEL} is a 5-bit control word that represents the subset of addressable rows (the user may operate only a portion of the display rows, this mode is usually referred to as *partial display mode*). FF_{SEL} is a 4-bit control word that represents the preferred frame frequency (required for instance to set the idle mode, or to allow frequency change with the temperature and LC mixture). Both are stored in the OTP and tuned by the module maker.

In conclusion, the above equation shows that the oscillation frequency can be set according to the addressable rows and frame frequency selected.

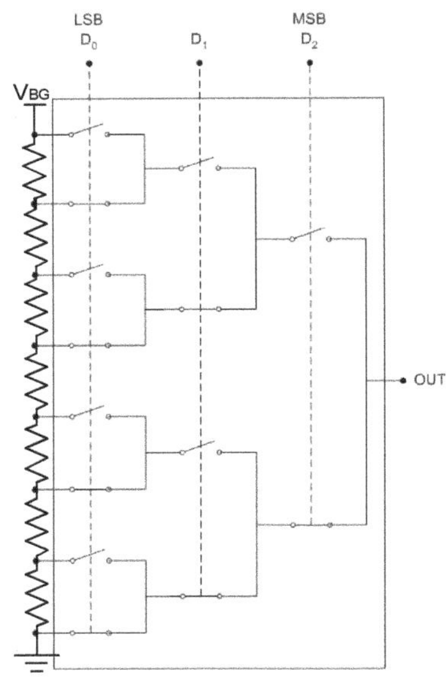

Fig. 4.15 String resistor DAC with three-bit multiplexer (MUX) scheme

4.4.3 Control Logic

The principal task of the Control Logic section is to decode the incoming command (Instruction Decoder) and to take the appropriate actions. It supervises the initialization phase of Timing Controller, Oscillator, DC/DC Converter and Temperature Compensation blocks. It also contains a (volatile) register where OTP parameters are latched at power on.

Advanced tasks offered by this section may include 2D graphic acceleration features to reduce the required processing power from the MCU and leaving it free for other features. For instance, graphic functions are implemented that facilitate the user with simple instructions to draw lines, rectangles, copy or move objects and even define pop-up windows, while speeding up the application system.

4.5 Image Processing

The incoming pixel dataflow is stored into the RAM and subsequently goes through a data elaboration pipeline that performs color processing.

4.5.1 Display Data RAM

The column data of each frame, before being displayed, are stored into a static random access memory, Video SRAM, that is included into the driver. In practice, each pixel of the display is mapped into the RAM. To display color images, each RAM location must be a C-word, where C is the number of color bits. For instance, a 262k-color display requires 18 bit to code the column data. For a 132×162 display the required RAM dimension is $N \cdot M \cdot C$ bit=132·162·18=384,912 bit.

The RAM is essential for power saving when displaying the same image for a long time (like in PMLCD applications). In this case the image data are stored into the RAM and are no longer transmitted by the host processor.

Usually, when data are written in the RAM several displaying modes are possible such as for instance: *normal horizontal*, *normal vertical*, *horizontally mirrored* and *vertically mirrored*.

It is worth noting that CMOS processes with progressively reduced channel lengths are needed to implement single-chip drivers able to manage 4k, 65k, 262k color data, due to the RAM requirements in terms of area occupation. However, finer processes tolerate reduced voltages and hence a trade-off between required performance and suitable technology may arise, which is solved by the development of specific processes, as discussed in Section 4.7.

4.5.2 Color Processing

Color processing includes gamma expansion, correction of nonlinear electro-optical transfer function and, in principle, color space conversion with black/white points adjustment, all performed in the digital domain.

(a) Gamma correction ensures accurate reproduction of color and image detail over a wide range of luminance. To fully understand the subject, we must refer first to the CRT performance, because gamma correction was originally applied to this class of displays. As illustrated in Fig. 4.16a, curve 1, a CRT does not respond to an input voltage linearly, its luminance is approximately proportional to the input voltage raised to the 2.5 power. This number is known as **gamma** (γ) [P1993]. In general any kind of display has a nonlinear T/V characteristic, so that the normalized luminance, L, is proportional to some power of the normalized signal amplitude, x, according to $L = x^{\gamma}$ where $L, x \in [0 \ldots 1]$ and $\gamma > 1$.

As a result, in a CRT, linear RGB data must be predistorted (gamma corrected) before being displayed to compensate for the nonlinearity of electro-optic transfer.

This correction effectively accomplishes a compression of the video signal through the inverse electro-optic transferfunction of the display, $x^{1/\gamma}$ (see curve 2 in Fig. 4.16a), so that the whole system (gamma-correction plus CRT display) response approaches the ideal straight line (see curve 3 in Fig. 4.16a). The input voltage is hence raised to the 0.4 power (i.e., 1/2.5). In reality, since also the human visual system (eyes to brain transferfunction) is nonlinear, such exponent is changed into 0.45 and video standards assume accordingly a display gamma of about 2.2 for CRT displays.[6]

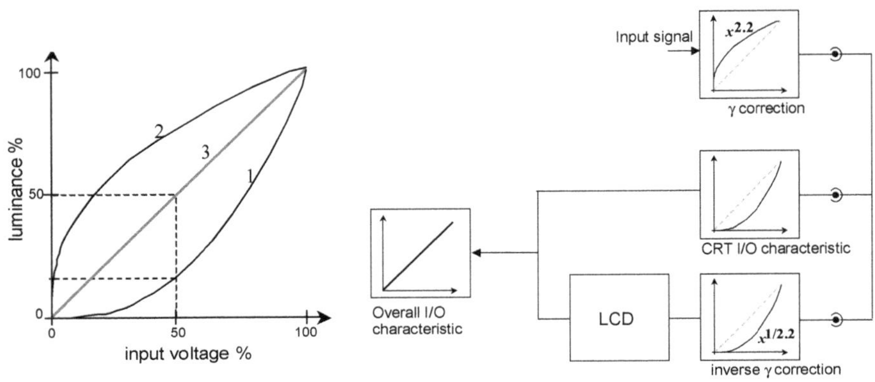

(a) (b)

Fig. 4.16 (**a**) CRT luminance versus input voltage: (1) actual curve, x^{γ}, (2) gamma correction curve, $x^{1/\gamma}$ and (3) ideal (linear) response, x. (**b**) TV broadcasting system. LCD is assumed linear but requires its own γ correction

In the early period of CRT TVs, it was decided to apply the correction directly at the video cameras; a less expensive solution than having to provide gamma correction to every TV set, see Fig. 4.16b.

This means that nowadays, if we want to display TV-coded (and also MPEG-coded) video on an LCD, we must take the gamma correction into account and introduce a transfer function that raises the associated data voltages to the power of 2.2 (x^{γ}, gamma expansion, or inverse gamma correction). This scenario is further complicated by the fact that there are several standards defining different gamma compression functions with slightly different γ values, such as ITU Rec. 709 (TV), CIE L*, sRGB (WEB), etc. [IEC1998, IEC1999].

Of course, gamma expansion must be matched with the standard used to encode incoming images or video.[7] As a consequence, an LCD driver IC cannot operate with a fixed gamma value.

Besides, an adjustable γ is needed for other two main reasons detailed below.

[6]Besides, human eyes are more sensitive to luminance changes in dark areas than to similar-sized changes in bright areas. Therefore, the compression should be reduced in low gray levels because that becomes much significant in dark areas. The solution is to use $\gamma = 1.8$ for $0 \le x \le 0.35$ and $\gamma = 2.2$ for $0.35 < x \le 1$.

[7]In addition, most commercial digital cameras dynamically vary the amount of gamma.

First, a specific gamma correction for the electro-optic nonlinearity of the LCD itself must be devised. This nonlinear behavior depends on variations in manufacturing and process parameters from a production line (LC mixture, polarizers, backlight, etc.). Thus, different panels have different T/V curves and require different gamma values (otherwise, different panels next to each other will have slightly different color response). It should be observed at this purpose that the T/V curves are not exact analytical functions but they are expressed as approximated interpolation of experimental data.

Second, factors such as ambient illumination can play an important role in our perception of gray scale and color. Without the ability to adjust gamma, the reproduced image will suffer from lack of detail in shadow areas (black crush) or oversaturated whites.

Considering a normally white LCD and its T/V nonlinearity, we represent in Fig. 4.17 the gamma correction curve required to display equally distributed gray levels, *GL*, by applying non-equally distributed voltage levels V_{GL}. A curve for positive and negative values is needed by polarity inversion driving techniques.

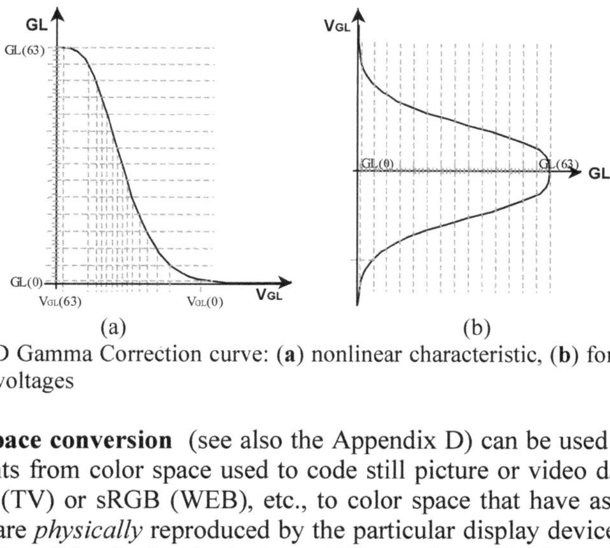

Fig. 4.17 LCD Gamma Correction curve: (**a**) nonlinear characteristic, (**b**) for positive and negative voltages

(b) Color space conversion (see also the Appendix D) can be used to translate RGB components from color space used to code still picture or video data, such as PAL, Rec. 709 (TV) or sRGB (WEB), etc., to color space that have as tristimulus base those that are *physically* reproduced by the particular display device. Indeed, if the RGB primaries of the display device have different chromaticities with respect to either interchange or transmission primaries, then an accurate color reproduction cannot be achieved. A 3×3 matrix is used to transform the color space from the interchange primaries to the display device primaries. This 3×3 matrix takes place in linear-light (tristimulus) space, after gamma expansion. The 3×3 matrix is unambiguously defined by primaries and white reference chromaticities of both interchange and display device color spaces. Hence, by suitably changing the matrix it is possible to change the white reference maintaining the same color primaries. Black and white points are the optical outputs of the display device when RGB inputs are respectively [0, 0, 0] and [255, 255, 255] (considering 24-bit RGB, that uses 8 bit per color component). But the color characteristics of these points depend on characteristics of display components such as backlight, liquid crystal cell, color filters, etc. It is then possible to adjust these two points by adjusting electrical minimum and maximum values for each RGB component.

Actually, in passive displays, in which color quality is not the main concern, this task is heavily simplified and $\mathbf{T}_{1\to2}$ is a straightforward diagonal matrix.

From the above considerations it emerges that the **pixel elaboration flow** has to be adaptable to different ambient conditions, to different encoding standards and to a wide range of display devices, and this has to be done by using a limited storage medium such as the OTP Memory previously described.

Figure 4.18 shows the three elaboration steps performed by a typical Driver IC. The incoming pixel data flow (R′, G′, B′) is described in perceptual domain. The first operation is gamma expansion; typical values of γ are 1.0, 1.8, 2.2 and 2.5. It is possible to store in the OTP memory a custom γ value. After gamma expansion, the pixel data flow (R_1, G_1, B_1) is described in physical (intensity) domain. Now it is possible to transform color tristimulus from source space (R_1, G_1, B_1) to destination space (R_2, G_2, B_2) by using the 3×3 conversion matrix $\mathbf{T}_{1\to2}$. In this step it is possible to adjust black point voltages by adding constants (c_R, c_G, c_B) to color components. White point voltages can by adjusted by proper choosing $\mathbf{T}_{1\to2}$ matrix. The last step is to correct the nonlinearity of the electro-optical transfer function (EOTF) of LCD. To do this a Look Up Table (**Gamma LUT**) is used. In this LUT the inverse LCD T-V function is stored. It is possible to use different inverse T-V functions for each color component ($\mathrm{EOTF}_R^{-1}, \mathrm{EOTF}_G^{-1}, \mathrm{EOTF}_B^{-1}$). The outputs are the voltages of elaborated pixel data flow (V_R, V_G, V_B), that is the RMS value driving the columns of the LCD allowing the desired pixel luminance (L_R, L_G, L_B).

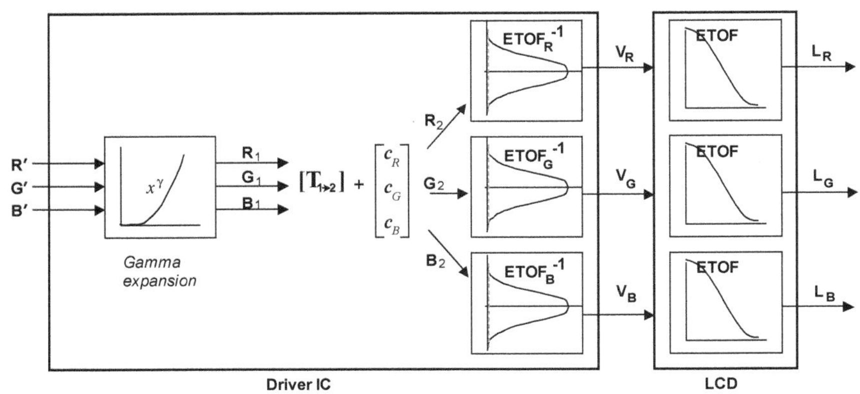

Fig. 4.18 Pixel elaboration steps

4.5.3 MLA and Gray Generation

Purpose of this block is to implement a **Column Logic** in order to latch Video RAM output, generate PWM/FRC control signals, calculate MLA waveforms and provide them to analog column drivers. Since these functions are tightly connected to the column (and row) drivers, their implementation issues will be discussed in Section 4.6.

4.5.4 Timing Controller (TCON)

This block generates the clock signals for (1) RAM access, (2) MLA and PWM/FRC elaboration and (3) n-line inversion. It synchronizes the column and row operations by selecting the proper rows with associated RAM data. These synchronization signals are all derived by division from the on-chip oscillator signal (described in Section 4.4.2). In the case of MLA, the orthogonal functions could be in principle memorized into a dedicated ROM. However, due to the small number of matrix elements usually needed (up to MLA-4 approach, at least), they are managed within the Timing Controller.

4.6 Output Drivers

The output drivers have the role to interface the driver IC with the LCD rows and the columns. These lines offer relatively large capacitive loads and should hence be driven with a sufficient current drive capability. This task is obtained by the analog buffers, directly connected at the output of the voltage generation section (DC/DC converters). Therefore, the architecture of the output drivers simplifies considerably and, ultimately, reduces to a multiplexer.

4.6.1 Row Drivers

It was described in Section 3.2 that both *single line* and *multiple line addressing* approaches are possible in PMLCDs.

In the former case, considering first the A&P approach, the row drivers are simple multiplexers designed to select one out of three voltages. As already illustrated in Fig. 3.9, the row voltage of a selected row can be $+V_R$ (or $-V_R$, for frames with reversed polarity), while unselected rows are addressed by a mid voltage (0 V).

The row driver for the IA&P technique must select one out of four voltages, i.e., V_R+V_C for the selected row and V_C for unselected rows in positive-polarity frames (and respectively 0 and V_R in negative-polarity frames). Note that the above voltages are those depicted in Fig. 3.10, but shifted of $+V_C$ to exploit positive values only.

A possible row driver architecture for IA&P is depicted in Fig. 4.19. The N-stage shift register (see Section 6.4 for shift register architectures) selects one row at a time through the bank of (row select) switches S_{RS}. For illustration, row 1 is the selected one. The voltages with the proper frame polarity are obtained through switches S_{FP}.

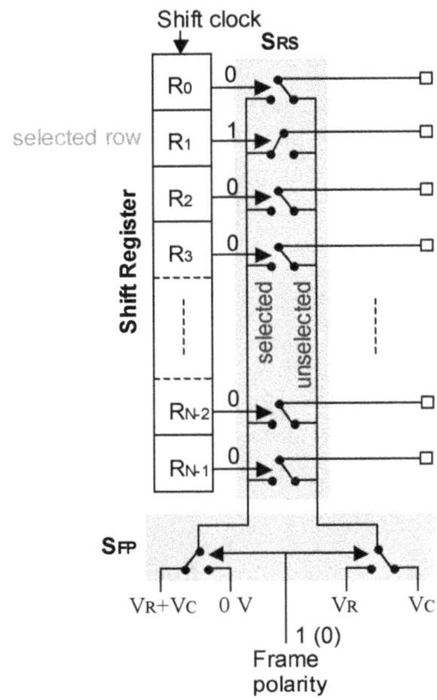

Fig. 4.19 Row driver block diagram for Improved Alt and Pleshko approach

Consider now the case of MLA-p addressing. We have seen in Section 3.2.2 that in this approach p lines at a time are addressed and the row signals are derived from orthogonal matrices, whose elements have just two values. Hardware complexity of the row (and column) driver depends on the number of voltage levels in the addressing waveforms, therefore those matrices minimizing the number of voltage levels in the scanning waveforms are preferred.

Also in this case, the row drivers are multiplexers that are designed to select one out of the three voltages i.e., $+V_R$ or $-V_R$, corresponding to elements +1 or −1 of matrix column (select pattern), and a mid voltage (0) corresponding to the unselected rows. The block diagram is very simple and can be deduced from that of a column driver for MLA described in the next section.

4.6.2 Column Drivers

Again, let us first consider single line addressing. In the A&P approach, the **column drivers** are multiplexers designed to select one of the two voltages i.e., $+V_C$ or $-V_C$, for ON and OFF pixels, respectively (the voltages are the same but with reversed meaning, for frames with reversed polarity). The column drivers used in the IA&P approach must instead select one out of four voltages, i.e., 0 or $2V_C$ for ON and OFF pixels, in positive frames, and V_R or V_R–$2V_C$ for ON and OFF pixels, in negative frames (see Fig. 3.10 after shifting all waveforms by $+V_C$).

Consider now the case of MLA-p. The column signal is given by the *dot product* between the row select pattern and the data pattern in the selected subgroup of rows (as explained in Section 3.2). Assume that p is the number of grouped lines and that p+1 is the number of voltage levels in the column waveforms. The number of data bits x must be consequently not lower than $\log_2(p+1)$. If M are the columns, there are hence M stages of shift register and latch and every stage has x bit each. After serially shifting the data corresponding to all column lines, the content of the shift registers are transferred in parallel to the latches and can be applied simultaneously to the column electrodes even though the dot products are computed sequentially. The schematic diagram of a column driver following this strategy is illustrated in Fig. 4.20 [RS2005].

This solution has the advantage of small area occupation, but requires high power consumption because the shift registers must work at the highest frequency (which hence coincides with f_{OSC} of the local oscillator) given by $f_F \cdot N \cdot M \cdot p = f_{OSC}$. For instance, if f_F= 60 Hz, N=160, M=128 and p=4, we get that the required local oscillator frequency is approximately f_{OSC} = 5 MHz.

An alternative low-power column driver scheme is depicted in Fig. 4.21.

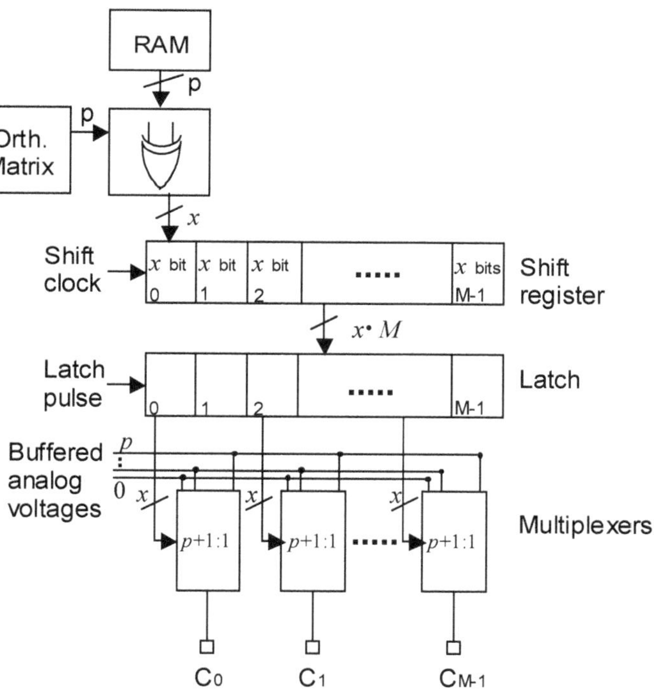

Fig. 4.20 Block diagram of a column driver used for MLA. (The operation performed by the XOR block is detailed in Fig. 4.21)

The column data patterns for each MLA row sub-group are stored into the Frame Buffer Memory and the XOR gates followed by the summer function perform the required product with the proper column of the orthogonal matrix. For instance, in Fig. 4.21 a MLA-4 approach is considered. The associated matrix is the 4×4 unity matrix shown at the left and the product with the first matrix column is performed. It is important to observe that to allow binary-logic computation, each −1 matrix entry is replaced by 0. In the figure it is also assumed that the data pattern for the considered row subgroup is 1011 (meaning that in the considered row subgroup the first, third and fourth pixels are ON while the second pixel is OFF). The result of the XOR and subsequent summing computation is associated (through multiplexers) with the analog voltage that will be applied to the column. More precisely, the values 000, 001, 010, 011 and 100 are respectively associated with $-V_C$, $-V_C/2$, 0, $+V_C/2$ and $+V_C$. In the case exemplified the computation gives 010 that means 0 V. Compared to the previous solution, this one requires a lower clock frequency expressed by $f_F \cdot N \cdot p$, i.e., M times lower. Indeed, with the same data of the previous example this one works with a frequency equal to 38.4 kHz. The lower consumption allows high-resolution PWM (up to 64 levels) to be implemented. Actually, the above considered column driver schemes are all for black and white displays. Gray levels through the PWM approach can be obtained with a modification of the scheme in Fig. 4.21, as illustrated in Fig. 4.22, where a 64-PWM technique is considered. Now, each cell of the frame buffer memory contains the number of bits encoding the gray levels (6 in our example). Besides, PWM generator blocks (implemented through comparators) are inserted at the output of the buffer. They compare the output of a 6-bit counter with the incoming 6-bit data word. The comparator output goes high only if the gray-level-encoding word is greater than the counter word.

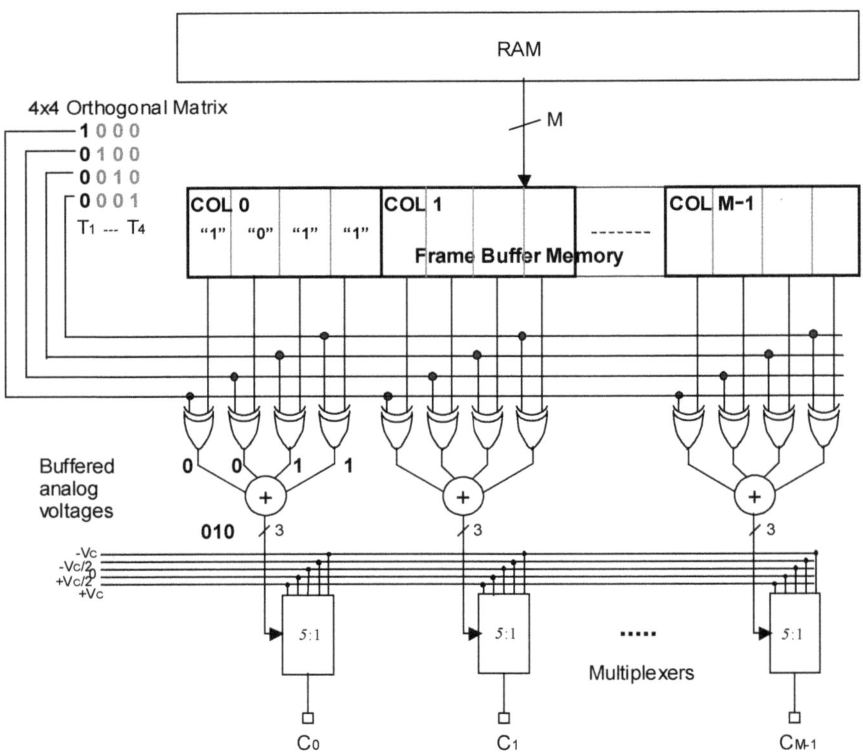

Fig. 4.21 Block diagram of another column-driver scheme for MLA-4

Considering again $f_F = 60$ Hz, N=160, M=128, p=4, n_{GL}=64(=2^6), we get approximately that the required maximum frequency of the local oscillator is $f_{OSC} = f_F \cdot N \cdot p \cdot n_{GL}$=2.46 MHz which is practical for mobile applications owing to the associated relatively low power dissipation.

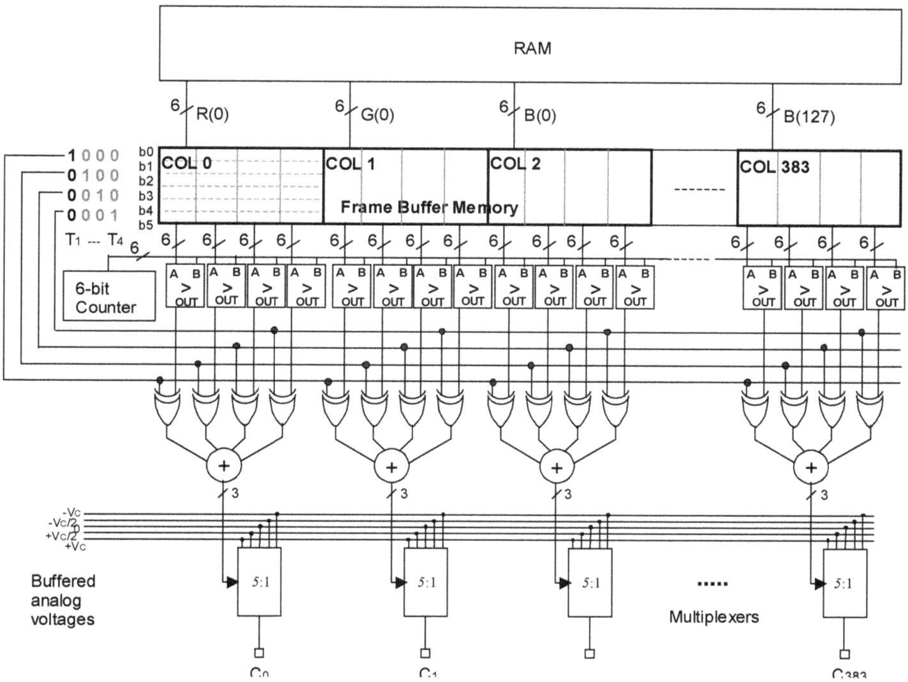

Fig. 4.22 Block diagram of column driver for color LCD adopting MLA-4 and PWM

4.7 Silicon Processes for Display Drivers

The silicon process required to implement single-chip driver ICs is quite challenging. Other than providing mixed-signal functionalities and allowing the integration of SRAM banks, it must posses also high-voltage capability (from 20 to 40 V).[8] In addiction, the demand for increased resolution and number of colors mandates for high gate density. Therefore, special technologies compatible with advanced CMOS families, but based on High Voltage Gate (HVG) CMOS devices have been developed.

Considering the specific case of mobile devices for consumer electronics, the market is very competitive and products have very short life (from 1 to 3 years).

[8]A process with these behaviors is of course developed by silicon foundries not only to cover passive-matrix and active-matrix LCDs (see Chapters 5 and 6) but also emissive (OLED) displays.

Keeping low the cost is thereby mandatory. This issue is addressed through the use of mature CMOS technologies (0.35 μm–0.13 μm), reduction of available process steps (e.g., limited number of metal layers) and efforts on IO area reduction including either *Circuit Under PAD* (CUP)[9] or *Bond over Active* (BOA)[10] approaches.

This section is largely based on [ABC2005], a paper describing the BCD-HVG8 (Bipolar, CMOS and DMOS) process family developed by STMicroelectronics. The HVG8 process exploits a 0.18-μm standard CMOS technology with several enhancements such as:

(1) Shallow Trench Isolation (STI) for CMOS isolation;

(2) Borderless Contacts;

(3) Dual Gate Oxide for 1.8-V/5-V CMOS devices (whose main electrical characteristics are summarized in Table 4.1). The low leakage current of the 1.8-V devices allows also the realization of compact SRAM banks based on the 6T cell (bit size 4.45 μm^2, in line with all available advanced 0.18-μm CMOS processes). Figure 4.22a shows the schematic of a SRAM cell and the micrograph of a SRAM bank is depicted in Fig. 4.23a. The vertical geometries are diffusions (transistor source/drain terminals) and the horizontal ones are polysilicon layers (transistor gate terminal). Metal contacts and paths are not shown in the photograph 4.22b. Symbol X means to row decoder;

(4) A number of 5 Al-Cu metal levels are allowed with the last one realized by either Al-Cu or pure Cu thick layer (though only three metal levels are utilized for LCD driver implementations, to keep fabrication costs low). The higher thickness of the upper metal allows long low-voltage-drop lines realization. This feature is very useful in display drivers because of their typical high aspect ratio.

The high voltage options offered by the HVG8 process are several and include:

(1) 20-V/32-V/40-V HVG MOS transistors through thick gate oxide;

(2) High voltage MOS transistors with low voltage (1.8/5 V) thin gate oxides (Drift MOS) for Antifuse One Time Programmable (OTP) memory blocks (OTP bit density in the range of 1 $kbit/mm^2$ has been achieved);

(3) LDMOS transistors for OLED drivers where large currents are needed.

---

[9]To save chip space, active circuitry is placed underneath the bonding pads. The CUP technique, which violates the rule of conventional layout for a chip, uses a variety of support structures that contain metals, dielectrics and their combinations designed to protect device components and to minimize damage caused by probe testing and subsequent assembly.

[10]Typical wirebond pad designs consist of a top-level metal that does not include any circuitry beneath the bonding region. The BOA technique was developed to utilize this region beneath wirebond pads in order to minimize die area.

Table 4.1 Main electrical parameters of HVG8 1.8-V and 5-V CMOS

	Parameter	1.8 V (L_{min}=0.18 μm)	5 V (L_{min}=0.6 μm)
nMOS	V_T (mV)	680	850
	I_{on} (μA/μm)	510	500
	I_{off} (pA/μm)	2	<1
pMOS	V_T (mV)	–770	–775
	I_{on} (μA/μm)	–220	–290
	I_{off} (pA/μm)	–3	–1

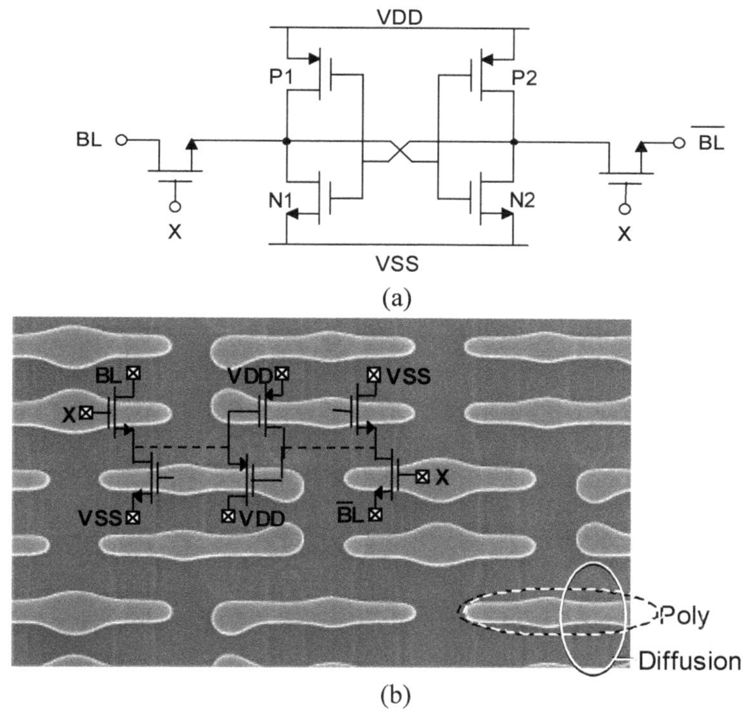

Fig. 4.22 (**a**) SRAM cell schematic, (**b**) micrograph of a SRAM bank in HVG8 technology

4.8 Packaging and Assembling Techniques

In this section we will describe the two main technologies that allow the electrical connection of the display to the driving circuits and of the driving circuits to the PCB. The same techniques described are not restricted to PMLCDs but are used also for Active Matrix LCDs.

Due to the great number of connections and the near impossibility of soldering ITO, both **Chip on Flex** (COF) and **Chip on Glass** (COG) approaches are widely used for driver IC attachment.

Anisotropically Conductive Adhesive Films (ACFs) are usually exploited to as an interconnection material for COG and COF [YP1998, S2004]. ACFs are composite materials consisting of polymer matrix, i.e., the adhesive and conductive particles, as illustrated in Fig. 4.23.

The fraction of conductive particles varies between 0.5 and 5% of the volume. The particle types vary from solid metal particles to polymer spheres coated with metal, the latter being most popular for high-density display applications [WT1999].

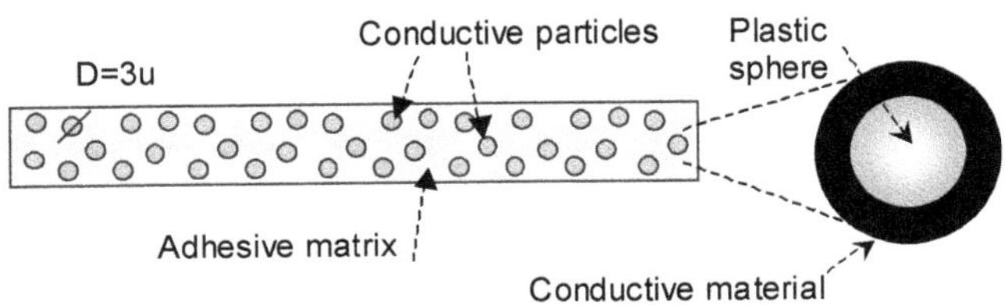

Fig. 4.23 Anisotropically Conductive Adhesive Films (ACFs) are used to electrically connect the Driver ICs to either the track on glass or tape automated bonding

4.8.1 Chip on Flex

In the original approach (first used by SHARP in 1985), the driver IC is packaged in a flexible **Tape Carrier Package** (TCP) and attached to the glass with **Tape Automated Bonding** (TAB) as illustrated in Fig. 4.24.

To allow flexibility the TCP has slits at regular distances (see Fig. 4.25a), and the tape can be bent only in a position corresponding to a slit. To improve performance and reduce costs, a new approach was developed called Chip on Film (1998).

The appearance of TCP or COF is similar, as both techniques rely in mounting the IC on a tape (see Fig. 4.25a, b). However, COF offers some substantial advantages: (1) it is free of slits and can be flexed in any position and with a greater freedom (oblique bend is possible), (2) it allows a finer bonding pitch (< 35 μm), (3) it allows high density packaging. Figure 4.26a shows the micrograph of a gold bump, while Fig. 4.26b shows the connections between the inner leads and bumps (in this case ACF is not used, but instead a gold-tin eutectic between the bump and the tin of the inner leads is formed [ML1986]).

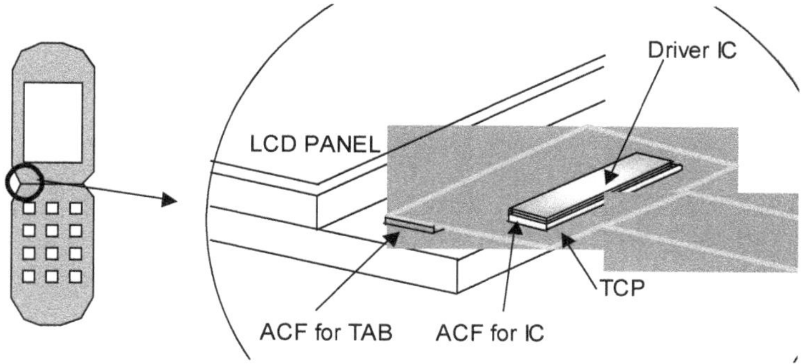

Fig. 4.24 TCP approach

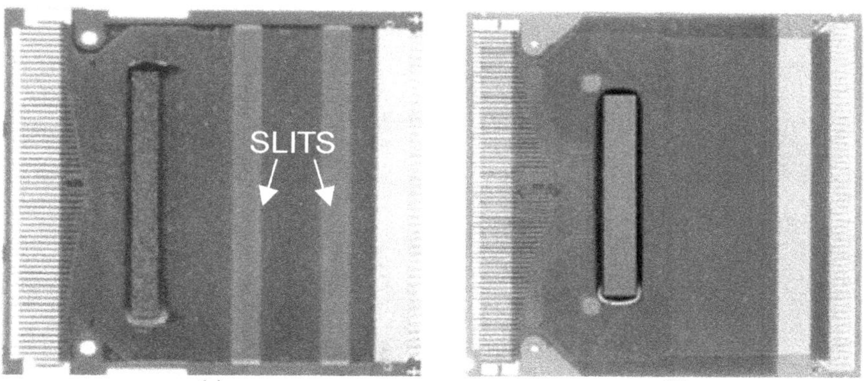

(a) (b)

Fig. 4.25 Examples of (**a**) COF and (**b**) TCP [TNF2003]

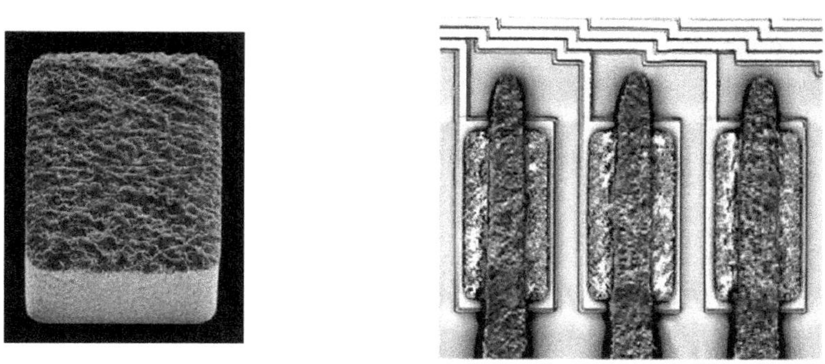

(a) (b)

Fig. 4.26 Micrographs of (**a**) a gold bump and (**b**) connection between bumps and inner leads in COF via gold-tin eutectic bonding [TNF2003]

4.8.2 Chip on Glass

COG was first introduced in 1983 by Citizen and since then the method diffused significantly [KL1999]. In this approach the IC is mounted directly on the glass substrate, avoiding the use of the flex and reducing costs. Figure 4.27 depicts an LCD panel with Driver IC connected via COG approach. The drive lines of an LCD are typically brought to a single glass bonding ledge, and routed along the inactive sides of the display to their intended destinations.

As illustrated in Fig. 4.28, conventional IC wire bonding is here replaced by the formation of gold bumps. After prebonding the ACF to the glass with mild heat and low pressure, the driver IC is aligned in such a way that its bump pattern matches the pad pattern on the glass. Then, the IC is pressed against the ACF and glass applying heat through the bonding tool. Conductive particles should be trapped between each bump-pad pair in order to ensure conductivity. The adhesive allows anisotropic conduction, only in the vertical direction. Extra polymer should be pressed out from under the IC.

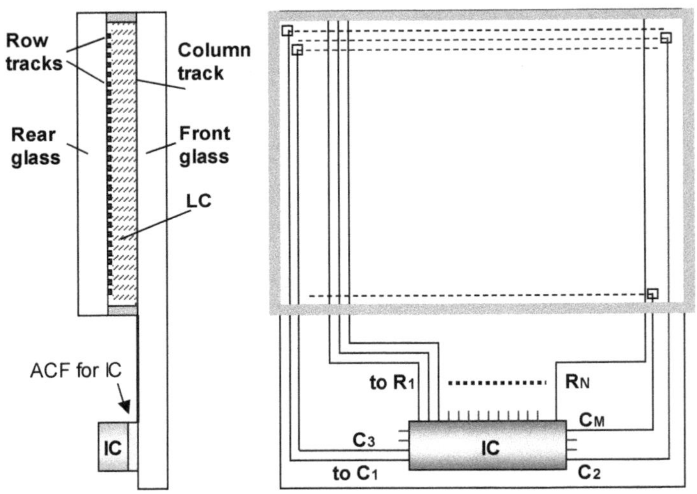

Fig. 4.27 LC panel with COG driver IC

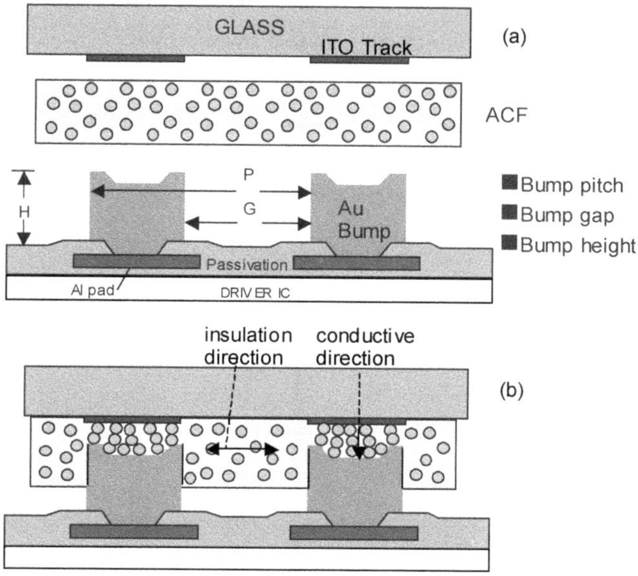

Fig. 4.28 COG technique. Cross section of a driver IC, Bumps, ACF, and Glass with typical dimensions: (**a**) before and (**b**) after electrical connection is formed

4.9 Concluding Remarks

We conclude this chapter by presenting an example of CSTN LCD ('C' stands for color) driver IC developed by STMicroelectronics, called the STE2028. It is designed for a 162×132-RGB LCD with 262k colors. It adopts the HVG technology and provides fully integrated booster (with the exception of the bulk capacitors), a 385-kbit SRAM and a 0.8-kbit OTP memory. It exploits MLA-4, n-line inversion and both PWM and FRC algorithms. It allows also partial display mode, enabling screen saver controls for power consumption reduction. The chip size is around 17 mm^2. Figure 4.29 shows the complete layout. Figure 4.30 illustrates the percentage of analog and digital component sections of the driver.

The complete block diagram is illustrated in Fig. 4.31. The reader will recognize many of the blocks although indicated with somewhat slightly different nomenclature (eg., MLS instead of MLA).

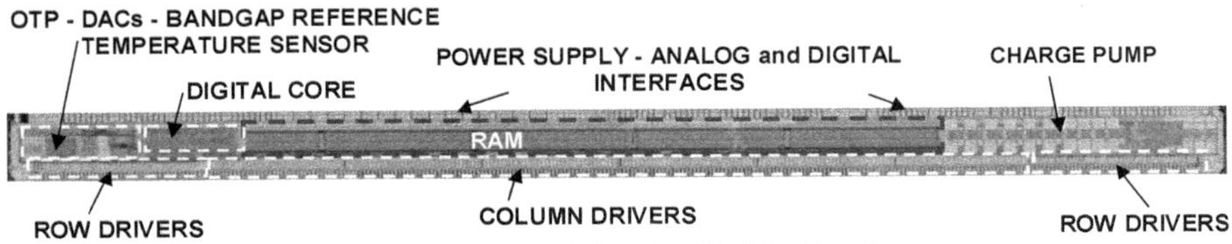

Fig. 4.29 Layout of the STE2028 LCD driver

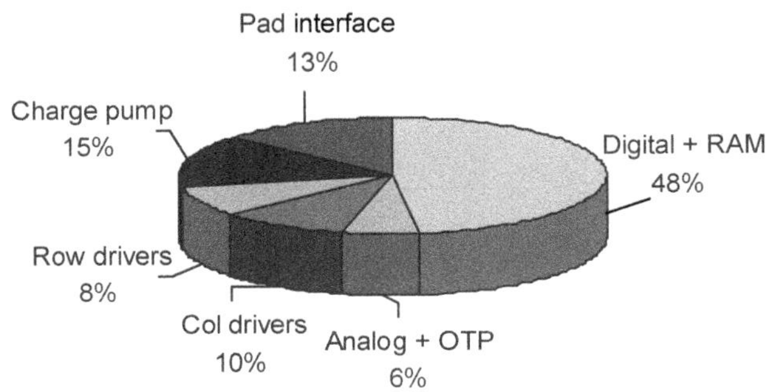

Fig. 4.30 Block partitioning of the STE2028 LCD driver

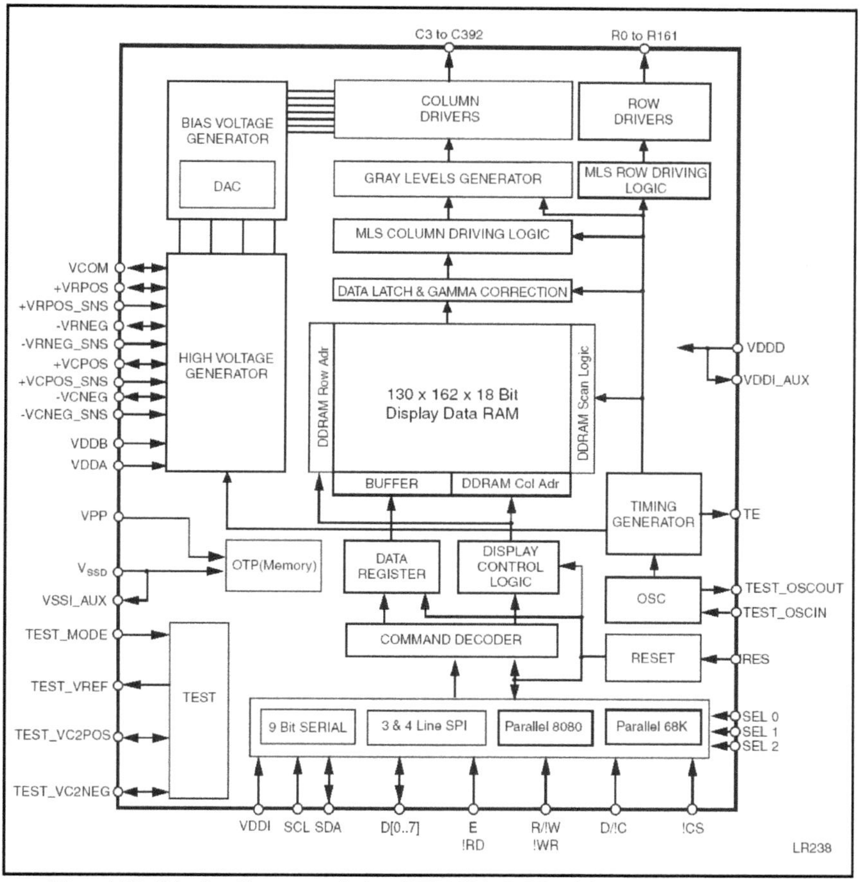

Fig. 4.31 Block diagram of the STE2028 LCD driver IC (Courtesy of STMicroelectronics)

References

[ABC2005] M. Annese, S. Bertaiola, G. Croce, A. Milani, R. Roggero, P. Galbiati, and C. Costiero, "0.18 µm BCD-High Voltage Gate (HVG) Process to address Advanced Display Drivers Roadmap," *Proceedings of Power Semiconductor Devices and ICs*, pp. 363–366, 2005.

[B1974] A.P. Brokaw, "A Simple Three-Terminal IC Bandgap Reference," *IEEE Journal of Solid-State Circuits*, Vol. 9, No. 6, pp. 388–393, 1974.

[BSU1999] H. Banba, H. Shiga, A. Umezawa, T. Miyaba, T. Tanzawa, S. Atsumi, and K. Sakui, "A CMOS Bandgap Reference Circuit with Sub-1-V Operation," *IEEE Journal of Solid-State Circuits*, Vol. 34, No. 5, pp. 670–674, 1999.

[D1976] J.F. Dickson, "On-Chip High-Voltage Generation in MNOS Integrated Circuits Using an Improved Voltage Multiplier Technique," *IEEE Journal of Solid-State Circuits*, Vol. 11, No. 3, pp. 374–378, 1976.

[D2006] J. Doutreloigne, "A Monolithic Low-Power High-Voltage Driver for Bistable LCDs," *Microelectronics Journal*, Vol. 37, pp. 1220–1230, 2006.

[DYY1992] J.W. Doane, D.K. Yang, and Z. Yaniv, "Front-Lit Flat Panel Display from Polymer Stabilized Cholesteric Textures," *Proceedings of 12th International Display Research Conference*, pp. 73–76, 1992.

[GP1996] R. Gariboldi and F. Pulvirenti, "A 70 mΩ Intelligent High Side Switch with Full Diagnostics," *IEEE Journal of Solid-State Circuits*, Vol. 31, No. 7, pp. 915–923, 1996.

[GPC2003] G. Giustolisi, G. Palumbo, M. Criscione, and F. Cutri, "A Low-Voltage Low-Power Voltage Reference Based on Subthreshold MOSFETs," *IEEE Journal of Solid-State Circuits*, Vol. 38, No. 1, pp. 151–154, 2003.

[IEC1998] IEC 61966-4, "Colour Measurement and Management in Multimedia Systems and Equipment – Part 4: Equipment Using Liquid Crystal Display Panel," 1998.

[IEC1999] IEC 61966-2-1, "Multimedia Systems and Equipment - Colour Measurement and Management – Part 2-1: Colour Management - Default RGB Colour Space – sRGB," 1999.

[J2001] K. Jack, *Video Demystified: A Handbook for the Digital Engineer,* LLH Technology Publishing, third edition, 2001.

[K1973] K.E. Kuijk, "A Precision Reference Voltage Source," *IEEE Journal of Solid-State Circuits*, Vol. 8, No. 3, pp. 222–226, 1973.

[KL1999] H. Kristiansen and J. Liu, "Overview of Conductive Adhesive Technologies for Display Applications," *Conductive Adhesives for Electronics Packaging*, J. Liu (editor), Electrochemical Publications, pp. 376–399, 1999.

[L2001] E. Lueder, *Liquid Crystal Displays: Addressing Schemes and Electro-Optical Effects*, Wiley, 2001.

[LCV2007] D.-U. Lee, R.C.C. Cheung, and J.D. Villasenor, "A Flexible Architecture for Precise Gamma Correction," *IEEE Transactions on Very Large Scale Integration Systems*", Vol. 15, No. 4, pp. 474–478, 2007.

[M1997] M.S. Makowski, "Realizability Conditions and Bounds on Synthesis of Switched-Capacitor DC/DC Voltage Multiplier Circuits," *IEEE Transactions on Circuits and Systems - part I*, Vol. 44, No. 8, pp. 684–691, 1997.

[ML1986] C. Mackay and S. Levine, "Solder Sealing Semiconductor Packages," *IEEE Transactions on Components, Hybrids, and Manufacturing Technology*, Vol. 9, No. 2, pp. 195–201, 1986.

[MMF2001] P. Malcovati, F. Maloberti, C. Fiochi, and M. Pruzzi, "Curvature-Compensated BiCMOS Bandgap with 1-V Supply Voltage," *IEEE Journal of Solid-State Circuits*, Vol. 36, No. 7, pp. 1076–1081, 2001.

[P1993] C. Poynton, "Gamma and its Disguises: The Nonlinear Mappings of Intensity in Perception, CRTs, Film and Video," *Society of Motion Picture and Television Engineers Journal*, Vol. 102, No. 12, pp. 1099–1108, 1993.

[P2002] F. Pulvirenti, "Low Supply Voltage Relaxation Oscillator having Current Mirror Transistors Supply for Capacitors," *U.S. Patent #6 362 697 B1*, issued 2002.

[RS2005] T.N. Ruckmongathan and A.R. Shashidhara, "Sparse Orthogonal Matrices for Scanning Liquid Crystal Displays," *Journal of Display Technology*, Vol. 1, No. 2, 2005.

[S2004]. P. Savolainen, "Display Driver Packaging: ACF Reaching the Limits?," *Proceedings of 9th International Symposium on Advanced Packaging Materials*, pp. 7–10, 2004.

[SK2008] F. Su and W.-H. Ki, "Component-Efficient Multiphase Switched-Capacitor DC-DC Converter with Configurable Conversion Ratios for LCD Driver Applications," *IEEE Transactions on Circuits and Systems - Part II*, Vol. 55, No. 8, pp. 753–757, 2008.

[TNF2003] K. Toyosawa, N. Nakamura, K. Fukuta, and Y. Chikawa, "COF (Chip-On-Film) Technology for LCD Driver ICs Using Reel-to-Reel System," *Sharp Technical Journal*, No. 3, 2001. (http://sharp-world.com/corporate/info/rd/index.html).

[TC2003] K. Toyosawa and Y. Chikawa, "Development of 35 μm Fine-Pitch SOF Technology for LCD Driver LSIs," *Sharp Technical Journal*, No. 4, 2003. (http://sharp-world.com/corporate/info/rd/index.html).

[W1971] R.J. Widlar, "New Developments in IC Voltage Regulators," *IEEE Journal of Solid-State Circuits*, Vol. 6, No. 1, pp. 2–7, 1971.

[WT1999] I. Watanabe and K. Takemura, "Anisotropic Conductive Adhesive Films for Flip-Chip Interconnection," *Conductive Adhesives for Electronics Packaging*, J. Liu (editor), Electrochemical Publications, pp. 256–271, 1999.

[WW1998] C.C. Wang and J.C. Wu, "Efficiency Improvement in Charge Pump Circuits," *IEEE Journal of Solid-State Circuits*, Vol. 32, No. 6, pp. 852–860, 1998.

[YKC2003] T.R. Ying, W.H. Ki, and M. Chan, "Area-Efficient CMOS Charge Pumps for LCD Drivers," *IEEE Journal of Solid-State Circuits*, Vol. 38, No. 10, pp. 1721–1725, 2003.

[YP1998] M.-J. Yim and K.-W. Paik, "Design and Understanding of Anisotropic Conductive Films (ACF) for LCD packaging," *IEEE Tranactions on. Components, Packaging and Manufacturing Technology - Part A*, Vol. 21, No. 2, pp. 226–234, 1998.

Chapter 5

Active Matrix LCDs and Their Addressing Techniques

We have seen that in PMLCDs the display contrast depends on the drive margin, but this parameter is in turn a function of the number of rows so that the more rows are added, the less contrast is obtained. **Active matrix LCDs** (**AMLCDs**, also called **TFT LCDs**) remove these multiplexing limitations by exploiting a switching element and a storage capacitor at every pixel of the display. These switches are usually implemented through transistors made up of deposited thin films, and are therefore called thin-film transistors,[1] TFTs. A TFT at each pixel allows the column voltages to be applied only to the row that is being addressed, while the storage capacitor maintains the pixel information for the whole frame also when the addressing signal is removed. Therefore, a high contrast is possible and a fast LC mixture can be used, since the pixel no longer has to respond to the average voltage over a whole frame period, as in PMLCDs. For the same reason the phenomenon of crosstalk is also minimized.

AMLCDs are hence suitable for full-motion video, requiring time responses below 20 ms. As main drawbacks, the TFT process increases production costs (at this purpose, a limit in the number of masks of the TFT process is from 4 to 6), reduces the yield and decreases the aperture ratio of the pixel: thus, to maintain an acceptable display luminance a brighter backlight is needed (by increasing power consumption).

In this chapter, after briefly describing the principal TFT devices and the structure of an AMLCD panel, addressing techniques for AMLCDs will be discussed. Related driver ICs and their implementation issues will be treated in Chapter 6.

---

[1]The majority of AMLCDs use TFTs as switching devices. An alternative commercially exploited solution is constituted by the Metal-Insulator-Metal (MIM) diode. The MIM low fabrication temperature is compatible with plastic substrates.

D.J.R. Cristaldi et al., *Liquid Crystal Display Drivers*, DOI 10.1007/978-90-481-2255-4_5,

5.1 Thin Film Transistors

In this section we will discuss the main characteristics of TFTs. Compared to conventional Metal Oxide Semiconductor Field Effect Transistors (MOSFETs) implemented in a crystalline Si substrate, TFTs have similar model equations but, due to the reduced carrier mobility, μ, they exhibit poorer performance both in terms of conductance and switching speed. Being implemented on an insulating substrate, TFTs lack the substrate electrode (called Body or Bulk in conventional MOSFET devices), and have thus only three terminals named **Drain** (D), **Gate** (G), and **Source** (S). The presence of the insulating substrate provides ideal isolation of each device and negligible parasitic capacitances [LLB1992].

Figure 5.1 shows the cross section and top view of a basic[2] TFT (with the layers materials typically adopted) along with the circuit symbol. The gate-source voltage, v_{GS}, determines the TFT ON/OFF condition as it causes the formation of the conducting **channel** between the source and the drain (i.e., charges are induced in the semiconductor layer).

One type of TFT, the *n*-channel TFT, operates similarly to the *n*-channel MOSFET. In particular, we define the threshold voltage (whose typical value is about 1–4 V) as

$$V_{TH} = -qn_0 d_{CH} / C_{ox} \tag{5.1}$$

where q is the charge of the electron, n_0 is the charge density in the channel, d_{CH} is the channel thickness and C_{ox} is the capacitance of the gate dielectric per unit area ($C_{ox} = \varepsilon_0 \varepsilon_r / t_g$, where t_g is the thickness of the gate dielectric, see Fig. 5.1). The channel is formed if $v_{GS} > V_{TH}$, in this case the TFT is turned on and a current, i_D, can flow through the source and drain terminals, provided that a drain-source voltage, v_{DS}, is applied. Conversely, $i_D = 0$ if $v_{GS} \leq V_{TH}$.

Assume now $v_{GS} > V_{TH}$. Denoting as W and L the channel width and length, respectively (as shown in Fig. 5.1), we can write the expression of the drain current as a function of the gate-source and drain-source voltages both in the triode and saturation operating conditions [S2006, MK2003].

The **triode region** (also termed the **linear** region) is characterized by $v_{DS} \leq v_{GS} - V_{TH}$ and the drain current expression is

$$i_D = \mu C_{ox} \frac{W}{L} \left[\left(v_{GS} - V_{TH} \right) v_{DS} - \frac{v_{DS}^2}{2} \right] \tag{5.2}$$

In this region, one of the most important parameters is the ON resistance offered by the TFT between the source and drain (when used as a switch), whose expression is

---

[2]The device depicted in Fig. 5.1 is a "bottom-gate" (or inverted staggered) TFT implemented in either amorphous silicon or polysilicon. Other alternatives are possible; see for instance [B2005, K2004].

$$R_{ON} = \left(\frac{\partial i_D}{\partial v_{DS}}\right)^{-1} = \frac{1}{\mu C_{ox}\frac{W}{L}(v_{GS} - V_{TH})} \tag{5.3}$$

The above relation shows that the TFT conductivity is increased by increasing the **aspect ratio**, W/L, and/or the **overdrive voltage** $v_{GS}-V_{TH}$.

In the **saturation region** we have $v_{DS} > v_{GS} - V_{TH}$ and the drain current is dependent to a first approximation only on v_{GS} (we will see that this assumption is not accurate in polysilicon TFTs, due to the so called *kink* effect)

$$i_D = \frac{\mu C_{ox}}{2}\frac{W}{L}(v_{GS} - V_{TH})^2 \tag{5.4}$$

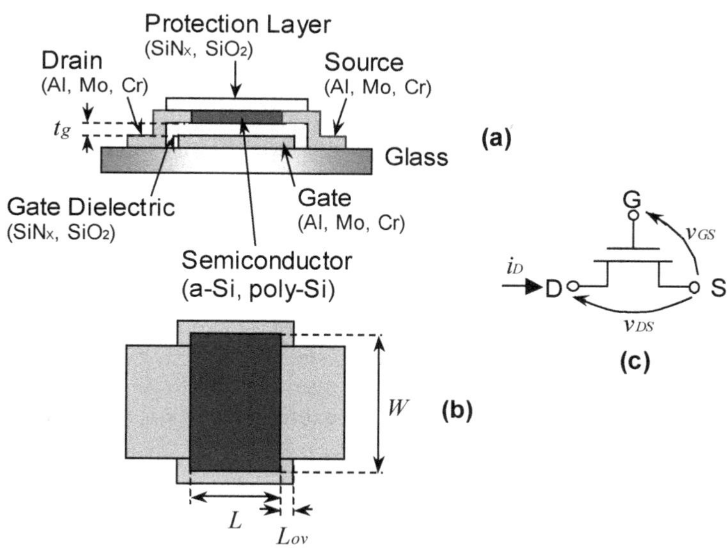

Fig. 5.1 Basic TFT: cross section (**a**), top view (**b**) and circuit symbol (**c**)

The output characteristics derived form (5.2) and (5.4) with v_{GS} as a parameter (and normalized to the factor $\mu C_{ox}W/L$), are plotted in Fig. 5.2. The transition region between the triode and saturation region (for $v_{DS} = v_{GS}-V_{TH}$) is represented by the dashed portion of a parabola whose equation is

$$i_D = \frac{\mu C_{ox}}{2}\frac{W}{L}v_{DS}^2 \tag{5.5}$$

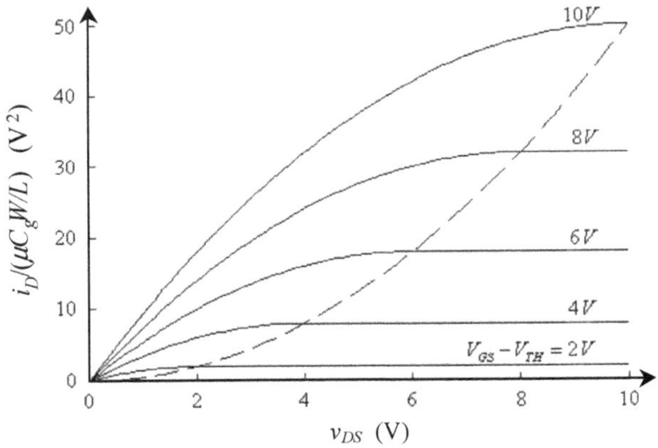

Fig. 5.2 Output characteristics (drain current versus drain-source voltage) of a generic TFT for $V_{GS} - V_{TH} = 2, 4, 6, 8, 10$ V

Ideally, i_D is equal to zero when $v_{GS} < V_{TH}$, in realty a **leakage conduction** is always present. Therefore, an important aspect regarding TFT addressing is the proper minimization of this current. Indeed, when the pixel is not selected, a high leakage current discharging the storage capacitor leads to a loss of information, but a larger storage capacitor results in a lower aperture-ratio and in a slowdown of the charging time.

Other important parameters for the dynamic behavior of a TFT are the **parasitic capacitances** between couples of terminals, C_{GS}, C_{GD} and C_{DS}. In particular, the gate-drain capacitance (or equivalently the gate-source capacitance, because these capacitances are equal in a transistor operating in triode) has two main contributions: one is due to the channel capacitance, the other is caused by the overlap region between the gate and drain. Assuming v_{DS} small in triode region, about half of the channel capacitance contributes to C_{GD} and the other half to C_{GS}. If the length of the overlap is labeled as L_{OV} (see Fig. 5.1) the expressions of maximum and minimum (when the TFT is near the cutoff region) of C_{GD} are

$$C_{GD,\max} = W(L/2 + L_{OV})C_{ox} \quad (5.6a)$$

$$C_{GD,\min} = WL_{OV}C_{ox} \quad (5.6b)$$

We will show in Section 5.3 that C_{GD} is the origin of performance degradation as it causes unwanted charge redistribution effects.

The above equations are useful for pencil-and-paper calculations. Circuit designers need more accurate but efficient models (time-domain simulations of a complete panel may be an onerous task in terms of required CPU time) to be used in their simulators (SPICE-like models are the most popular) [HS1989, SSJ1997].

5.1.1 a-Si TFT

The most common semiconductor layer used for the integration of TFTs on glass substrates is realized with idrogenated **amorphous silicon** (a-Si:H, in the following a-Si for simplicity). It allows large-area fabrication in a low-temperature process (300°C–400°C) and is the dominant technology in computer and television LCDs. The main difference between a-Si TFT and a conventional Si MOSFET is the lower electron mobility (in the range of 0.2–1.5 cm²/Vs, while it is greater than 1800 cm²/Vs in crystalline Si), due to the presence of electron traps in amorphous silicon. Hence, a-Si TFTs are inadequate to be used in analog and high-speed digital processing directly on glass, but they are instead suitable for the implementation of ON/OFF pixel switches with refresh frequencies around 60 Hz. However, even for this simple scope, the low mobility would require extremely large aspect ratios, thereby reducing the **aperture-ratio** (i.e., only a portion of the pixel's area is transparent to light when the pixel is ON) and ultimately decreasing the display's brightness and contrast. To mitigate this problem, since the transconductance depends also on the gate-source overdrive, the gate of a TFT is driven by a high voltage, typically around 20 V.

Because a-Si materials exhibit poor stability and their structure can be modified by strong illumination and charge carrier injections (in particular its OFF current increases with incident light), the a-Si TFT must be shielded from light. Besides, this technology allows only *n*-channel TFTs to be realized. The conductivity of a-Si *p*-channel TFTs would not be sufficiently high, even as switches for LCDs.

Another disadvantage of a-Si TFTs is the **threshold voltage shift** (by several volts over a few hours with a DC voltage stress [GW2001, KNH2004]). It can be expressed by $\Delta V_{TH}(t) = A(V_{G,ST} - V_{TH,i})t^{\beta}$, where A and β are two temperature dependent parameters, $V_{G,ST}$ is the stress gate voltage and $V_{TH,\mathrm{i}}$ is the threshold voltage before stress which has duration t [PBH1989]. It is caused by two mechanisms. At higher gate electric fields, electrons are injected into the gate dielectric [P1983, PD1993]; at lower electric fields, charges are trapped in the interface a-Si and gates [HMM1986]. The threshold voltage increase is larger for lower temperature processes [GW2001]. The shift is not a severe problem for switch operations, but the function of digital circuits is impaired when the static noise margin decreases considerably (becoming zero when V_{TH} increases to $V_{DD}/2$) [LVS2006]. The shift has different polarity for ON and OFF state, therefore a partial cancellation occurs. In any case, a practical driving scheme must take into account a threshold tolerance in the TFTs.

5.1.2 Poly-Si TFT

Polysilicon represents an alternative material appropriate to realize the TFT silicon layer, adopted in recent years for small-area displays. It is a silicon-based material, which contains numerous Si grains with sizes ranging from 0.1 to several microns. In semiconductor manufacturing, polysilicon is usually deposited as amorphous films by LPCVD (Low Pressure Chemical Vapor Deposition) and then crystallized by laser annealing above 900°C. This technology requires also expensive quartz substrates. The same approach cannot be exploited by the LCD industry since the strain temperature of glass is only about 650°C. **Low Temperature Polysilicon**

(**LTPS**) technologies, built on low-cost glass, were then developed based on either excimer laser annealing or by combined solid state phase crystallization [H1986, TDM1999, MS1999, JSH1996].

The preparation of LTPS film is more complicated than a-Si, and LTPS is more exposed to process variations over the substrate area. However, LTPS TFTs have a higher mobility than a-Si TFT (up to two orders of magnitude for electrons, 30 to 400 cm^2/Vs, and up to 150 cm^2/Vs for holes) and the implementation of both *n*-channel and *p*-channel transistors is possible, therefore they allow to increase the pixel aperture ratio (higher mobility means smaller TFT area for a given transconductance) and are suitable for the integration of CMOS-like circuits directly on the glass substrate (ultimately leading to a complete System on Glass, SOG). The latter is a remarkable advantage that allows the number of electrical connections on the LCD panel to be drastically reduced, thereby increasing reliability and compactness. The connections are made in the thin film technology using a mask which is in any case needed for processing the display.

A schematic cross section of a poly-Si TFT is shown in Fig. 5.3a. For high drain-source voltages, poly-Si TFTs exhibit the so called **kink** effect consisting in an anomalous increase of the drain current, as illustrated in Fig. 5.3b. This effect is mainly originated by the high voltage at the drain end of the channel that accelerates charges causing in turn an avalanche current (impact ionization). This current is able to flow to ground in monocristalline Si ICs, being the body electrically connected to a bias voltage. For TFTs the substrate is insulating glass, therefore this current increases the drain current. Of course, to contrast this problem, low supply voltages should be adopted. A Lateral Body Terminal (i.e., the extension of the channel area serving as collector of the charges) was also employed [Y1999].

Recently, Sharp in conjunction with Semiconductor Energy Laboratory Co. Ltd., has developed a process called **Continuous Grain Silicon** (CG Silicon), a further evolution of the LTPS technology, leading to the commercial introduction of high-resolution, high-definition system LCDs for mobile terminal devices in 2002. A complete CPU on a glass substrate is shown in Fig. 5.4 [N2003].

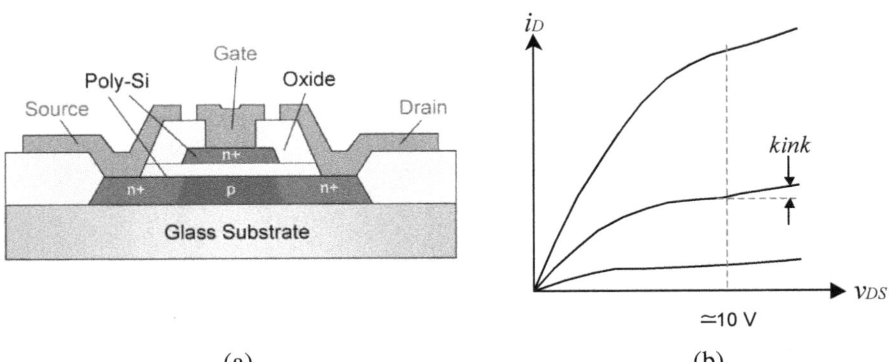

Fig. 5.3 Poly-Si TFT, (**a**) cross section and (**b**) drain current versus drain-source voltage putting in evidence the kink effect

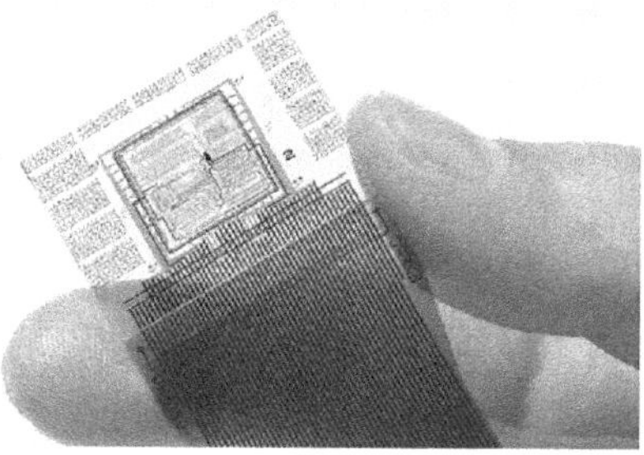

Fig. 5.4 CG Silicon CPU formed on a glass substrate (Semiconductor Energy Laboratory) [N2003]

5.1.3 Further Comparison of a-Si and Poly-Si TFT Performance

Figure 5.5 shows the a-Si and poly-Si TFT characteristics for V_{DS} = 5 V. The poly-Si TFT size is only 1/10 of its a-Si counterpart, to provide the same drain current corresponding to V_{GS} = 20V. The smaller size of transistors (in consequence of the greater mobility electron) can increase both resolution (pixels per inch) and aperture-ratio improving the image definition. For this reason and for the large process tolerances of poly-Si, displays with a high pixel density and pixel size of 15 μm×15 μm are best addressed by poly-Si TFTs. Application of this kind are small high-definition displays for view finders in camcorder, or projection displays [B2005].

Compared to a-Si TFTs, poly-Si TFTs exhibit higher OFF currents, caused by the electric field at the highly doped drain and the imperfect crystalline structure of the poly-Si material. Methods currently adopted against this limitation are the dual gate structure or adding another process step to realize a lightly doped drain (LDD) [L2001].

From the driving voltage point of view, a negative gate-source voltage around –5 to –8 V (not just 0 V) is adopted for a minimum leakage condition below 10 pA. On the other hand, to compensate for the low electron mobility the gate-source voltage is set at around 20 V for switch-on, to provide a drain current in the order of 100 μA.

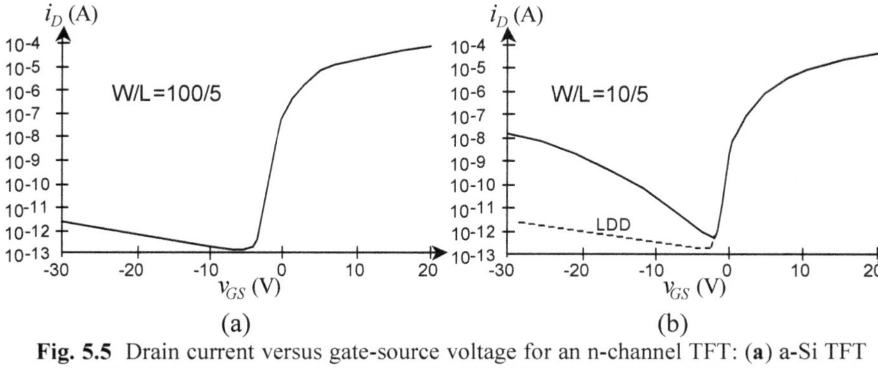

(a) (b)

Fig. 5.5 Drain current versus gate-source voltage for an n-channel TFT: (**a**) a-Si TFT with W/L=100 μm/5 μm and (**b**) poly-Si with W/L=10 μm/5 μm. Voltage V_{DS} is 5 V in both cases. The *dashed curve* in (b) represents the OFF current obtained with lightly doped drain (LDD) technology

5.2 Structure of an AMLCD Panel

Like a passive-matrix color LCD, an AMLCD has a sandwich-like structure with liquid crystal filled between two glass plates kept to a controlled distance by spacers and exploits backlight, polarizers and color filters (the latter fitting the upper substrate). The cross section of an AMLCD is shown in Fig. 5.6.

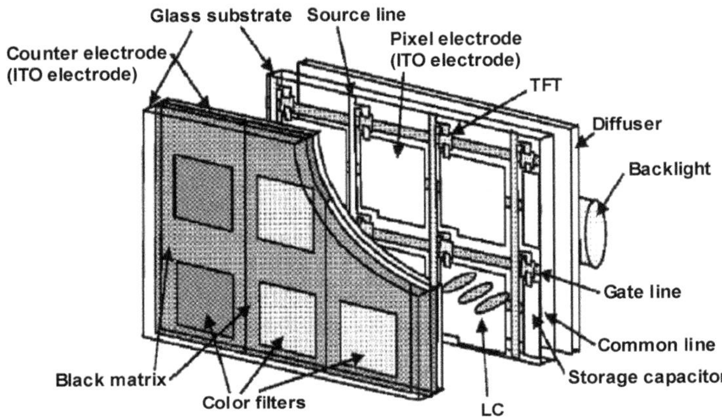

Fig. 5.6 Cross section of an AMLCD [U2006]

Each pixel of a color panel is subdivided into three RGB subpixels. TFTs are integrated on the back glass substrate, along with storage capacitors, ITO electrodes and interconnect wiring, as shown in Fig. 5.7a. In an AMLCD panel, the row and column electrodes are called gate and source electrodes (with reference to the associated TFT terminals). By scanning the gate lines sequentially, and by applying signal voltages to all source lines in parallel, we can address all pixels. As a result, the addressing of an AMLCD is very simple and performed line by line. Unlike passive-matrix addressing which required orthogonal row and column signals, the use of TFTs in AMLCDs makes each row independent from the others. Indeed, the thin-film storage capacitor (C_S), accessed through the associated TFT, memorizes the source (pixel data) voltage and maintains it even after the gate (selection) signal is removed. Note that C_S is added in parallel to the intrinsic LC capacitor, as the value of C_{LC} is too low for adequate charge retention. Between the color filters is placed a **black matrix**, made of an opaque metal, such as chromium (Cr), which shields the (a-Si) TFTs from room light and prevents light leakage between pixels to minimize photo-generated leakage currents in the TFTs. A double layer of Cr and CrOx is used to minimize reflection from the black matrix.

Figure 5.7b shows the electrical model of a subpixel with the storage capacitor C_S in parallel to C_{LC} connected as a load of the TFT. It is worth noting that C_S can be connected either to an adjacent gate line (**Cap-on-Gate**) or to a dedicated common storage bus (**Cap-on-Common**), see Fig. 5.7b, c. As main drawbacks, the former solution increases the capacitive load seen by the gate drivers while the latter reduces the pixel aperture.[3] The former approach is thus unsuitable for panels with a diagonal size larger than about 20 in.

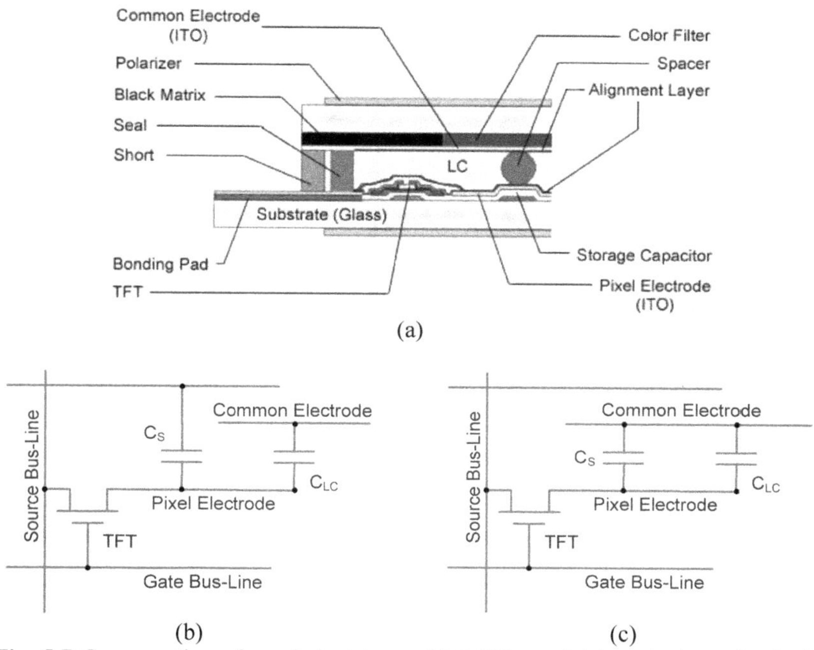

Fig. 5.7 Cross section of a subpixel in an AMLCD panel (**a**) equivalent circuits in Cap-on-Gate structure and (**b**) in Cap-on-Common structure (**c**)

In the unit cell, as shown in Fig. 5.8, TFT electrodes, storage-capacitor electrodes, signal bus-lines, and the black-matrix material constitute opaque areas. The combined areas of these elements, along with the area of the pixel aperture through which light can pass, determine the **aperture ratio** of the pixel: the aperture ratio is given by the area of the pixel aperture divided by the total pixel area (aperture area plus the area of the opaque elements). To increase this value as much as possible, the size of the opaque elements must be made as small as possible, while maintaining a design that maximizes the size of the pixel-electrode area. The Source Line (ITO) overlaps the bottom plate of the storage capacitor and the Gate Line at each pixel in the same column. These overlapping areas cause coupling capacitances.

A complete view of the TFT, C_S and C_{LC} is provided in Fig. 5.9a. The relative areas of C_S and C_{LC} are clearly evidenced. A top view showing the RGB color filters is depicted in Fig. 5.9b.

[3]To increase the pixel luminance for outdoor use, the opaque electrode of C_S can be covered with a reflective material, as suggested in [FNK2003].

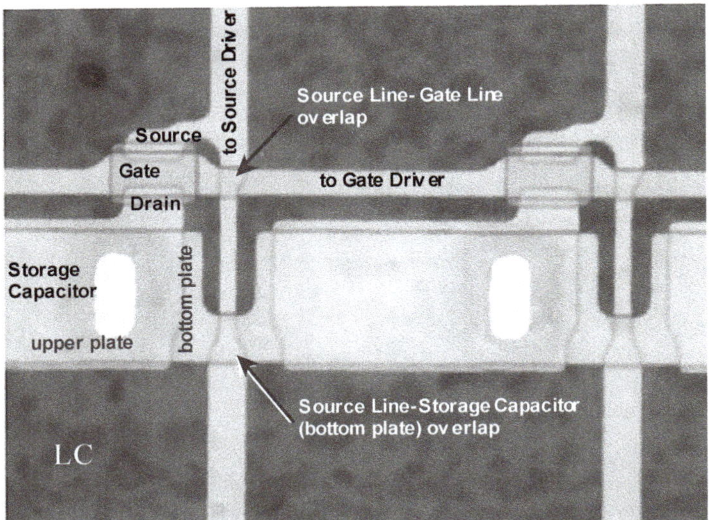

Fig. 5.8 Micrograph of an AMLCD panel showing the TFT, Storage Capacitor and interconnections

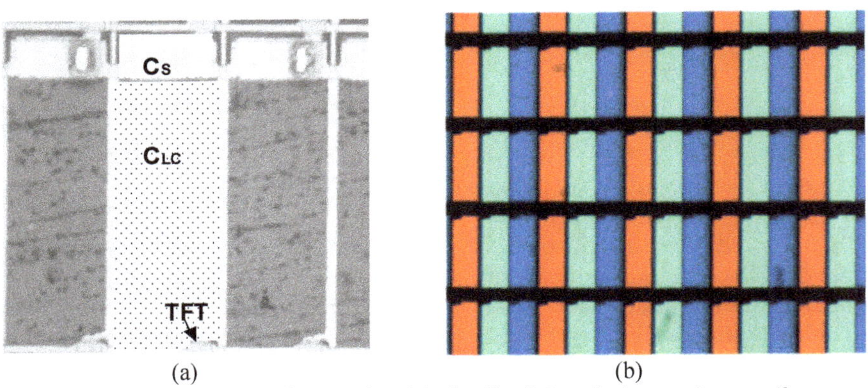

Fig. 5.9 AMLCD panel micrographs: (**a**) detail of the pixel capacitances (*bottom view*); Fig.5.8 is a magnified version of this figure. (**b**) detail of the RGB color filters (*top view*)

5.3 General Considerations

In contrast to PMLCDs that need some form of elaboration technique (such as PWM and/or FRC) for gray level generation, AMLCDs are inherently designed to produce gray levels depending on the voltage magnitude applied to the pixel from the source line. With reference to Fig. 5.7, applying a suitable positive pulse voltage through a specific gate line has the effect of turning on all the TFTs whose gates are connected to that bus line. Capacitors C_{LC} and C_S are then charged at the voltage level applied through the (data) source line. This charged state (and consequently the pixel voltage level) can be maintained as the gate voltage goes to a negative value, at which time the TFTs nominally turn off. The main function of C_S is then to maintain the voltage on the pixel electrode until the next source voltage is applied. An example of AMLCD addressing waveforms is illustrated in Fig. 5.10.

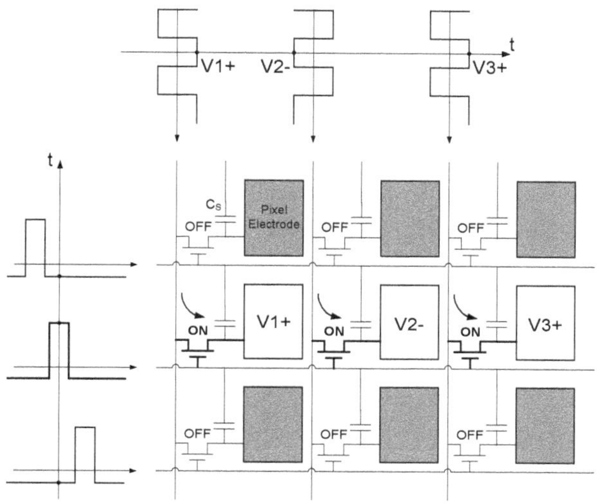

Fig. 5.10 Addressing example of an Active Matrix LCD

As already discussed, the scanning frequency should be at least 30 Hz to avoid flickering. A refresh frequency of 60 Hz is the result of a trade-off between elimination of flicker and low power dissipation. If $f_F = 1/T_F$ is the frame (scanning) frequency and N is the number of rows of display, the maximum access time to pixel (row time), T_R, will be:

$$T_R = \frac{T_F}{N} \tag{5.7}$$

When switched on, the resistance offered by the TFT, R_{ON}, should be adequate to charge the pixel capacitance at the maximum source line voltage within the row time. The ON time constant guaranteeing that, within $0.5T_R$, the voltage stored on C_S is only 1% below the desired value is given by (we consider that the source line voltage polarity is inverted in two consecutive addressing of the same pixel):

$$R_{ON}(C_S + C_{LC}) \le \frac{T_R}{2\ln(2\cdot 100)} = \frac{T_R}{10.6} \tag{5.8}$$

On the other hand, from (5.3), the expression of R_{ON} is

$$R_{ON} = \frac{1}{\mu C_g \frac{W}{L}(v_G - v_{DATA} - V_{TH})} \tag{5.9}$$

where the TFT is assumed in the triode region. To ensure this, voltage v_G is made quite high, and v_{DS} is small, being zero at the steady state (i.e., no current flows in the TFT at the steady state). Hence, from (5.8) and (5.9) we find the minimum aspect ratio W/L of the TFT

$$\frac{W}{L} \geq \frac{1}{\mu C_g \left(v_G - v_{DATA} - V_{TH}\right) R_{ON}} = \frac{10.6(C_S + C_{LC})}{\mu C_g \left(v_G - v_{DATA} - V_{TH}\right) T_R} \tag{5.10}$$

When the TFT is OFF, several causes concur to discharge the pixel capacitance, modifying gray scale uniformity. The TFT leakage current is the most relevant one, but also resistive leakage through the LC material (R_{LC}) should be considered. Considering only the TFT leakage, I_{OFF}, this should be low enough to retain the charge on the pixel capacitance during the frame time. Assuming the parallel of C_S and C_{LC} be discharged by the constant current I_{OFF} during T_F, we get

$$I_{OFF} \leq \frac{C_S + C_{LC}}{T_F} \Delta V_{LK} \tag{5.11}$$

where ΔV_{LK} is limited by the maximum acceptable voltage loss (referred to one gray level).

Tables 5.1 and 5.2 show some technology and display parameters for a very large (56.3-in.) monolithic a-Si AMLCD [KSG2002].

Table 5.1 Technology parameters of a 56.3-in. a-Si AMLCD [KSG2002]

AMLCD Technology Parameter	Value	Unit	Symbol
Row (Gate) Bus Resistance (Width 30 μm)	18.8	Ω/mm	R_R
Row Bus Capacitance (Width 30 μm)	226	fF/mm	C_R
Column (Source) Bus Resistance (Width 10 μm)	45.0	Ω/mm	R_C
Column (Source) Bus Capacitance (Width 10 μm)	82.4	fF/mm	C_C
Row-to-Column Bus Overlap Capacitance	15.3	fF	C_{XC}
Column-to-Row Bus Overlap Capacitance	15.3	fF	C_{XR}
Sub-Pixel Liquid Crystal Capacitance	8.24	pF/mm^2	C_{LC}/A
Sub-Pixel Storage Capacitance	204	pF/mm^2	C_S/A
TFT Channel On-Resistance (Width/Length 100 μm/4.5 μm)	0.0634	MΩ-mm	R_{TFT}^{On}-W
TFT Channel Off-Resistance (Width/Length 100 μm/4.5 μm)	0.607	TΩ-mm	R_{TFT}^{OFF}-W
TFT Gate-to-Source Capacitance	689	fF/mm	C_{GS}/W
TFT Gate-to-Drain Capacitance	919	fF/mm	C_{GD}/W
Row Driver Resistance	1.00	kΩ	R_{RS}
Row Driver Capacitance	1.00	pF	C_{RS}
Column Driver Resistance	1.40	kΩ	R_{CS}
Column Driver Capacitance	1.40	pF	C_{CS}

Table 5.2 Display parameters of a 56.3-in. a-Si AMLCD [KSG2002]

AMLCD Display Parameters	Value	Unit
Number of Horizontal Pixel	1280	1
Number of Vertical Pixel	720	1
Number of Column Lines	3840	1
Number of Row Lines	720	1
Aspect Ratio	16 x 9	1
Horizontal Pixel Pitch	0.974	mm
Vertical Pixel Pitch	0.974	mm
Pixel Array Width	1,246	mm
Pixel Array Height	701	mm
Display Size	1,429 (56.3)	mm (in)
Worst Case Sub-Pixel Charging Time	26.8	µs

As a numerical example, consider an XGA display with 768 rows and 60-Hz frame frequency (N=768, T_F=16.6 ms and T_R=21 µs) in which 256 gray levels are generated by a voltage swing (v_{DATA}) of 8 V (typical for large size panels). Thus, 31.2 mV defines one gray level. Assume also that $C_{LC,min}$=150 fF, $C_{LC,max}$=300 fF, C_S= 400 fF, V_{ROW}=v_G=20 V, C_g=1.06·10^{-8} F/cm, µ=0.45 cm^2/V·s, V_{TH}=3.5 V. From (5.8) we get that $R_{ON} \leq 2.83$ MΩ, yielding $W/L \geq 7.4$ from (5.10). Assuming also as tolerable a ΔV_{LK} lower than 1/6 one gray level (about 5 mV), we get from (5.11) $I_{OFF} \leq 0.16$ pA. The values of W/L and I_{OFF} we have found are compatible with usual TFT technologies.

5.3.1 Kickback

The gate-drain parasitic capacitor C_{GD} is also the cause of charge redistribution with the pixel capacitor C_S. This phenomenon is referred to as the **kickback effect** (it is the same effect known as the clock-feedthrough experienced by designers of digital and switched-capacitor circuits) and causes a decrement, ΔV_{KB}, in the LC voltage when the TFT is turned OFF at the end of a select period (the same effect holds also when the TFT is turned ON, but it is not cause of error since it occurs at the beginning of a select period). Considering Fig. 5.11, if ΔV_G is the voltage change at the TFT gate when the row is deselected and assuming V_{COM} constant, the voltage shift of V_{LC} is expressed by

$$\Delta V_{KB} = \frac{C_{GD}}{C_{GD} + C_S + C_{LC}} \Delta V_G \tag{5.12}$$

As a result of both kickback and leakage effects, the behavior of a typical V_{LC} waveform appears as qualitatively illustrated in Fig. 5.12. The kickback causes a modification of the data voltage and the generation of an unwanted DC component (if the common electrode is held at the middle of data signal voltage swing). It is thus the cause of gray-scale errors [KTK1982], shadings [KKN1990] and flickering [K1989].

From (5.12) we see that the error is nullified for C_{GD} ideally equal to zero. Besides, ΔV_{KB} decreases by increasing C_S+C_{LC}, but this decreases also the ON charging time constant. We recall that C_{GD} for a TFT in triode is not a constant as it depends on v_{GS}. A conservative choice in order to evaluate (5.12) is to take the maximum value (5.6). Also C_{LC} is not a constant, as already discussed in Section 2.2.3. Referring to Fig. 2.12, we can take the minimum value $C_{LC,\min}$.

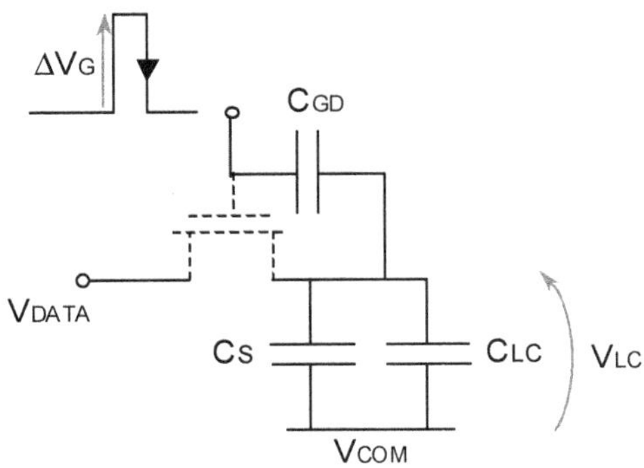

Fig. 5.11 Pixel electrical model for evaluating the kickback effect

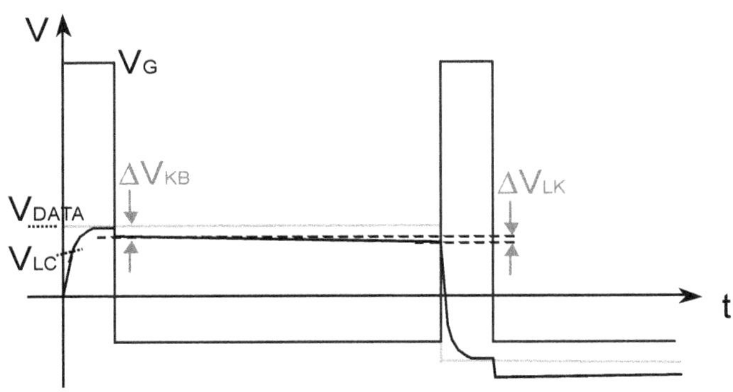

Fig. 5.12 Qualitative behavior of the pixel voltage due to kickback and leakage

As a numerical example, with the same data given before, assuming $\Delta V_G = 20-(-5) = 25$ V, $C_{GD,\max}$=50 fF and again C_S+$C_{LC,\min}$ = 550 fF, we get the maximum ΔV_{KB}= 2.08 V, corresponding to about 67 gray levels. In addition, since the voltage shift is always negative, a DC component is added to V_{LC} that, as said in many occasions, must be avoided (limited to below 100 mV). This demonstrates that a compensation of the kickback is mandatory. To be more precise, we can write

$$C_{LC} = C_{LC,\min} + \Delta C_{LC} \tag{5.13}$$

where ΔC_{LC} depends on V_{LC}. Since also C_{GD} is dependent on the applied voltage, we can rewrite (5.12) as follows

$$\Delta V_{KB} = \Delta V_G \cdot \left[\underbrace{\frac{C_{GD,\max}}{C_{GD,\max} + C_S + C_{LC,\min}}}_{\text{STATIC}} - \underbrace{\frac{C_{GD,\max} \cdot \Delta C_{LC}}{\left(C_{GD,\max} + C_S + C_{LC,\min}\right) \cdot \left(C_{GD,\max} + C_S + C_{LC,\min} + \Delta C_{LC}\right)}}_{\text{DYNAMIC}} \right] \quad (5.14)$$

We recognize a first static term, independent of the LC polarization, and a second dynamic term which depends on the LC capacitance variation (function of the gray level) and the TFT gate-drain capacitance variation. The sign of the dynamic term is opposite to that of the static term.

We will describe techniques for kickback compensation in Section 5.5.

5.3.2 RC Delay and Pre-emphasis Driving Method

As the size and resolution of AMLCD panels increase, the large RC delay introduced by the rows and columns causes a smoothing of the ideal square waveforms and may cause an incomplete settling of the gate and column voltages. In other words, the rate of transferring the analog video signal voltage to the pixel in one row time becomes insufficient. The effect is illustrated in Fig. 5.13a. This limitation is particularly evident when driving large voltage increments onto pixel located at the farthest end of a column and of a row (with respect to the driving ends) and is particularly critical for moving pictures causing motion blur (leading to the appearance of a smear behind any moving image)[M2004]. Besides, the delay of the gate pulse waveforms produces also different kickback moving from the nearest driver end to the farthest end. To alleviate these problems, various approaches have been developed. One is technological, through the adoption of materials with low resistance for gate and source lines: for example, copper [TLT2006] or aluminium alloy [AIH1998, TCC1998]. The other approaches rely on modifying the driving method.

Source and gate drivers can be assembled at both ends of columns and rows for simultaneous driving in **Both-Side Drive** or **Dual Driving** (see Fig. 5.14). In this case the worst-case driving occurs at the center of the panel. Compared to the single-side drive, this method reduces the RC delay time up to a factor of 4, but doubles the number of gate and source drivers, thereby increasing fabrication costs. Source drivers are more critical in terms of area and power dissipation, therefore this approach is used only for the gate drivers.

A fast single-side drive technique exploits **Pre Emphasis** (or **Overdrive)**. Here the row time interval is divided into a "pre-emphasising sub-period" and a "source-follower sub-period", as illustrated in Fig. 5.13b. During the former interval, the source data waveforms use pre-emphasis amplitudes that, as we will see, vary according to the magnitude of the gray-level shift to be driven [KSK2003]. During the source-follower period, the image data voltage is presented to the source line. Figure 5.13b illustrates qualitatively the improvement offered by this technique compared to the conventional one [NIM2007].

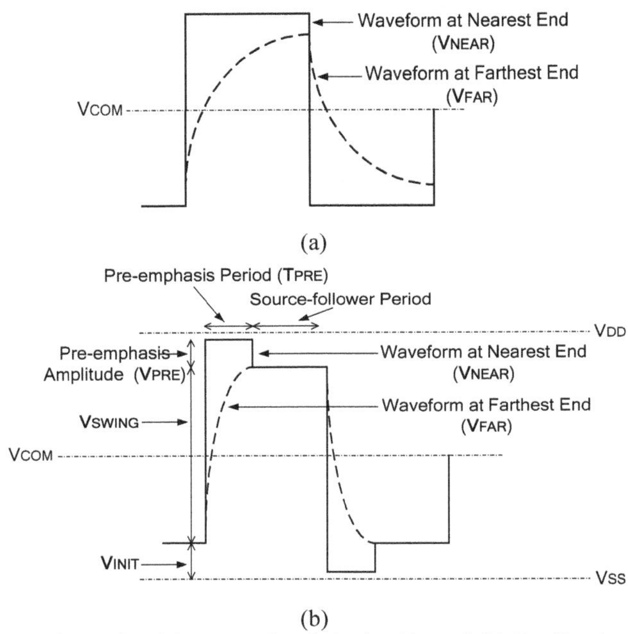

Fig. 5.13 Waveforms for: (**a**) conventional single side and (**b**) Pre-Emphasis driving. The curves in solid (*dashed*) line represent the waveform near (far from) the driver end

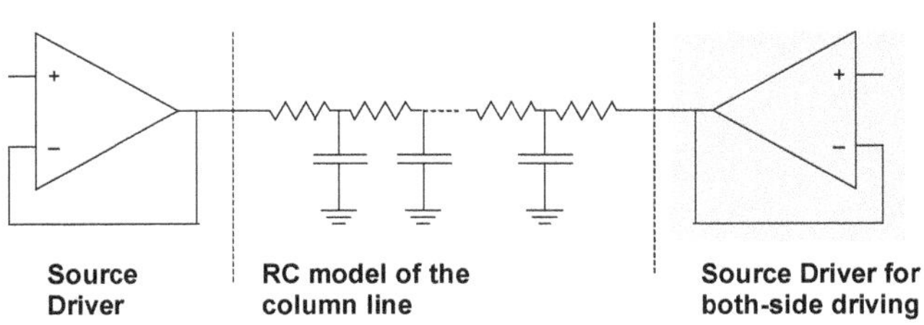

Fig. 5.14 Illustration of the Both-side Drive approach with RC-chain model of the source line and unity-gain amplifier(s) of the source driver

Let us evaluate analytically the required pre-emphasis voltage according to the data voltage swing and allowed pre-emphasis time T_{PRE}. During this time interval, the nearest end of the source line is driven at the voltage swing that is the sum of the target (nominal) voltage swing and the pre-emphasis voltage swing expressed by

$$V'_{SWING} = V_{SWING} + V_{PRE} \tag{5.15}$$

The farthest end source line voltage is expressed by

$$V_{FAR} = V_{INIT} + V'_{SWING}\left(1 - e^{-\frac{t}{R_{DATA}C_{DATA}}}\right) \tag{5.16}$$

where V_{INIT} is the initial pixel data voltage. The product R_{DATA} C_{DATA} can be evaluated through the Elmore's delay model [E1948] and, for an M-length RC-chain, it is

$$R_{DATA}C_{DATA} = \sum_{i=1}^{M}\left(C_i \sum_{j=1}^{i} R_j\right) \tag{5.17}$$

After imposing that at time T_{PRE} the far-end voltage of the source line must have settled to the target voltage level, $V_{INIT}+V_{SWING}$, we get [SK2007]

$$V_{PRE} = \frac{e^{-\frac{T_{PRE}}{R_{DATA}C_{DATA}}}}{1 - e^{-\frac{T_{PRE}}{R_{DATA}C_{DATA}}}} V_{SWING} \tag{5.18}$$

The above relation shows that the pre-emphasis voltage must be proportional to the required voltage swing. Observe that to maintain the advantages deriving form the method, pre-emphasis voltages need to change according to T_{PRE} and the display panel load conditions. Besides, to manage the overdrive voltages required by the approach, a higher-voltage IC fabrication process than that used for conventional source drivers must be adopted (i.e., V'_{SWING} may result higher than the ordinary supply).

Based on the fact that extremely white or black regions have insignificant effects on image quality, a modification of the approach that varies the duration of the pre-emphasis period, rather than its amplitude (this last can be maintained only in the mid-swing regions) was proposed in [YHY2007]. This method does not require higher-voltage IC processes.

5.4 Crosstalk Reduction and Polarity Inversion Techniques

5.4.1 Crosstalk in AMLCDs

As explained in Section 3.3.3 crosstalk is mainly due to capacitive coupling effects. These effects worsen with the scaling of lithographic dimensions and increasing number of gray levels. In addition, crosstalk in AMLCDs can be caused by the following mechanisms [LL1998]:

(1) incomplete pixel charging (due to high TFT ON resistance, unequal data driver loading, line delay);

(2) leakage current from the pixel through the TFT.

Remedies to these problems are based on material selection, fabrication processes, device design and layout. For our scopes, we will address only the capacitive coupling effects.

Consider the layout of a pixel and his crosstalk capacitance shown in Fig. 5.15.

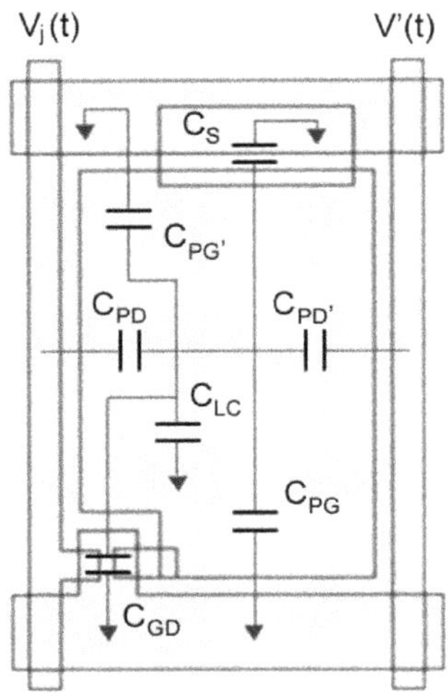

Fig. 5.15 Top view of a typical pixel layout showing the data-line-to-LC-electrode parasitic capacitance regions

The ratio of parasitic capacitances coupling the pixel-to-data line (α) and pixel-to-adjacent-data line (β) can be expressed as

$$\alpha = \frac{C_{PD}}{C_{LC} + C_S + C_{PG} + C_{PG'} + C_{GD} + C_{PD} + C_{PD'}} \tag{5.19a}$$

$$\beta = \frac{C_{PD'}}{C_{LC} + C_S + C_{PG} + C_{PG'} + C_{GD} + C_{PD} + C_{PD'}} \tag{5.19b}$$

where C_{PD} and $C_{PD'}$ are the data-line- and adjacent-data-line-to-pixel-electrode coupling capacitances, respectively, while C_{PG} and $C_{PG'}$ are the gate-line- and adjacent-gate-line-to-pixel-electrode coupling capacitances, respectively.

Note that all capacitances except C_S in (1) and (2) are functions of the liquid crystal dielectric constant. Both parameters α and β must be minimized to minimize the effects of crosstalk.[4] One method relies on increasing C_S. Another one requires a ground-plane shielding beneath the data line. Both methods need added pixel area, at the expense of a smaller aperture ratio and larger gate-line delay.[5] A third method is based on an appropriate data-driving scheme. It leaves unaltered the pixel layout and fabrication technology, thereby providing the least expensive and most straightforward approach to further pixel scaling. Note that this approach does not change α and β, but relies on incorporating some form of positive and negative data-line pulse-coupling cancellation. As a drawback, higher voltage drivers and/or higher driver power consumption may result.

We recall that any driving method must provide a nominally zero DC component of the pixel voltage. Actually a DC residual always exists and to avoid LC mixture damage, all present driving methods include some signal voltage polarity inversion in a time scale shorter than that needed for the mobile LC ions to move substantially.

According to the type of polarity inversion (in which the signs of both the row and column signals are periodically inverted) there are four possible driving methods. These are illustrated in Fig. 5.16 and are called **Frame Inversion, Column (Data-Line) Inversion, Row (Gate-Line) Inversion** and **Dot Inversion**.

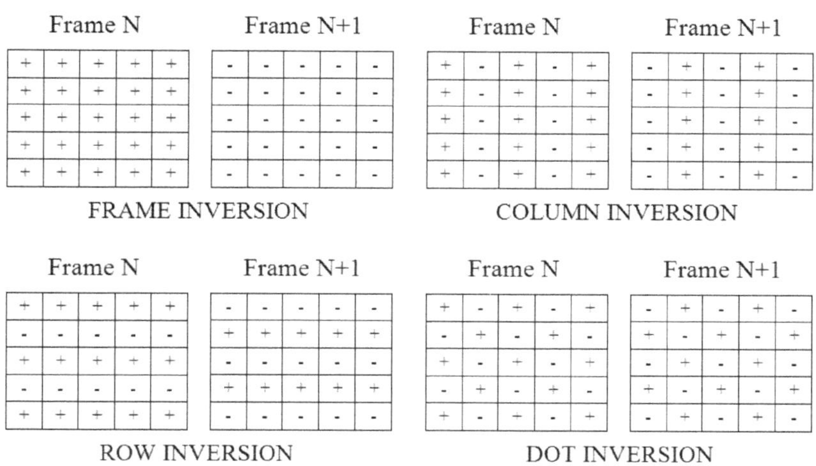

Fig. 5.16 Polarity inversion methods

[4] $\alpha = 0.08$ and $\beta = 0.03$ were extracted for a VGA display in [LL1998].

[5] The IPS-Pro pixel structure has large C_S capacitance without reducing the aperture ratio because the common electrode is made of transparent material under the pixel electrodes [EOO2006].

The principal trade-off among the four mentioned driving methods is the degree of crosstalk cancellation versus driver chip power dissipation, and sensitivity to flicker.

In the **Frame Inversion** method, each source signal has the same data polarity and the data polarity is inverted once per frame time. The value of the instantaneous voltage across the liquid crystal capacitor in the *i*th row position is given by:

$$V_{pi}(t) = V_i + \alpha \cdot [V_j(t) - V_i] + \beta \cdot [V_j'(t) - V_i'] \tag{5.20}$$

where V_i and V_i' are the data voltages initially written onto the pixel and the adjacent pixel through pixel selection, respectively, through their corresponding TFTs. $V_j(t)$ and $V_j'(t)$ are the data-line and adjacent-data-line voltages with their respective parasitic coupling factors to the pixel, α and β.

In the Frame Inversion method, the RMS voltage at the i_{th} row position in a N-row display can be derived from (5.20) as

$$\overline{V}_{pi}^2 = \frac{1}{N}\left\{ V_i^2 + \sum_{j>i}^{N}\left[V_i + \alpha\left(V_j - V_i\right) + \beta\left(V_j' - V_i'\right)\right]^2 + \sum_{j=1}^{i-1}\left[-V_i + \alpha\left(V_j - (-V_i)\right) + \beta\left(V_j' - (-V_i')\right)\right]^2 \right\} \tag{5.21}$$

where the first term (V_i^2) represents the value of voltage pixel during the TFT switching-on (line *i-th* selected), the second term represents the value of voltage pixel during the switching-on of all subsequent rows, while the third is the voltage pixel during the switching-on of the previous rows (reversed polarity).

The simplified expression, obtained by neglecting the second-order crosstalk terms (proportional to α^2, β^2, and $\alpha\beta$) in (5.21), is

$$\overline{V}_{pi}^2 \cong \left(V_i - \alpha V_i + \beta V_i'\right)^2 + \frac{2}{N} V_i \left[\sum_{j\geq i}^{N}\left(\alpha V_j + \beta V_j'\right) - \sum_{j=1}^{i-1}\left(\alpha V_j + \beta V_j'\right)\right] \tag{5.22a}$$

then

$$\overline{V}_{pi}^2 \cong \left(V_i - \alpha V_i + \beta V_i'\right)^2 + \frac{2}{N}\alpha V_i \left[\sum_{j\geq i}^{N} V_j - \sum_{j=1}^{i-1} V_j\right] + \frac{2}{N}\beta V_i \left[\sum_{j\geq i}^{N} V_j' - \sum_{j=1}^{i-1} V_j'\right] \tag{5.22b}$$

Since each pixel is made up of three (RGB) sub-pixels, in general it is $V_j(t) \neq V'_j(t)$. This condition guarantees that the first-order term in (5.22b) (linear in α and β) is dependent on the data-line voltage for one frame. The Frame Inversion is sensitive to flickering due to the absence of spatial averaging.

In a similar way, a general expression can be written for the case of the **Column Inversion** method, for which each consecutive signal line has alternating positive and negative data polarity and the data polarity is inverted once per frame time. By neglecting the second-order crosstalk terms, we obtain:

$$\overline{V}_{pi}^2 \cong \left(V_i - \alpha V_i + \beta V_i'\right)^2 + \frac{2}{N}\alpha V_i \left[\sum_{j \geq i}^{N} V_j - \sum_{j=1}^{i-1} V_j\right] - \frac{2}{N}\beta V_i \left[\sum_{j \geq i}^{N} V_j' - \sum_{j=1}^{i-1} V_j'\right] \tag{5.23}$$

Note that although $V_j(t) \neq V'_j(t)$, the Column Inversion method provides a better crosstalk compensation compared to the Frame Inversion owing to the polarity inversion between two adjacent columns that gives a *negative* contribution in the expression of RMS pixel voltage. This method is less sensitive to horizontal crosstalk caused by capacitive coupling between adjacent columns and can produce better crosstalk compensation to a larger number of pattern images compared to Frame Inversion. It is also less exposed to flicker due to the presence of spatial averaging.

For the **Row Inversion** driving method, where each signal line has the same data polarity and the data polarity is inverted once per scan line time, we obtain the expression in (5.24a) and by neglecting the second-order crosstalk terms we obtain the simplified expression in (5.24b).

$$\begin{aligned}\overline{V}_{pi}^2 = \frac{1}{N}\Bigg\{ & V_i^2 + \sum_{j(even)>i}^{N} \left[-V_i + \alpha\left(V_j - (-V_i)\right) + \beta\left(V_j' - (-V_i')\right)\right]^2 \\ & + \sum_{j(even)=2}^{i-1} \left[V_i + \alpha\left(V_j - V_i\right) + \beta\left(V_j' - V_i'\right)\right]^2 \\ & + \sum_{j(odd)>i}^{N} \left[V_i + \alpha\left(V_j - V_i\right) + \beta\left(V_j' - V_i'\right)\right]^2 \\ & + \sum_{j(odd)=1}^{i-1} \left[-V_i + \alpha\left(V_j - (-V_i)\right) + \beta\left(V_j' - (-V_i')\right)\right]^2 \Bigg\}\end{aligned} \tag{5.24a}$$

$$\overline{V}_{pi}^{2} \cong \left(V_i - \alpha V_i + \beta V_i'\right)^2 + \frac{2}{N}\alpha V_i \left[\sum_{j(odd)\geq i}^{N} V_j - \sum_{j(odd)=1}^{i-1} V_j - \sum_{j(even)>i}^{N} V_j + \sum_{j(even)>1}^{i} V_j\right] + \frac{2}{N}\beta V_i \left[\sum_{j(odd)\geq i}^{N} V_j' - \sum_{j(odd)=1}^{i-1} V_j' - \sum_{j(even)>i}^{N} V_j' + \sum_{j(even)>1}^{i} V_j'\right] \quad (5.24b)$$

Compared to the Frame Inversion driving method described by (5.23), the Row Inversion one contains an additional polarity reversal, as it is evident from the separation of even and odd data-line values (V_j) and adjacent data-line (V_j') values. For this reason, the Row Inversion driving method, in general, provides better crosstalk compensation than the Frame Inversion driving method, but this is not guaranteed, because the magnitude of crosstalk is dependent upon the screen image pattern. For instance, if an alternating image pattern of black to white on the spatial periodicity of a scan line were present, the result is a worst-case adding of all coefficients of α and β for the case of the Row Inversion method. In general, image patterns tend to be uniform over a screen area of at least several gate-line spatial dimensions [LL1998]. In contrast to the Column Inversion method, this one provides a better compensation of the vertical crosstalk produced by capacitive coupling between the rows, but it is more sensitive to the horizontal crosstalk. As the Column Inversion, it is also less exposed to flicker. As above, a general expression can be written for the RMS pixel voltage in the case of the **Dot Inversion** method, where the column and the row inversion approaches are both applied to the data so that each consecutive signal line (column), as well as each consecutive gate line (row), has an alternating positive and negative data polarity, and the data polarity is inverted once per frame time. By neglecting the second-order crosstalk terms, the resulting expression is expressed in (5.25) and includes an additional polarity reversal compared to (5.24b).

$$\overline{V}_{pi}^{2} \cong \left(V_i - \alpha V_i + \beta V_i'\right)^2 + \frac{2}{N}\alpha V_i \left[\sum_{j(odd)\geq i}^{N} V_j - \sum_{j(odd)=1}^{i-1} V_j - \sum_{j(even)>i}^{N} V_j + \sum_{j(even)>1}^{i} V_j\right] - \frac{2}{N}\beta V_i \left[\sum_{j(odd)\geq i}^{N} V_j' - \sum_{j(odd)=1}^{i-1} V_j' - \sum_{j(even)>i}^{N} V_j' + \sum_{j(even)>1}^{i} V_j'\right] \quad (5.25)$$

This method has virtually no flicker.

Equations (5.22b), (5.23), (5.24b), and (5.25) describing all four crosstalk-reduction driving methods show a gain-correction term $(V_i - \alpha V_i + \beta V_i')^2$ as the first term. For a simplified case, if $V_i = V_i'$, the gain correction is $(1 - \alpha + \beta)^2$, and the effect is only a parallel shift in the T/V curve which depends on the values of α and β. Any DC gain in the required RMS operation of the LC can easily be offset. In general, however, $V_i \neq V_i'$, which makes the gain term data dependent on the voltage or image pattern, and it cannot be easily offset.

Among the previous techniques, the Dot Inversion gives the best quality. Row and Column Inversion are suitable for the reduction of vertical and horizontal crosstalk effects, respectively. Since Column and Dot inversion needs source drivers with high voltage swings, a good trade-off between area/power consumption and image quality is given by the Row Inversion. Another approach that was not discussed and that is a further simplification of the Row Inversion is called **n-Line Inversion.** In this case polarity of the addressing waveforms is reversed at every n rows, where n is an integer less than the number N of matrix rows. Recently, **Vertical n-dot Inversion** has been adopted as a polarity reversal method to reduce power consumption while maintaining image quality [SKC2007]. The Vertical 2-dot Inversion is exemplified in Fig. 5.17.

Frame N

+	-	+	-	+
-	+	-	+	-
-	+	-	+	-
+	-	+	-	+
+	-	+	-	+

Frame N+1

-	+	-	+	-
+	-	+	-	+
+	-	+	-	+
-	+	-	+	-
-	+	-	+	-

Fig. 5.17 Vertical 2-Dot Inversion driving method

In another technique, referred to as **Data Complement**, the row addressing time is partitioned into two intervals. During the first interval the data is applied to the source line, as usual. During the second interval all the row TFTs are turned OFF and the data complement (voltages with opposite sign) is applied to the source lines. If the two time intervals are equal, it can be easily shown that first-order crosstalk is eliminated [HAW1989]. The simplest case of equal subperiods halves however the net row time (it is now $T/2N$ instead of T/N) and hence requires an LC mixture with doubled switching speed. To avoid this drawback, the compensation interval can be a fraction $\delta T/N$ of the row time (with $\delta<0.5$), and the amplitude of the complement voltage must be increased by a $\sqrt{\delta/(1-\delta)}$ factor.

5.4.2 Power Analysis of Polarity-Inversion Techniques

In this section we will evaluate the power consumption of the above polarity inversion methods. We will consider a display with N rows and M columns and assume that the load capacitance of each source driver is constant and that the common electrode voltage V_{COM} is 0 V.

(a) Frame Inversion. In this case the lowest power consumption is obtained when all the pixel voltages of the same column are equal during a given frame time (white pixels are considered, otherwise for all black pixels the power consumption is theoretically zero). Indeed, in this case the Source Drivers have the minimum number of commutations. The power dissipated by a single column is hence:

$$P_{COL} = \frac{1}{2} \cdot C_{TOT} \cdot \left[V_{SAT} - (-V_{SAT})\right]^2 \cdot f_F = 2 \cdot C_{TOT} \cdot V_{SAT}^2 \cdot f_F \tag{5.25}$$

where C_{TOT} is the total capacitance seen by the Source Drivers, whose largest output voltages in the positive and negative frames are $+V_{SAT}$ and $-V_{SAT}$, respectively and f_F is the frame frequency.

The overall power consumption in this best-case condition is

$$P_{\min} = 2 \cdot M \cdot C_{TOT} \cdot V_{SAT}^2 \cdot f_F \tag{5.26}$$

The worst case occurs when we have an alternating succession of black and white rows. In this case each Source Driver must provide N commutations in each frame.

$$P_{\max} = \frac{1}{2} \cdot N \cdot M \cdot C_{TOT} \cdot V_{SAT}^2 \cdot f_F \tag{5.27}$$

$P_{\max}$ is $N/4$ times greater than $P_{\min}$ and then power consumption in Frame Inversion is limited by

$$P_{\min} \leq P_{FI} \leq \frac{M}{4} \cdot P_{\min} \tag{5.28}$$

(b) Column Inversion. The power consumption is exactly the same as in the previous case, because the pixels of the same column are subject to the same voltage. Hence, P_{CI} equals P_{FI}.

(c) Row and Dot Inversion. Dually to the previous approaches, here the maximum power consumption occurs when all the pixel voltages of the same column are equal during a given frame time. Indeed, in this case the Source Drivers have the maximum number of commutations.

$$P_{\max} = 2 \cdot N \cdot M \cdot C_{TOT} \cdot V_{SAT}^2 \cdot f_F \tag{5.29}$$

and is 4 times greater than P_{MAX} of Frame e Column Inversion approaches. Besides from (5.26) it is N times greater than P_{MIN} obtained with Frame and Column Inversion. Therefore, we get

$$P_{min} \leq P_{RI} = P_{DI} \leq M \cdot P_{min} \qquad (5.30)$$

The power consumption is greater in those driving methods where row polarity is switched from frame to frame.

For simplicity, we have assumed V_{COM} equal to 0 V. Actually, V_{COM} can be different form 0 and even a variable voltage (V_{COM} switching) in order to meet the constraints set by the adopted CMOS technology, area occupation, and/or low voltage demand. The V_{COM} switching approach increases the power dissipation and causes additional charge injection effects through parasitic capacitances. We will see in Section 5.5.2 that V_{COM} switching requires the same polarity for pixel of the same row and therefore cannot be adopted in both Column and Dot Inversion techniques.

5.5 Kickback Compensation Methods

We have seen in (5.14) that the error due to the kickback includes two contributions: a static one, independent of the gray level, and a dynamic one

$$\Delta V_{KB} = \Delta V_{KB,ST} + \Delta V_{KB,DYN} \qquad (5.31)$$

The dynamic contribution, being dependent on the actual data, must be compensated by the source driver. The static part may be compensated through different approaches: we can shift V_{COM} (as illustrated in Fig. 5.18), or through the subsequent gate voltage for Cap-on-Common architectures or other techniques that will be described in the following.

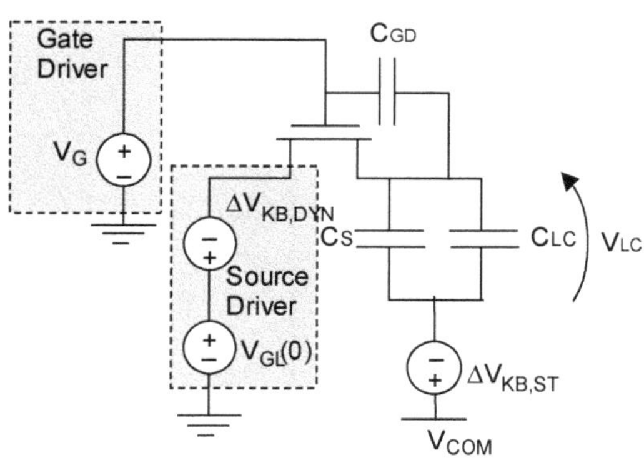

Fig. 5.18 Kickback compensation through the Source Driver (dynamic part) and V_{COM} (static part). The static part can be also compensated via the subsequent gate line, for cap-on-gate architectures

The gray-level voltage, V_{GL}, that through the source drivers must be applied to each pixel is hence

$$V_{GL} = V_{GL}(0) - \Delta V_{KB,DYN} \tag{5.32}$$

where $V_{GL}(0)$ represents the nominal gray-level voltage without kickback correction ("0" stands for zero kickback). Using (5.14), (5.32) can be rewritten as

$$V_{GL} = V_{GL}(0) + \frac{C_{GD,\max} \cdot \Delta C_{LC}}{\left(C_{GD,\max} + C_S + C_{LC,\min}\right) \cdot \left(C_{GD,\max} + C_S + C_{LC,\min} + \Delta C_{LC}\right)} \tag{5.33}$$

In practice, the nonlinear compensation required by (5.33) is implemented in the same circuit block that operates the Gamma Correction. Considering only the nonlinearity of the electrooptic T/V behavior, the Gamma Correction curve should be symmetrical in both positive and negative values, as already described in Section 4.5.2. This symmetry is lost if we introduce also the compensation of the dynamic part of kickback, as illustrated in Fig. 5.19. Note that due to the nonlinear dependence of C_{LC} on the applied voltage, curve (b) is not obtained by a simple shifting of curve (a). This implies that two different gamma curves must be constructed for both positive and negative polarity.

The kickback compensation for both the static and dynamic contributions is closely related to the driving scheme adopted [N1986, YKK1986, S1987, T1989, TNM1990]. In the following we will examine this issue with reference to the four polarity inversion techniques discussed before.

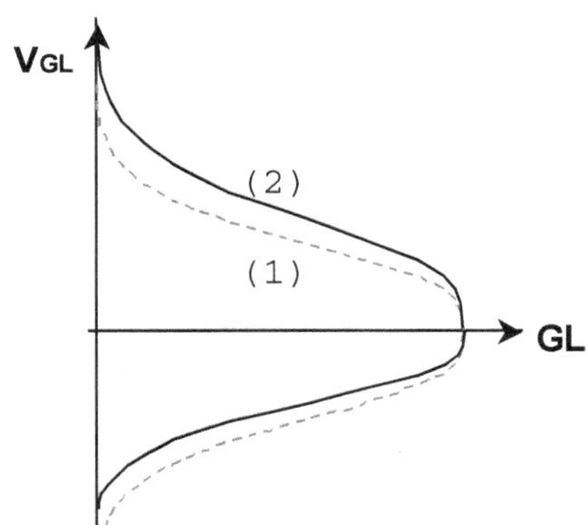

Fig. 5.19 Gamma correction curves: ideal symmetrical (1) and asymmetrical including dynamic kickback compensation (2)

5.5.1 2-Level Driving

In this scheme the Gate Driver provides a two-level output voltage, V_{ON} and V_{OFF}, for selected and unselected rows, as illustrated in Fig. 5.20. Voltage V_{COM} is held constant and consequently the Source Drivers signal swing must be $2V_{SAT}$ ($\approx$ 10V) to allow polarity inversion and thus covering the maximum excursion range given by $V_{COM}-V_{SAT}$ to $V_{COM}+V_{SAT}$. Nominally, V_{COM} can be either equal to 0 V or to V_{SAT}.

The dynamic kickback contribution is compensated via the asymmetrical Gamma Correction curve described before. Considering the static contribution and (5.14), its compensation is achieved by shifting the nominal value of V_{COM} by a quantity ΔV_{COM}

$$\Delta V_{COM} = \Delta V_{ST} = \left(V_{OFF} - V_{ON}\right) \cdot \frac{C_{GD,\max}}{C_{GD,\max} + C_S + C_{LC,\min}} \tag{5.34}$$

Figure 5.21 illustrates the driving voltages in the case of 2-Level Driving for a Cap-on-Common LCD panel structure. The compensation effect is apparent. After the charge injection, the magnitude of the V_{LC} voltages in both positive and negative polarity are equal.

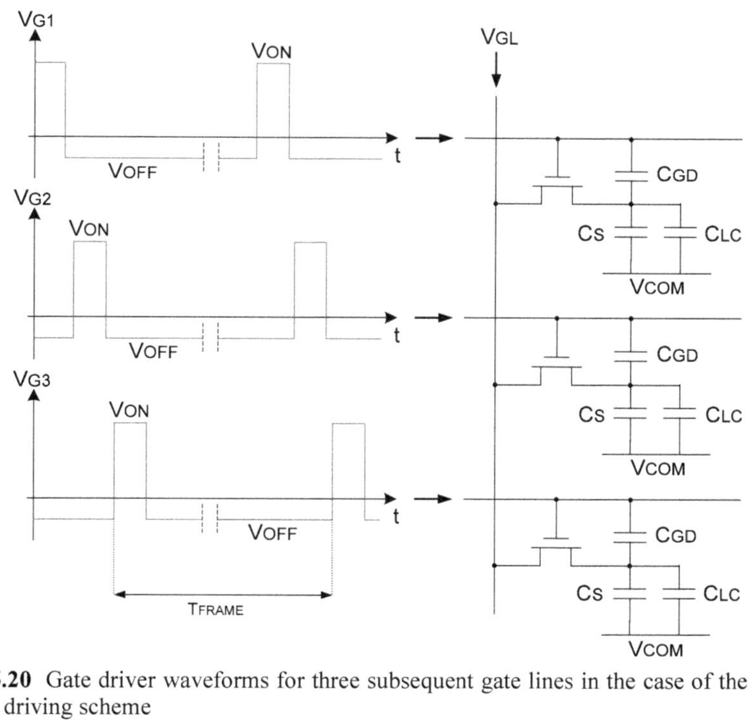

Fig. 5.20 Gate driver waveforms for three subsequent gate lines in the case of the 2-Level driving scheme

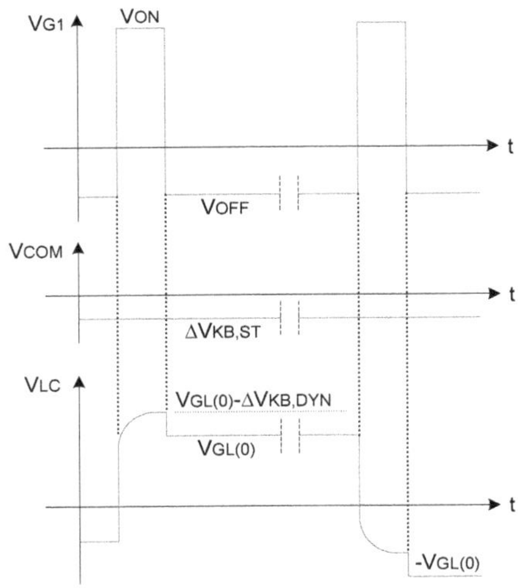

Fig. 5.21 Driving voltages in the case of 2-Level approach for a Cap-on-Common LCD panel structure

5.5.2 V_{COM} Switching

This technique operates the inversion polarity through the modulation of voltage V_{COM}, as illustrated in Fig. 5.22, where V_{GL} is the data voltage to be applied (in both polarities) to the LC pixel and V_{SD} is the output voltage of the Source Driver. Compared to the two-level approach, a Source Driver with a halved output swing is required. The approach is thereby suitable to be employed with low-voltage CMOS technologies.

Unfortunately, an additional error is introduced due to the charge injection from the common electrode when the TFT is OFF and V_{COM} is switching from 0 V to V_{SAT}, in order to allow the polarity inversion of the other rows (see Fig. 5.23). This problem is simply eliminated by introducing the same voltage change onto the gate voltage. Therefore, voltage V_{OFF} at the output of the Gate Driver can be either $V_{OFF,H}$ or $V_{OFF,L}$ so that

$$V_{OFF,H} - V_{OFF,L} = V_{COM,H} - V_{COM,L} \tag{5.35}$$

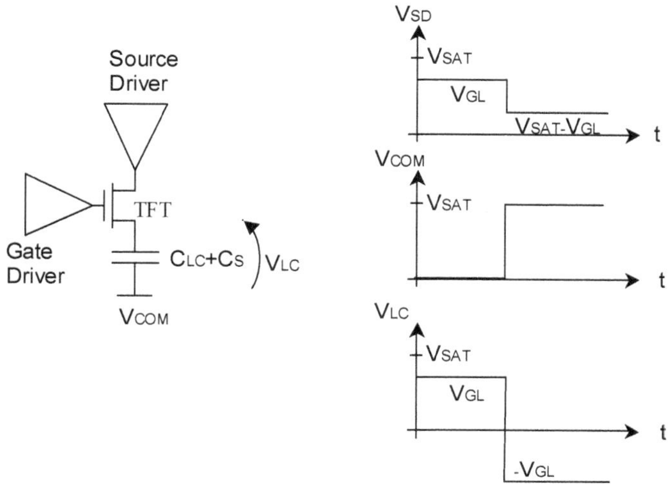

Fig. 5.22 Source Driver and V_{COM} waveforms for V_{COM}-switching driving scheme

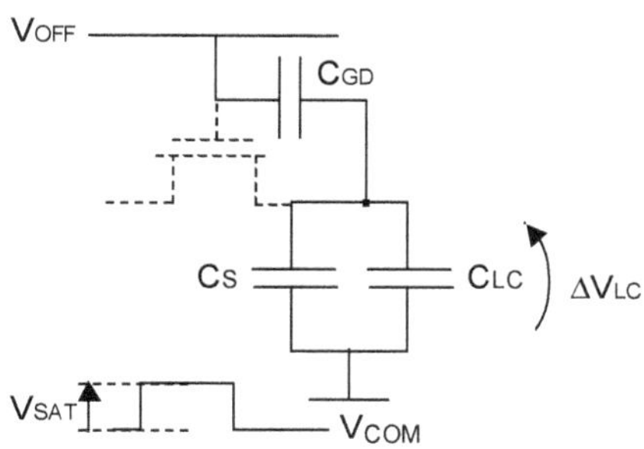

Fig. 5.23 Illustration of the error voltage (ΔV_{LC}) caused by the switching of V_{COM} when the row is unselected

The gate driving waveforms required in the V_{COM}-switching technique are exemplified in Fig. 5.24. To better illustrate the behavior, a Row Inversion method is assumed.

As far as the kickback is concerned, in this case the amount of kickback differs from positive to negative polarity, owing to the different C_{LC} and C_{GD} and also to the different step amplitude, that is $V_{ON} - V_{OFF,L}$ for negative polarity and $V_{ON} - V_{OFF,H}$ for positive polarity.

Consequently, the kickback amount is

$$\Delta V_{KB}^{+} = (V_{ON} - V_{OFF,H}) \cdot \left[\frac{C_{GD,\max}}{C_{GD,\max} + C_S + C_{LC,\min}} + \right.$$
$$\left. - \frac{C_{GD,\max} \cdot \Delta C_{LC}}{\left(C_{GD,\max} + C_S + C_{LC,\min}\right) \cdot \left(C_{GD,\max} + C_S + C_{LC,\min} + \Delta C_{LC}\right)} \right] \tag{5.36a}$$
$$= \Delta V_{KB,ST}^{+} + \Delta V_{KB,DYN}^{+}$$

$$\Delta V_{KB}^{-} = (V_{ON} - V_{OFF,L}) \cdot \left[\frac{C_{GD,\max}}{C_{GD,\max} + C_S + C_{LC,\max}} + \right.$$
$$\left. - \frac{C_{GD,\max} \cdot \Delta C_{LC}}{\left(C_{GD,\max} + C_S + C_{LC,\min}\right) \cdot \left(C_{GD,\max} + C_S + C_{LC,\min} + \Delta C_{LC}\right)} \right] \tag{5.36b}$$
$$= \Delta V_{KB,ST}^{-} + \Delta V_{KB,DYN}^{-}$$

The compensation of the static contribution is performed by shifting the value of V_{COM} by $\Delta V_{KB,ST}^{+}$ and $\Delta V_{KB,ST}^{-}$ and in turn the values $V_{OFF,H}$ and $V_{OFF,L}$. The expression of the pixel voltages in both polarities are

$$V_{LC}^{+} = V_{GL}^{+} + \Delta V_{KB}^{+} - \Delta V_{COM,L} \tag{5.37a}$$

$$V_{LC}^{-} = V_{GL}^{-} + \Delta V_{KB}^{-} - \left(V_{SAT} + \Delta V_{COM,H}\right) \tag{5.37b}$$

where $\Delta V_{COM,L}$ and $\Delta V_{COM,H}$ are the correction values of the low and high V_{COM} level, respectively. As stated before, 0 and V_{SAT} are the low and high V_{COM} ideal levels (neglecting any kickback).

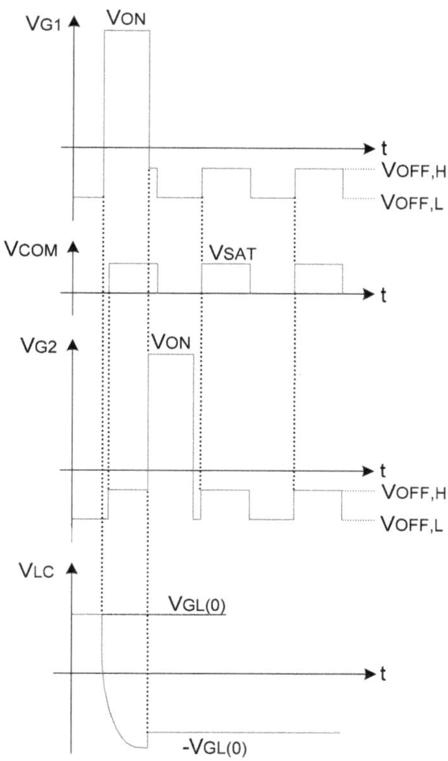

Fig. 5.24 Waveforms of two adjacent gate lines, common electrode, and pixel for V_{COM}-switching driving scheme

The dynamic contribution is compensated as usual via the Gamma Correction curve. Observe that for negative polarity the Gamma curve extends to negative voltage values. This should be avoided if source drivers with positive output swing are to be adopted. To obtain a completely positive Gamma curve it must be shifted by a constant value equal to the dynamic kickback of a *black* pixel. Rearranging (5.37) we get:

$$V_{LC}^{+} = V_{GL}^{+}(0) + \Delta V_{KB,ST}^{+} - (0 + \Delta V_{COM,L}) \tag{5.38a}$$

$$V_{LC}^{-} = V_{GL}^{-}(0) + \Delta V_{KB,DYN}^{-}(\text{Black}) + \Delta V_{KB,ST}^{-} - (V_{SAT} + \Delta V_{COM,H}) \tag{5.38b}$$

To obtain the correct pixel voltages $V_{LC}^{+} = V_{GL}^{+}(0)$ and $V_{LC}^{-} = V_{GL}^{-}(0) - V_{SAT}$, the required $\Delta V_{COM,L}$ and $\Delta V_{COM,H}$ are:

$$\Delta V_{COM,H} = \Delta V_{KB,ST}^{-} + \Delta V_{KB,DYN}^{-}(\text{Black}) \tag{5.39a}$$

$$\Delta V_{COM,L} = \Delta V_{KB,ST}^{+} \tag{5.39b}$$

Finally, from (5.35) and (5.39) we get:

$$V_{COM,H} = V_{SAT} + \left(V_1 - V_{OFF,H}\right) \cdot \frac{-C_{GD,\max}}{C_{GD,\max} + C_S + C_{LC,\max}} \tag{5.40}$$

$$V_{COM,L} = \frac{-C_{GD,\max} \cdot \left[C_{GD,\max} \cdot V_{SAT} + \left(C_{LC,\max} + C_S\right) \cdot \left(V_1 - V_{OFF,H} + V_{SAT}\right)\right]}{\left(C_{LC,\min} + C_S\right) \cdot \left(C_{GD,\max} + C_{LC,\max} + C_S\right)} \tag{5.41}$$

$$V_{OFF,L} = V_{OFF,H} - V_{SAT} - \Delta V_{COM,H} + \Delta V_{COM,L} \tag{5.42}$$

Considering the negative polarity, the gamma correction amount is

$$V_{GL}^{-} = V_{GL}^{-}(0) - \Delta V_{KB,DYN}^{-} + \Delta V_{KB,DYN}^{-}(\text{Black}) \tag{5.43}$$

It should also be observed that the numerous commutations of the common electrode and Gate Lines, increase, in the case of Row Inversion, the power consumption. This is a clear limit of this approach for mobile applications. Besides, owing to the commutation of V_{COM}, techniques such as Column e Dot Inversion cannot be implemented. This scheme is however compatible with both Cap-on-Common and Cap-on-Gate structures.
As a numerical example, consider the typical values listed below:

$V_{SAT} = 5$ V, $V_1 = 18$ V, $V_{OFF,H} = -8$ V, $C_{GD,\max} = 50\text{fF}$, $C_{LC,\min} = 150\text{fF}$, $C_{LC,\max} = 300\text{fF}$, $C_S = 400\text{fF}$

Using (5.40)–(542) we get the following voltages:

$V_{COM,H} = 3.267$ V, $V_{COM,L} = -2.661$ V, $V_{OFF,L} = -13.928$ V

5.5.3 3-Level Driving

Like in the 2-Level approach, a constant V_{COM} is here employed (a high-voltage Source Driver is hence needed). This approach exploits the Cap-on-Gate panel structure, where the storage capacitor C_S is connected to the adjacent Gate Line, as depicted in Fig. 5.25, showing also the driving waveforms. V_1 is the TFT ON voltage, V_3 is the OFF voltage, and V_2 is called a pre-pulse voltage.

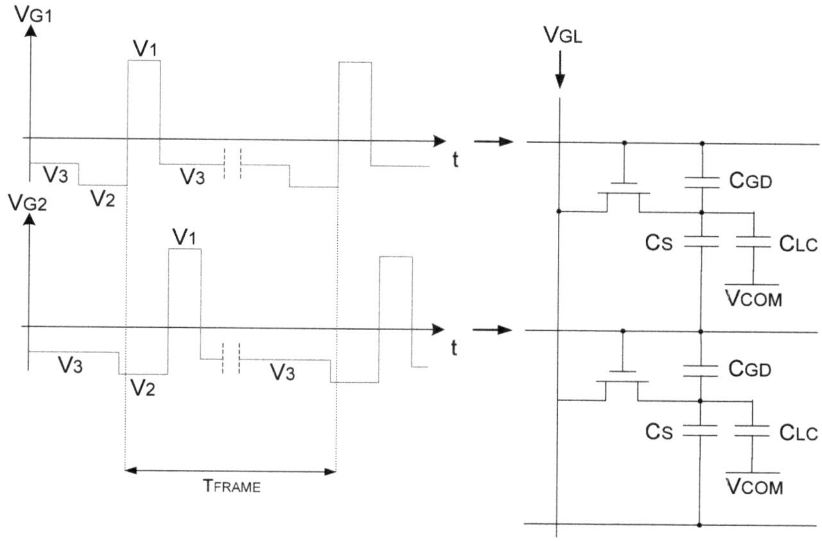

Fig. 5.25 Gate voltages in the case of 3-Level driving scheme

Separating the static and dynamic kickback contributions we get:

$$\Delta V_{KB,ST} = (V_3 - V_1)\cdot\frac{C_{GD,\max}}{C_{GD,\max}+C_{LC,\min}+C_S} - (V_3 - V_2)\cdot\frac{C_S}{C_{GD,\min}+C_{LC,\min}+C_S} \tag{5.44a}$$

$$\Delta V_{KB,DYN} = (V_3 - V_1)\cdot\frac{-C_{GD,\max}\cdot\Delta C_{LC}}{(C_{GD,\max}+C_S+C_{LC,\min})(C_{GD,\max}+C_S+C_{LC,\min}+\Delta C_{LC})}$$

$$+(V_3 - V_2)\cdot\frac{C_S\cdot\Delta C_{LC}}{(C_{GD,\min}+C_{LC,\min}+C_S)(C_{GD,\min}+C_{LC,\min}+C_S+\Delta C_{LC})} \tag{5.44b}$$

After setting $\Delta V_{KB,ST} = 0$ we get

$$V_2 = V_3 + (V_3 - V_1)\cdot\frac{C_{GD,\max}}{C_S}\cdot\frac{C_{GD,\min}+C_{LC,\min}+C_S}{C_{GD,\max}+C_{LC,\min}+C_S}$$
$$\approx V_3 + (V_3 - V_1)\cdot\frac{C_{GD,\max}}{C_S} \tag{5.45}$$

Substituting (5.45) into (5.44b) we get the dynamic kickback component

$$\Delta V_{KB,DYN} \approx (V_3 - V_1) \cdot \frac{-C_{GD,\max} \cdot \Delta C_{LC}}{C_{GD,\max} + C_S + C_{LC,\min}} \cdot \left(\frac{1}{C_{GD,\max} + C_S + C_{LC,\min} + \Delta C_{LC}} - \frac{1}{C_{GD,\max} + C_S + C_{LC,\min} + \Delta C_{LC}} \right) = 0 \tag{5.46}$$

As an important result, we see that with the aid of a pre-pulse voltage both static and dynamic kickback components are ideally eliminated, without requiring the use of Gamma correction. The Gamma curve can be therefore symmetrical, with an advantage both for circuit complexity and calibration task. Actually, due to the fact that C_{GD} is not constant, we cannot achieve a perfect cancellation of kickback. The driving voltages are illustrated in Fig. 5.26.

Assuming V_1=18 V, V_3= –8 V, $C_{GD,\max}$=50 fF, $C_{GD,\min}$=30 fF and C_S=400 fF, we get from (5.45) V_2= –11.142 V (and with the approximation –11.250 V).

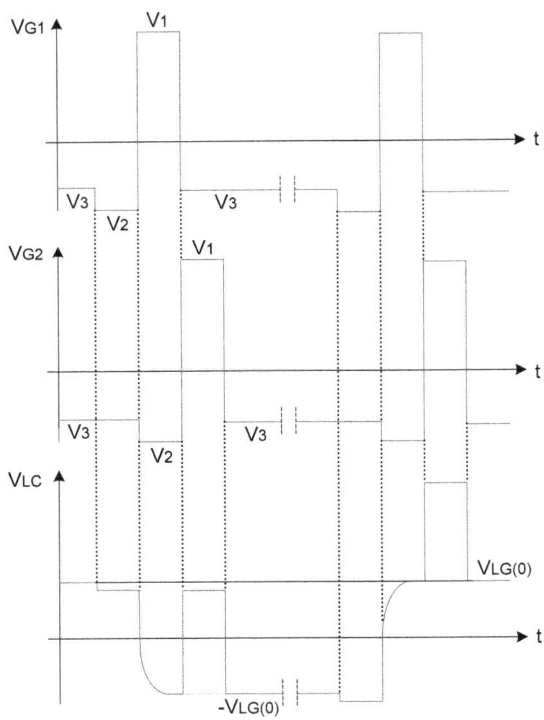

Fig. 5.26 Driving and pixel voltages in the case of 3-Level driving scheme

5.5.4 4-Level Driving

As for the 3-Level approach, also this driving scheme exploits the Cap-on-Gate structure. However, in this case the aim is to use a low-voltage source driver and a constant V_{COM}. The polarity inversion is here obtained by using four voltage levels in the gate line. These waveforms are shown in Fig. 5.27 (Frame Inversion is assumed).

Denoting as ΔV_{KB}^{+} and ΔV_{KB}^{-} the kickback in a positive and negative frame polarity, respectively, and separating the static and dynamic components we get:

$$\Delta V_{KB,ST}^{+} = \left(V_3 - V_1\right)\cdot\frac{C_{GD,\max}}{C_{GD,\max}+C_S+C_{LC,\min}} + \left(V_3 - V_4\right)\cdot\frac{C_S}{C_{GD,\min}+C_S+C_{LC,\min}} \tag{5.47a}$$

$$\Delta V_{KB,ST}^{-} = \left(V_3 - V_1\right)\cdot\frac{C_{GD,\max}}{C_{GD,\max}+C_S+C_{LC,\min}} + \left(V_3 - V_2\right)\cdot\frac{C_S}{C_{GD,\min}+C_S+C_{LC,\min}} \tag{5.47b}$$

$$\Delta V_{KB,DYN}^{+} = \left(V_3 - V_1\right)\cdot\frac{-C_{GD,\max}\cdot\Delta C_{LC}}{\left(C_{GD,\max}+C_S+C_{LC,\min}\right)\left(C_{GD,\max}+C_S+C_{LC,min}+\Delta C_{LC}\right)}+$$

$$-\left(V_3 - V_4\right)\cdot\frac{C_S\cdot\Delta C_{LC}}{\left(C_{GD,\min}+C_S+C_{LC,min}\right)\left(C_{GD,\min}+C_S+C_{LC,\min}+\Delta C_{LC}\right)} \tag{5.48a}$$

$$\Delta V_{KB,DYN}^{-} = \left(V_3 - V_1\right)\cdot\frac{-C_{GD,\max}\cdot\Delta C_{LC}}{\left(C_{GD,\max}+C_S+C_{LC,\min}\right)\left(C_{GD,\max}+C_S+C_{LC,\min}+\Delta C_{LC}\right)}+$$

$$-\left(V_3 - V_2\right)\cdot\frac{C_S\cdot\Delta C_{LC}}{\left(C_{GS,\min}+C_S+C_{LC,\min}\right)\left(C_{GS,\min}+C_S+C_{LC,\min}+\Delta C_{LC}\right)} \tag{5.48b}$$

The pixel voltage is hence

$$V_{LC}^{+} = V_{GL}^{+} + \Delta V_{KB,ST}^{+} + \Delta V_{KB,DYN}^{+} - V_{COM} \tag{5.49a}$$

$$V_{LC}^{-} = V_{GL}^{-} + \Delta V_{KB,ST}^{-} + \Delta V_{KB,DYN}^{-} - V_{COM} \tag{5.49b}$$

where:

$$V_{GL}^{+}=V_{GL}^{+}(0)+\Delta V_{GAMMA}^{+},\ V_{GL}^{-}=\left[V_{SAT}-V_{GL}^{+}(0)\right]+\Delta V_{GAMMA}^{-},\ V_{LC}^{+}=-V_{LC}^{-}$$

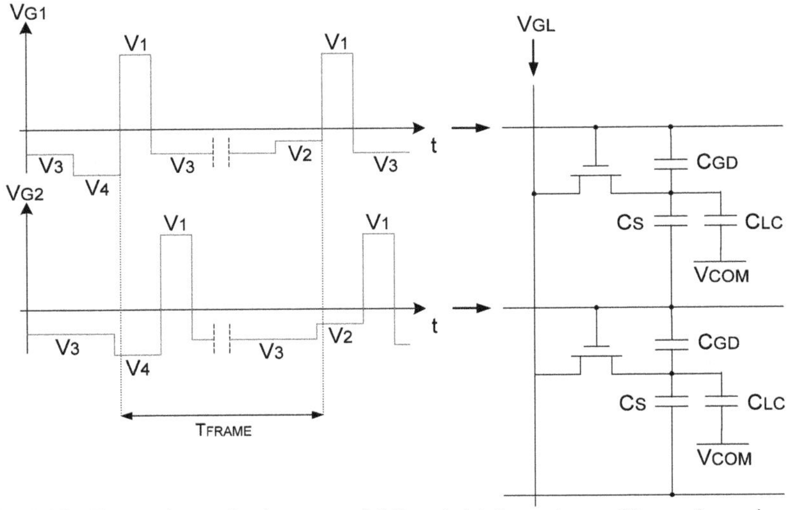

Fig. 5.27 Gate voltages in the case of 4-Level driving scheme (Frame Inversion is assumed)

Similarly to the previous case of V_{COM} modulation, in order to have a positive gamma voltage for both polarities we must set:

$$\Delta V_{GAMMA}^{+}=-\Delta V_{KB,DYN}^{+}\ \text{and}\ \Delta V_{GAMMA}^{-}=-\Delta V_{KB,DYN}^{-}+\Delta V_{KB,DYN}^{-}(\text{Black})$$

$$\Delta V_{KB,ST}^{+}=V_{COM}\ \text{and}\ \Delta V_{KB,ST}^{-}=V_{COM}-V_{SAT}-\left|\Delta V_{KB,DYN}^{-}(\text{Black})\right|$$

After setting $V_{COM}=V_{SAT}/2$, we get the prepulse voltages:

$$\begin{aligned}V_2&=V_3+(V_3-V_1)\cdot\frac{C_{GD,\max}}{C_S}\cdot\frac{C_{GD,\min}+C_{LC,\max}+C_S}{C_{GD,\max}+C_{LC,\max}+C_S}\\&\quad+\frac{V_{SAT}}{2}\cdot\frac{C_{GD,\min}+C_{LC,\max}+C_S}{C_S}\\&\approx V_3+(V_3-V_1)\cdot\frac{C_{GD,\max}}{C_S}+\frac{V_{SAT}}{2}\cdot\frac{C_{GD,\min}+C_{LC,\max}+C_S}{C_S}\end{aligned}\tag{5.50a}$$

$$V_4 = V_3 + (V_3 - V_1) \cdot \frac{C_{GD,\max}}{C_S} \cdot \frac{C_{GD,\min} + C_{LC,\min} + C_S}{C_{GD,\max} + C_{LC,\min} + C_S}$$

$$- \frac{V_{SAT}}{2} \cdot \frac{C_{GD,\min} + C_{LC,\min} + C_S}{C_S} \quad (5.50b)$$

$$\approx V_3 + (V_3 - V_1) \cdot \frac{C_{GD,\max}}{C_S} - \frac{V_{SAT}}{2} \cdot \frac{C_{GD,\min} + C_{LC,\min} + C_S}{C_S}$$

We show in Fig. 5.28 the driving and pixel voltages in the case of Frame Inversion.

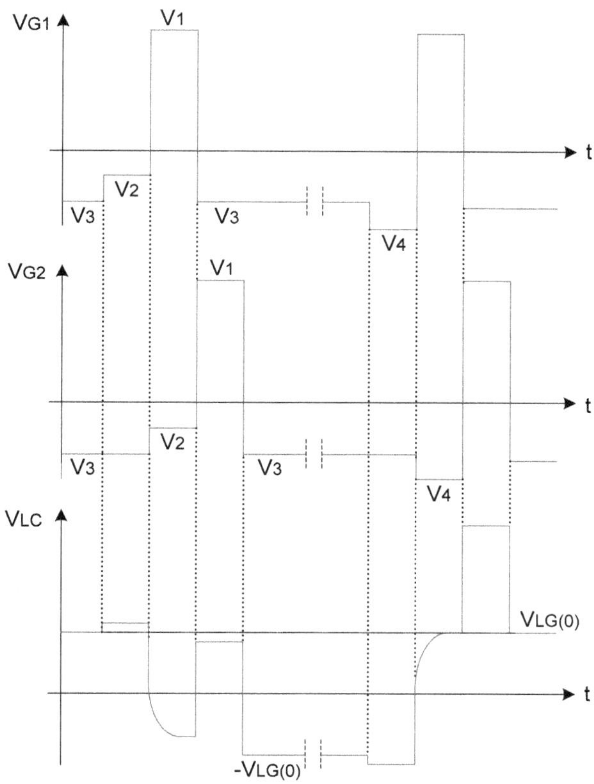

Fig. 5.28 Driving and pixel voltages in the case of 4-level driving scheme. Frame Inversion is considered

The advantage of the 4-level approach derives from the use of a low-voltage source driver and a constant V_{COM}, but a complex gamma correction to eliminate the kickback dynamic component. The approach is compatible only with Cap-on-Gate panels. Finally, either by increasing (for positive-polarity frames) or decreasing (for negative-polarity frames) by the same amount all the ON pixels, it is possible to implement only those inversion methods in which the pixels of a given row have the same polarity, that is Row and Frame Inversion.

5.6 Concluding Remarks

The AMLCD industry has rapidly attracted a great interest by producers and originated a vast excitement in the consumers, so that now it has become an upstream dominant industry. As first investments were made in Japan and subsequently in Korea, Taiwan and recently in China, virtually all AMLCD panel fabrication lines are located in Asia. At the time of writing AMLCDs have penetrated completely the PC notebook market and by more than 50% that of PC monitors, while the TV sector represents the greatest potential of growth (in 2007, the number of LCD-TVs sold worldwide overcome that of CRT-TVs). This impressive and successful expansion has been possible thanks to the efforts of all divisions of the industry, from the manufacturers to component and equipment suppliers.

In this chapter we summarized the a-Si and poly-Si TFT electrical behaviors, we described the AMLCD panel structure and focused our attention on the dominant driving approaches. Conventional inversion polarity techniques were analyzed in detail with their related power consumption issues. Pre-emphasis and kickback compensation methods were addressed as well.

Based on these considerations, we can anticipate the first main differences between drivers for PMLCDs and AMLCDs:

(1) The pixels do not respond to average voltages, frame response effect is absent and MLA techniques are no longer required;

(2) Gray levels are generated by means of the sole source voltage (and not on both row and column voltages, as for PMLCDs), therefore an accurate Digital-to-Analog conversion is mandatory to obtain a source voltage from a digital RGB information. PWM techniques are no longer required;

(3) Due to the higher data transfer rates, low-speed serial interfaces cannot be used because are too slow.

To conclude the discussion, we briefly describe some further methods that have been devised to improve AMLCD picture quality. They are based on the human vision perception system[6] [YW2000, SAK2006, SL2007]. Of course, most of these techniques are meant for LCD TVs, but it is expected that they will be rapidly transferred to high-performance portable multimedia devices as well.

The first two approaches we treat are driving methods aimed to reduce the *motion blur* and in general to improve the quality of moving pictures [M2004].

[6]It is now recognized that optimized display performance can be only achieved by the understanding of the human color perception (see Appendix D), preferences, perceptible maximum contrast level (in general viewing circumstances and not only in dark rooms), and so on.

Studies on the human visual system show that the conventional display-and-hold driving method of LCDs, which maintains the previous frame image information for about 16.7 ms (=1/60 Hz) until the subsequent frame image, is the cause of *afterimage formation* on the retina and, in turn, the origin of motion blur perception [K2001, IYT2003]. This drawback is absent in CRTs with their inherent impulsive driving behaviour. Note that motion blur cannot be eliminated even if the LC response time approaches zero. This explains why, despite the use of progressively faster LCs and pre-emphasis techniques, we have now reached a limit in reducing motion blurring. The amount of afterimage on the human retina can be further reduced only by decreasing the frame time duration.

Fast Frame Rate (FFR) driving (i.e., the use of a frame frequency of 120 Hz) is at this purpose the most straightforward method [LSP2005, OIM2005], though there are issues to resolve in order to contrast the temperature increase in the data driver ICs without affecting the system cost.

Impulsive Driving is an attempt to recreate the CRT behavior through either *Black Data Insertion* [YKI2005, YTI2006] or *Backlight Blinking* [FNC2001, HST2001]. The former is based on the insertion of black data subframes between two consecutive video frames and is implemented via software. It takes advantage of the fast gray to black response of recent LC mixtures. For slower LCs the same effect can be simulated by pulsing the backlight. The technique is implemented via hardware through the backlight inverter and was proposed already in 1998 by a group of Japan Broadcasting Corporation (NHK) headed by Taiichiro Kurita [KSY1998]. To be effective, residual light leakage during the backlight turning OFF must be minimized. As a main drawback, both approaches reduce the overall display luminance.

Research has been carried out to improve also contrast and color reproduction.

LCDs provide reduced contrast when displaying very bright or very dark images. **Dynamic gamma** can be used to overcome this drawback. In this approach the gamma curve is adjusted on a frame by frame basis depending on the image content.

Another promising approach for contrast improvement is the **Backlight Dimming**, [SHS2004, SKT2005, SM2006], in which the whole screen is divided into some areas and LEDs are assembled on each one with independent brightness control function. The backlight intensity of different areas is hence separately controlled based on input video signal levels. A 100 000:1 contrast ratio was reported using this method [SHS2004], which takes advantage of the LEDs fast response. This approach allows reducing the power dissipation of the backlight as well. The better LED color property[7] will also improve color gamut in LCD TVs. To be completely exploited, this would require the introduction of an additional primary color (most probably Yellow) together with the traditional three RGB primaries (**Multi-Primary** color displays) [ELS2005, SL2007].

[7]A typical CCFL backlight will enable an LCD display to reproduce 60% of the colors that can be transmitted using an NTSC signal (see Appendix D). RGB LED backlight has the better color coverage spectrum (color gamut), which is about 110% NTSC (that of White LED is about 70–80% NTSC). Unfortunately, RGB LED backlight has the highest power consumption [CCC2007].

All these techniques are very important for the rapidly growing markets of LCD-TVs and home PC multi-media stations, where image quality is going to become a key factor. On the other side, the market pressure to reduce cost for mainstream desktop monitors (adopted massively by business companies for text and graphics usage) will cause a limited adoption of the above techniques into low-end displays. However, several improvements are foreseen also in the area of mobile AMLCD drivers. For example, Solomon Systech Ltd. (Hong-Kong), outlined the coming advances summarized in Table 5.3 already at the end of 2006.

Table 5.3 Some improvements expected in mobile LCD driver ICs [S2006a]

Features	Present	Coming advance
Resolution	QVGA	VGA
DAC	6-bit gray scale	8-bit gray scale
Motion	Pre-emphasis	Pre-emphasis and Black Data Insertion
Interface	Parallel and SPI	Custom parallel and high-speed serial*
Backlight	No control	LED Backlight Dimming
Contrast	No adjustment	Automatic adjustment according to ambient light level.
Touch panel	High-end feature	Popular feature

*High-speed serial interfaces are discussed in the next chapter.

References

[AIH1998] T. Arai, H. Iiyori, Y. Hiromasu, M. Atsumi, S. Ioku, and K. Furuta, "Aluminium-Based Gate Structure for Active Matrix Liquid Crystal Displays," *IBM Journal of research and Development*, Vol. 42, No. 3/4, pp. 491–499, 1998.

[B2005] W. den Boer, *Active Matrix Liquid Crystal Displays*, Newnes, 2005.

[CCC2007] Y.-F. Chen, C.-C. Chen, and K.-H. Chen "Mixed Color Sequential Technique for Reducing Color Breakup and Motion Blur Effects," *IEEE Journal of Display Technology*, Vol. 3, No. 4, pp. 377–385, Dec. 2007.

[E1948] W.C. Elmore, "The Transient Response of Damped Linear Networks with Particular Regard to Wideband Amplifiers," *Journal of Applied Physics*, Vol. 19, No. 1, pp. 55–63, 1948.

[ELS2005] D. Eliav, E.H.A. Langendijk, S. Swinkels, I. Baruchi, "Suppression of Color Breakup in Color-Sequential Multiple-Primary Projection Displays," *SID Digest*, Vol. 36, No. 1, pp. 1510–1513, 2005.

[EOO2006] K. Endo, R. Oke, and K. Ono, "Column Line Inversion Driving for IPS-Pro LCD-TVs," *Proceedings of International Display Workshops*, pp. 255–256, 2006.

[FNC2001] N. Fisekovic, T. Nauta, N.J. Cornelissen, and J. Bruinink, "Improved Motion-Picture Quality of AM-LCDs Using Scanning Backlight," *Proceedings of International Display Workshops*, pp. 1637–1640, 2001.

[FNK2003] K. Fujimori, Y. Narutaki, and N. Kimura, "High-Transmissive Advanced TFT LCD Technology," *Sharp Technical Journal*, No. 4, 2003. (http://sharp-world.com/corporate/info/rd/index.html).

[GW2001] H. Gleskova and S. Wagner, "DC-Bias Stressing of a-Si:H TFTs Fabricated at 150°C on Polyimide Foil," *IEEE Transactions on Electron Devices*, Vol. 48, No. 8, p. 1667, 2001.

[H1986] W.G. Hawkins, "Polycrystalline-Silicon Device Technology for Large Area Electronics," *IEEE Transactions on Electron Devices*, Vol. 33, No. 4, pp. 447–450, 1986.

[HMM1986] A. Hepburn, J.M. Marshall, C. Main, M.J. Powell, and C. van Berkel, "Metastable Defects in Amorphous Silicon Thin Film Transistors," *Physical Review Letters*, Vol. 56, No. 20, pp. 2215–2218, 1986.

[HS1989] M. Hack and M.S. Shur, "Physicals Models for Amorphous-Silicon Thin Film Transistors and their Implementation in a Circuit Simulation Program," *IEEE Transactions on Electron Devices*, Vol. 36, No. 12, pp. 2764–2769, 1989.

[HST2001] J. Hirakata, A. Shingai, Y. Tanaka, K. Ono, and T. Furuhashi, "Super-TFT-LCD for Moving Picture Images with the Blink Backlight System," *SID Digest*, Vol. 32, No. 1, pp. 990–993, 2001.

[IYT2003] Y. Igarashi, T. Yamamoto, Y. Tanaka, J. Someya, Y. Nakamura, M. Yamakawa, S. Hasegawa, Y. Nishida, and T. Kurita, "Proposal of The Perceptive Parameter Motion Picture Response Time (MPRT)," *SID Digest*, Vol. 34, No. 1, pp. 1039–1041, 2003.

[JSH1996] M.D. Jacunski, M.S. Shur, and M. Hack, "Threshold Voltage, Field Effect Mobility and Gate-to-Channel Capacitance in Polysilicon TFT's," *IEEE Transactions on Electron Devices*, Vol. 43, No. 9, pp. 1433–1440, 1996.

[K1989] A.G. Knapp and M.J. Powell, "Display Issues for a-Si Active Matrix LCTV," *Japan Display Conference Digest*, p. 494, 1989.

[K2001] T. Kurita, "Moving Picture Quality Improvement for Hold-Type AM-LCDs," *SID Digest*, Vol. 32, No. 1, pp. 986–989, 2001.

[K2004] Y. Kuo, "Amorphous Silicon Thin Film Transistors," *Thin Film Transistors: Materials and Processes*, Vol. 1: Kluwer Academic Publishers, pp. 186–188, 2004.

[KDH2005] M.-D. Ker, C.-K. Deng, and J.-L. Huang, "On-Panel Design Technique of Threshold Voltage Compensation for Output Buffer in LTPS Technology," *SID Digest*, Vol. 36, No. 1, pp. 288–291, 2005.

[KKN1990] Y. Kanemori, M. Katayama, N. Nakazawa, H. Kato, K. Yano, and Y. Fukuoka, "10.4 in.-Diagonal Color TFT-LCDs without Residual Images," *SID Digest*, pp. 408–411, 1990.

[KNH2004] K. Karim, A. Nathan, M. Hack, and W. Milne, "Drain Bias Dependence of Threshold Voltage Stability of Amorphous Silicon TFTs," *IEEE Electron Device Letters*, Vol. 25, No. 4, pp. 188–190, 2004.

[KSG2002] J.P. Krusius, D.P. Seraphim, R.G. Greene, D.S. Skinner, and B. Yost, "Approaches Toward Ultralarge FPDs," *Proceedings of IEEE*, Vol. 90, No. 4, pp. 559–580, 2002.

[KSK2003] S.-J. Kim, Y.-C. Sung, and O.-K. Kwon, "Pre-Emphasis Driving Method for Large Size and High Resolution TFT-LCDs," *SID Digest*, Vol. 34, No. 1, pp. 1354–1357, 2003.

[KSY1998] T. Krita, A. Saito, and I. Yuyama, "Consideration on perceived MTF of hold type display for moving images," *Proceedings of International Display Workshops*, pp. 823–826, 1998.

[KTK1982] S. Kawai, N. Takagi, T. Kodama, K. Asama, and S. Yanagisawa, "Amorphous Silicon Thin-Film Transistor for Liquid Crystal Display Panel," *SID Digest*, pp. 42–43, 1982.

[L2001] E. Lueder, *Liquid Crystal Displays: Addressing Schemes and Electro-Optical Effects*, Wiley, 2001.

[LL1998] F.R. Libsch and S.-C.A. Lien, "Understanding Crosstalk in High-Resolution Color Thin-Transistor Liquid Crystal Displays," *IBM Journal of Research and Development*, Vol. 42, pp. 467–479, 1998.

[LLB1992] A.G. Lewis, D.D. Lee, and R.H. Bruce "Polysilicon TFT Circuit Design and Performance," *IEEE Journal of Solid-State Circuits*, Vol. 27, No. 12, pp. 1833–1842, 1992.

[LSP2005] B.-W. Lee, K. Song, D.-J. Park, Y. Yang, U. Min, S. Hong, C. Park, M. Hong, and K. Chung, "Mastering the Moving Image: Refreshing TFT-LCDs at 120 Hz," *SID Digest*, Vol. 36, No. 1, pp. 1583–1585, 2005.

[LVS2006] Z. Li, S. Venugopal, R. Shringarpure, D.R. Allee, and L.T. Clark, "An Analytical Lifetime Model for Digital a-Si:H Circuits," *SID Digest,* Vol. 37, No.1, pp. 270–273, 2006.

[M2004] J. Miseli, "Motion Artifacts," *SID Digest*, Vol. 35, No. 1, pp. 86–89, 2004.

[MK2003] R. Muller and T. Kamins, *Device Electronics for Integrated Circuits*, Wiley, third edition, 2003.

[MS1999] M. Miyasaka and J. Stoemenos, "Excimer Laser Annealing of Amorphous and Solid-Phase-Crystallized Silicon Films," *Journal of Applied Physics*, Vol. 86, pp. 5556–5565, 1999.

[N1986] S. Nagata, "Capacitively Coupled Driving Method of TFT LCD," *SID Digest*, p. 242, 1986.

[N2003] H. Nakagawa, "Current Trends of Flat Panel Displays Viewed from Applications," *Sharp Technical Journal*, No. 4, 2003. (http://sharp-world.com/corporate/info/rd/index.html).

[OIM2005] K. Ono, Y. Imajo, I. Mori, R. Oke, S. Kato, K. Endo, H. Ishino, and Y. Ooishi, "New IPS Technology Suitable for LCD-TVs," *SID Digest*, Vol. 36, No. 1, pp. 1848–1851, 2005.

[P1983] M.J. Powell, "Charge Trapping Instabilities in Amorphous Silicon-Silicon Nitride Thin Film Transistors," *Applied Physics Letters*, Vol. 43, No. 6, pp. 597–599, 1983.

[PBH1989] M.J. Powell, C.V. Berkel, and J.R. Hughes, "Time and Temperature Dependence of Instability Mechanisms in Amorphous Silicon Thin-Film Transistors," *Applied Physics Letters,* Vol. 54, No. 14, pp. 1323–1325, 1989.

[PD1993] M.J. Powell and S.C. Deane, "Improved Defect-Pool Model for Charged Defects in Amorphous-Silicon," *Physical Review B*, Vol. 48, No. 15, pp. 10815–10827, 1993.

[S1987] K. Suzuki, "Compensative Addressing for Switching Distortion in a-Si TFTLCD," *Proceedings of 7th International Display Research Conference*, pp. 107–110, 1987.

[S2006] S. Sze, *Physics of Semiconductor Devices*, Wiley, third edition, 2006.

[SAK2006] K.H. Shin, J.Y. Ahn, K.D. Kim, H.H. Shin, and I.J. Chung, "Acceptable Motion Blur Levels of a LCD TV Based on Human Visual System," *SID Digest*, Vol. 37, No. 1, pp. 287–290, 2006.

[SHS2004] H. Seetzen, W. Heidrich, W. Stuerzlinger, G. Ward, L.A. Whitehead, M. Trentacoste, A. Gosh, and A. Vorozcovs, "High Dynamic Range Display Systems," *ACM Transactions on Graphics*, Vol. 23, pp. 760–768, 2004.

[SK2007] Y.-C. Sung and O.-K. Kwon, "Low-Cost TFT-LCDs with Pre-Emphasis Driving Method for Large-Size and High-Definition TVs," *IEEE Transactions on Consumer Electronics*, Vol. 53, No. 4, pp. 1674–1681, 2007.

[SKC2007] Y. Son, J. Kim, H. Cho, J. Hong, J. Na, D. Kim, D. Han, J. Hong, Y. Jeon, and G. Cho, "A Column Driver with Low-Power Area-Efficient Push-Pull Buffer Amplifiers for Active-Matrix LCDs," *Proceedings of IEEE International Solid-State Circuits Conference*, pp. 142–143, 2007.

[SKT2005] T. Shiga, S. Kuwahara, N. Takeo, and S. Mikoshiba, "Adaptive Dimming Technique with Optically Isolated Lamp Groups," *SID Digest*, Vol. 36, No. 1, pp. 992–995, 2005.

[SL2007] J.H. Souk and J. Lee, "Recent Picture Quality Enhancement Technology Based on Human Visual Perception in LCD TVs," *IEEE Journal of Display Technology*, Vol. 3, No. 4, pp. 371–376, 2007.

[SM2006] J.H. Stessen and J.G.R. vanMourik, "Algorithm for Contrast Reserve, Backlight Dimming, and Backlight Boosting on LCD," *SID Digest*, Vol. 37, No. 1, pp. 1249–1252, 2006.

[SSJ1997] M.S. Shur, H.C. Slade, M.D. Jacunski, A.A. Owusu, and T. Ytterdal, "Spice Models for Amorphous Silicon and Polysilicon Thin Film Transistors," *Journal of Electrochemical Society*, Vol. 144, No. 8, pp. 2833–2839, 1997.

[T1989] E. Takeda, "Simplified Method of Capacitively Coupled Driving for TFT-LCD," *Proceedings of 9th International Display Research Conference*, pp. 580–583, 1989.

[TCC1998] H. Takatsuji, E.G. Colgan, C. Cabral, Jr., and J.M.E. Harper, "Evaluation of Al(Nd)-alloy Films for Application to Thin-Film-Transistor Liquid Crystal Displays," *IBM Journal of Research and Development*, Vol. 42, No. 3/4, pp. 501–508, 1998.

[TDM1999] C.T. Angelis, C.A. Dimitriaids, M. Miyasaka, F.V. Farmakis, G. Kamarinos, J. Brini, and J. Stoemenos, "Effect of Excimer Laser Annealing on the Structural and Electrical Properties of Polycrystalline Silicon Thin-Film Transistors," *Journal of Applied Physics*, Vol. 86, No. 8, pp. 4600–4606, 1999.

[TLT2006] W.-C. Tsai, J.-Y. Lai, K.-Y. Tu, H.-T. Lin, and F.-Y. Gan, "A Novel Barrierless Cu Gate for TFT-LCD," *SID Digest*, Vol. 37, No. 1, pp. 1181–1184, 2006.

[TNM1990] E. Takeda, Y. Nan-no, Y. Mino, A. Otsuka, S. Ishihara, and S. Nagata, "Capacitively Coupled TFT-LCD Driving Method," *SID Digest*, Vol. 31, No. 2, pp. 87–94, 1990.

[U2006] S. Uchikoga, "Future Trend of Flat Panel Displays and Comparison of its Driving Methods," *Proceedings of 18th International Symposium on Power Semiconductor Devices & IC's*, pp. 1–5, 2006.

[Y1999] J.-S. Yoo, "A Novel Polysilicon TFT with Lateral Body Terminal for Driving Circuit Application of AMLCDs," *ITE Technical Report*, Vol. 23, No. 67, p. 27, 1999.

[YHY2007] Y.-S. Son, H.-H. Son, Y.-S. Kim, H.-S. Oh, and G.-H. Cho, "A Gray-Level Dependent Pre-Emphasis Column Driver with Fast Settling for Active-Matrix LCD Applications," *IEEE Transactions on Circuits and Systems - Part II*, pp. 1057–1061, 2007.

[YKI2005] Y. Yamada, K. Kimura, and Y. Ishii, "Technology Trend for High Quality Display Image of LC-TV," *Proceedings of International Display Workshops*, Vol. 12, No. 1, pp. 227–230, 2005.

[YKK1986] T. Yanagisawa, K. Kasahara, and M. Kajimura, "Compensation of the Display Electrode Voltage Distortion," *Proceedings of International Display Research Conference*, pp. 192–199, 1986.

[YTI2006] N. Yamagishi, Y. Tanabe, H. Ishibashi, and H. Yamagami, "Moving Picture Simulations with OCB for Image-Blurring-Free LC TVs," *SID Digest*, Vol. 37, No. 2, pp. 1804–1807, 2006.

[YW2000] Z. Yu and H.R. Wu, "Human Visual System Based Objective Digital Video Quality Metrics," *Proceedings of International Conference on Signal Processing*, Vol. 2, pp. 1088–1095, 2000.

Chapter 6

Drivers for Active-Matrix LCDs

AMLCDs were originally conceived for *large*-screen full-motion applications, but the demand for multimedia handheld equipments with embedded high-quality (*small*-screen) displays has recently pushed their utilization also in this market segment.[1] Nevertheless, the driver circuits for these two LCD panel categories are subject to different trade-offs:

(1) Small-area LCDs are typically battery operated and are cost driven and area driven. Thus, the evolution of the driver architecture has been aimed to the reduction of the number of IC components ultimately converging to a monolithic IC (single chip, or *combo* chip), conceptually similar to that studied in Chapter 4 for PMLCDs.

(2) Owing to the greater number of rows and columns, large-area high-resolution panels cannot be driven by a single chip, which would require an unfeasible number of pins. Besides, large-area LCDs are not strictly focused on reducing power consumption and dimensions. Their main concern is on performance. Their driver design is hence characterized by a multichip architecture where the main functionalities are implemented by separate specialized ICs.

(3) As already discussed in the previous chapter, small-area AMLCDs can be realized either in a-Si and LTPS technologies, whereas large-area AMLCDs exploit only a-Si because of the inherent wide process variations of LTPS over the substrate area. Research is however performed to overcome this limitation.

(4) Column voltages of small-area AMLCDs must have a reduced voltage swing (of about 5 V and V_{COM}-switching is adopted) to employ thin-gate-oxide transistors in the implementation of the Source Drivers, choice driven by size and cost reduction. In contrast, Source Drivers for large-size AMLCDs require a higher voltage swing in the range of 16 V, because this supports the increasing number of gray levels required (10 bit per RGB color, at least). Therefore, 20-V thick-gate transistors are adopted. In this case V_{COM} is held constant to $V_{LCD}/2(\approx 8\text{ V})$ and V_{COM}-switching is not exploited. Dot Inversion or column and n-Line inversion is generally used in large-area AMLCDs for its better performance.

---

[1]The demarcation between small-area and large-area AMLCDs can be roughly set to 5 in., despite this value is dependent on the application (e.g., mobile phones, smart phones, MP3/MP4 players, PDAs, etc.). This differentiation is becoming less and less evident with the increasing of the resolution.

D.J.R. Cristaldi et al., *Liquid Crystal Display Drivers*, DOI 10.1007/978-90-481-2255-4_6,

(5) High voltage boosters must be integrated on silicon in small-size AMLCD drivers and are consequently implemented through charge pumps. In contrast, large-size panel can adopt more standard DC-DC converters which are realized in a centralized fashion and are not implemented within each source driver and gate driver IC because of the unavoidable mismatches.

Other differences between large-size and small-size AMLCD drivers will emerge in the remaining of the chapter.

Due to the strategic commercial significance of AMLCDs, their driving circuits have been the subject of intensive study and prototyping from the most important manufacturers. In this chapter we shall consider some of the main accomplishments. Section 6.1 deals with the driver architectures suitable for small-area AMLCDs (either for a-Si or LTPS technologies) and for large-area AMLCDs. Video interfaces are discussed in Section 6.2. The implementation issues of Source Drivers for small- and large-size AMLCDs, sharing many common characteristics, will be studied in Section 6.3, whereas Gate Drivers will be treated in Section 6.4.

6.1 Amlcd Driver Architectures

We shall start our discussion with small-area LCDs which are typically used in mobile equipments. The basic requirement is here low power consumption but, in addition, continuous efforts are also on enhancing image quality to provide entertainment. In this context driver ICs are key components because they greatly affect the display performance from the point of view of power consumption, image quality, and module cost.

6.1.1 Driver Architecture for Small-area a-Si Panels

The first driver architecture for small-area a-Si AMLCDs was based on several specialized ICs (see Fig. 6.1a where IC interconnections are not shown for simplicity). In order to reduce power consumption, area, and costs, the functions were progressively integrated into only two main chips (Fig. 6.1b). The first is a high-speed and low-voltage IC, which includes Timing Control logic, and DC/DC converters (a RAM block may or not be included as we will shortly see), and that drives all the source lines on the panel using an output buffer/DAC per channel. The second is a high-voltage IC that implements the gate drivers.

Presently, the monolithic IC solution, supporting both high-voltage and low-voltage transistors, is adopted (Fig. 6.1c). It implements all functions in a similar way seen for PMLCDs;

An alternative architecture under feasibility study, which integrates the gate drivers directly on the a-Si TFT backplane is shown in Fig. 6.1d, [JCL2004, KSS2004, EWK2006]. In this case the single monolithic IC supports only low-voltage transistors.

The latter design approach reduces the number of interconnect lines to the display compared to that with external gate drivers, minimizing costs and improving reliability. Unfortunately, the limited a-Si carrier mobility makes difficult the realization of more complex digital and analog circuits directly on glass.

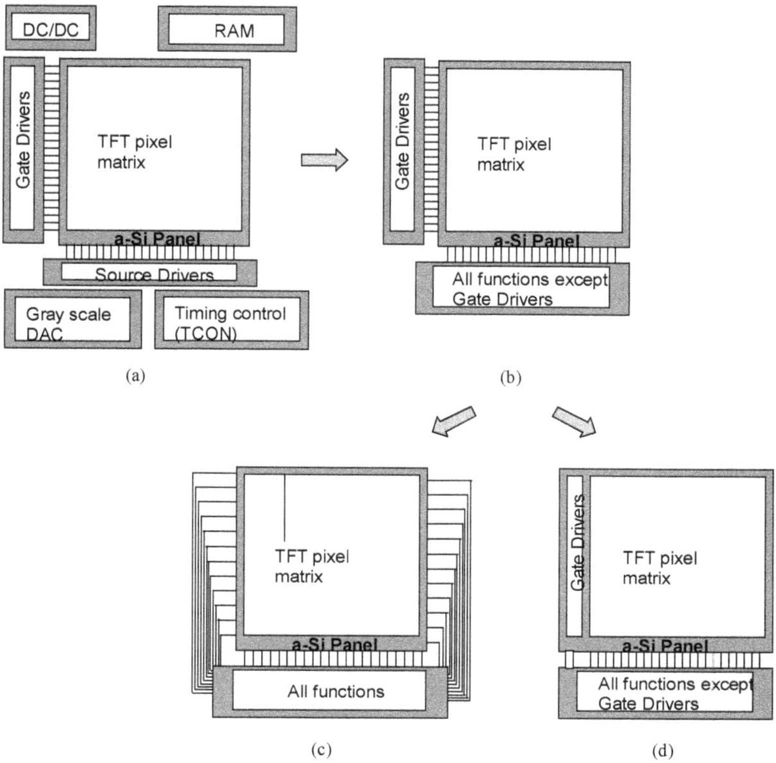

Fig. 6.1 Chipset evolution for a-Si small-size AMLCD drivers: multichip architecture (**a**), two-chip architecture (**b**), single-chip architecture (**c**), single-chip architecture with gate drivers integrated on the TFT backplane (**d**)

Since the driver IC in Fig. 6.1c is the most complete one, it will be used as a reference for the following discussion. Figure 6.2 depicts the block diagram of a possible driver IC for small a-Si AMLCDs. The architecture closely resembles that of a PMLCD already depicted in Fig. 4.1 and deserves no further explanation. The main difference is in the video interfaces (see Section 6.2), the presence of the VCOM driver and a larger non-volatile memory. In most of the cases, when a memory larger than 1-kbit is required, an E^2PROM is used instead of the OTP. When the E^2PROM is external, an E^2PROM interface must be included.

The above architecture includes a video RAM block. More frequently, a RAM-less type driver IC is adopted. The presence or the absence of the RAM is related to the intended application. A video RAM is useful when displaying fixed pictures or low-rate-refresh images. In this case the data frame are stored into the RAM and not transmitted at the frame rate by the baseband processor to the driver, thereby reducing power consumption. In contrast, if the LCD must primarily display full motion video, storing each frame into the RAM and then retransmitting it to the source driver not only is useless, but it even increases power dissipation (without mentioning the other drawback of the augmented silicon area overhead). As a rule, drivers for panels up to QVGA resolution include a full video RAM. For higher resolutions the driver IC does not include the RAM. Some implementation may have a partial RAM which stores some low-color-depth low-rate-refresh data. For instance, in a mobile videophone, these data may be either the clock information (which refreshes once a second or even once a minute) or the name of the calling person. They are usually written in a small portion of the display using 1 bit per RGB color.

The alternative RAM-less architecture differs from that shown in Fig. 6.2 for the absence, of course, of the RAM, as well as of the LUT/Dithering and the Oscillator. Video data presented to the driver interface are indeed in the RGB format that includes a clock signal. Therefore the internal oscillator is avoided.

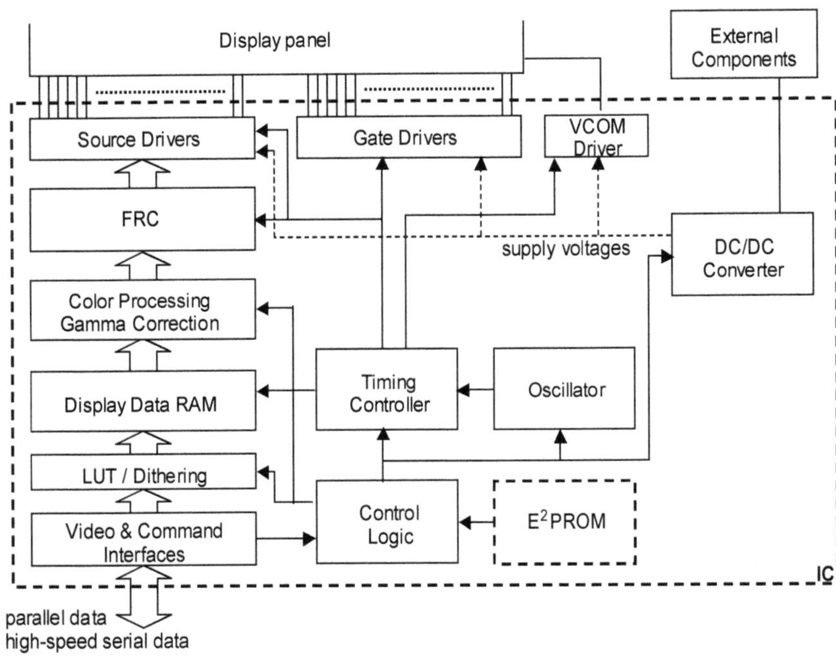

Fig. 6.2 Internal architecture of a-Si small-size AMLCD drivers. RAM-less types are for display resolutions higher than QVGA

6.1.2 Driver Architecture for LTPS Panels

With the increase of display resolution in smaller areas, the interconnections between the driver IC and the panel become more critical and the probability of electrical breaks increases, reducing in turn the yield. In addiction, the cost of a flexible PCB increases with the number of I/O pins. To minimize the number of bonding connections to the panel, LTPS technologies offer a substantial advantage because they allow integration of digital and analog electronics directly on glass (approach denoted alternatively as System on Glass, SOG, or System on Panel, SOP, [NKM2004]).

For instance, if M is the number of columns, LTPS can be exploited to multiplex a number (say m) of the source lines (by implementing the demultiplexers on glass, as shown in Fig. 6.3, showing a driver IC for an LTPS panel), thereby decreasing the number of external source lines as well as and the required number of DACs plus Buffers from $3M$ to $3M/m$. This relaxes the area requirements, but mandates for circuits with faster settling, because now only the fraction $1/m$ of the row scanning time is dedicated to each pixel.

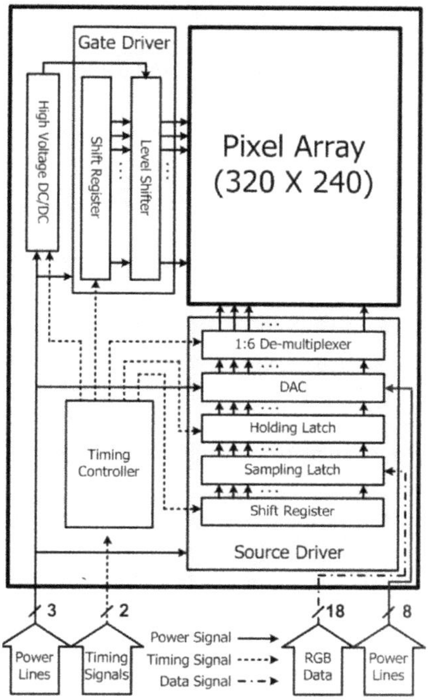

Fig. 6.3 Architecture of a 2.2-in. QVGA AMLCD LTPS panel [SCK2007]. Note the absence of a RAM and Oscillator and Interface since the incoming data are in the RGB format

LTPS TFT technology was initially adopted in high-end video projectors and digital cameras [SMH2000, K2001] in which the reduction of peripheral bonding was of outmost importance. Prototypes have been also recently developed for the consumer market of mobile devices, in the attempt to reduce costs and complexity required by the flexible PCB and to improve reliability. For instance, source/gate drivers, timing controller, DC-DC converters, interface circuits, and V_{COM} driver have been fully integrated in 2.2-in. QVGA displays [TSL2005, SCK2007].

The RAM-less architecture reported in the last referenced paper is illustrated in Fig. 6.3. Prototypes of 10.4-in. XGA and 10.2-in. VGA displays with integration of gate driver and demultiplexers for source driver have been also presented [PLY2005, YCL2007]. The recent trend is to use pMOS LTPS TFT circuits especially for medium panel sizes due to the less number of masks (four) compared to those of CMOS (six) [PLY2005]. Indeed, for small-area panels the CMOS LTPS retains its advantages even if the number of masks is higher than a-Si. However, the benefits will be lost with increasing panel sizes due to higher manufacturing costs and lower yield, unless the integration of the whole system is possible. Therefore, to overcome these disadvantages, the approach of photomask reduction has been followed.

The main shortcoming of LTPS technologies is the high variability of process parameters, mainly threshold voltage and mobility, presenting no spatial uniformity [KDH2006]. This problem impacts especially analog circuits. For instance, if analog buffers are sensitive to threshold voltage variations, the source voltage level will be affected by a (random) offset error, and the LCD will exhibit nonuniform brightness and grey levels. Techniques to combat the effect of process tolerances are well known [ET1996] and many of these can be exploited to the design of LTPS drivers [CLH2001, JPH2004, YCS2004, LLY2005, YKL2005, KDH2006, SCK2007]. An example is shown in Fig. 6.4 representing a voltage buffer compensated to the offset (which is modelled as a concentrate voltage at the input of the ideal amplifier).

The circuit adopts a well-known autozero technique that uses three switches driven by two nonoverlapping phases ϕ_1 and ϕ_2. The hold capacitor, C_H, samples the offset voltage during phase ϕ_1 high (T2 and T1 act as closed switches and the amplifier is in unity gain configuration). The resulting circuit is shown in Fig. 6.4b and the voltage across the capacitor is V_{OS} (also V_{out} equals V_{OS}, but this voltage is not applied to the columns). When ϕ_2 is high, the circuit becomes that shown in Fig. 6.4c, the voltage on C_H subtracts to the amplifier offset and the output voltage is V_{in} which is applied to the columns.

The main sources of error are now due to the finite gain of the amplifier causing the voltage stored on C_H to be $V_{OS}\,A/(1+A)$ and clock-feedthorugh and charge injection effects of the switches. They can be limited by adopting large C_H values (larger than the parasitic C_{gd} of the transistors implementing the swithces), and compelemtary switches (tansmission gates in place of simple pass transistors) or even more complex structures including dummy transistors.

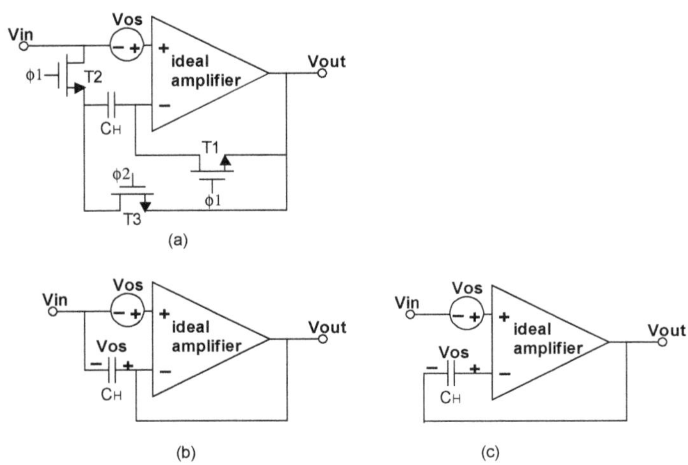

Fig. 6.4 Example of buffer amplifier with offset compensation [TSL2005]

6.1.3 Driver Architecture for Large-Area Panels

The development of improved multi domain IPS and VA pixel structures (discussed in Section 2.4.4) has been the key point to obtain wide viewing angles for large-area LCDs. However, every new mode of operation must provide its improvements while maintaining the same simple pixel driving scheme, with one gate line, one source line and one common line, as well as a standard timing controller.

We already mentioned in Chapter 5 that large-size high-resolution displays cannot take advantage of the LTPS technology, because of the large spread in the electrical characteristics of individual LTPS TFTs over the substrate area and kink effect. In contrast, the same advantages deriving form the integration of a-Si gate drivers could be in principle obtained also on large-area AMLCDs, but two main technological issues must be solved. One is the instability of a-Si TFTs, due to bias stress from an applied-voltage bias during circuit operation, causing a change in the TFT characteristics. The other is the required large width of the a-Si TFTs, extended to thousands of micrometers due to the low electrical mobility and high capacitive load of large-area panels.

As already anticipated, large-area AMLCDs adopt typically a multi-IC driver architecture. Figure 6.5 shows a block diagram where a Controller Board, including I/O Interfaces, Timing Controller (TCON), DC/DC converter, and Parallel-to-Serial decoders, is connected via a parallel bus to the Gate Driver and Source Driver sections without the need, as early observed, of a video RAM. The TCON PCB, Gate Driver PCB, and Source Driver PCB are connected together through a Flexible Printed Circuit (FPC).

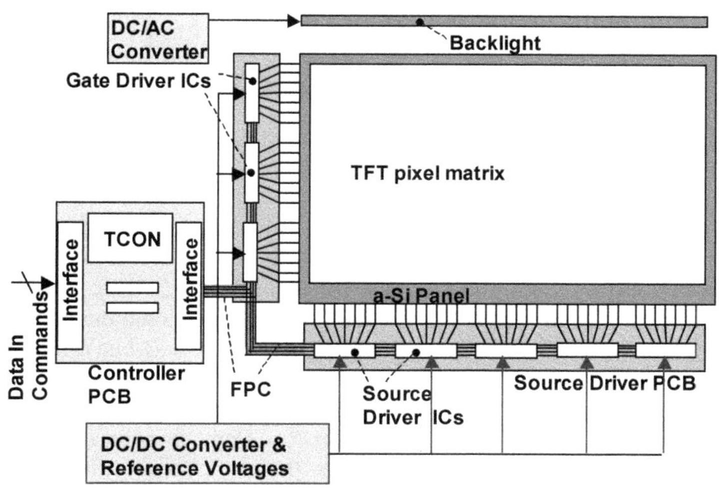

Fig. 6.5 Typical multichip driver architecture for large-area AMLCDs

The TCON of an AMLCD is conceptually simpler than that of a PMLCD. It is fully digital and operating at low voltages. The LCD gamma correction curve is stored in look-up tables within the timing controller. There may be three separate

LUTs, one for each color. The LUTs are loaded from an external E^2PROM. LUT data can be digitally manipulated to compensate for temperature, ambient light, and LCD response.

In the past TCON ICs have been implemented as full custom ASIC (Application Specific Integrated Circuit) devices, with limited possibility of reuse in different LCD panel systems e.g., characterized by different resolutions, color depths, etc.(see Fig. 6.6). Recently, highly functional TCON ICs (programmable through high-level description languages such as VHDL and Verilog) have been developed supporting even non-standard resolutions and different LCD panel configurations from different manufacturers. These TCON ICs result in smaller PCBs and lower power consumption by integrating also on chip I/O data Interfaces (see also Section 6.2).

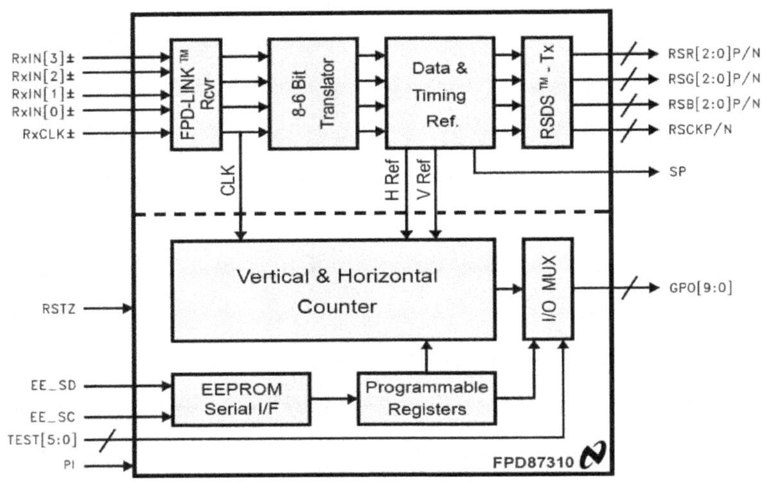

Fig. 6.6 Example of fully integrated TCON IC from National Semiconductor Corp [LL2000]

One of the main problems associated with the multi IC architecture is due to the length of the wiring between the TCON IC and the Source Driver ICs (for a 45-in. display it commonly exceeds 50 cm, being the manufacturing limit equal to 60 cm [K2005]) thus making communication difficult via a conventional bus interface. One practical solution is to use 2–4 timing controller ICs (Fig. 6.7) but it requires more components (including peripheral FPCs and connectors) and results in higher cost.

In a conventional bus interface (Fig. 6.8a) the number of wires remains the same even if we reduce the number of Source Driver ICs. Point-to-point connection (Fig. 6.8b) minimizes the length of wires and reduces their number for a lesser number of source drivers used. Besides, thanks to the decreased loading, higher speeds can be achieved. As detailed in Section 6.2, the point-to-point connection supports high-speed serial data transfer.

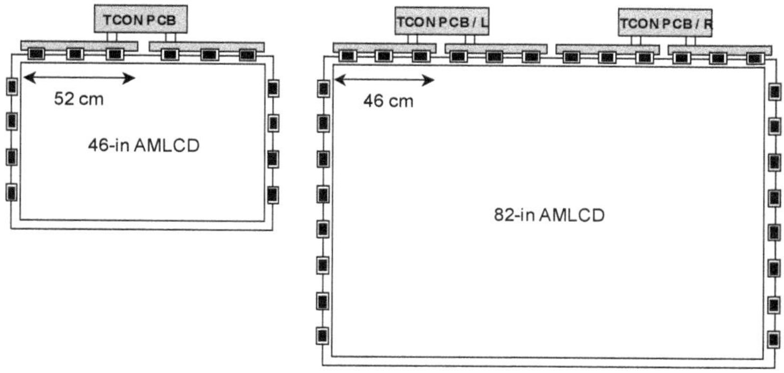

Fig. 6.7 82-in. AMLCD with two TCON PCBs

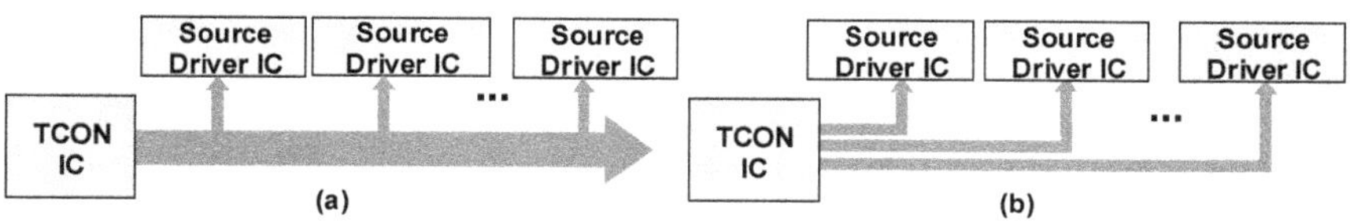

Fig. 6.8 Connection between TCON IC Source Driver ICs: (**a**) Parallel style (a) and (**b**) point-to-point style

While LCD drivers for TV application have substantially maintained the multi-IC architecture,[2] notebook computers (where reduction of size, weight, cost, and driving efficiency are more important objectives) have changed somewhat radically their LCD driver architecture to follow almost the same evolution of small a-Si LCDs described in Fig. 6.1, in order to reduce the chip count and flexible PCBs. Figure 6.9 shows the photograph of a conventional 15-in. AMLCD for notebook PCs. The constituting elements are clearly evidenced.

[2] LCD TVs main accomplishments are oriented towards improvements in color and image quality. For instance, the gray level distribution is expanded to make the level utilization more efficient and improve *contrast*. At this purpose, histogram equalization is usually adopted by adjusting the entire image frame according to a global statistics regardless of the spatial information. Such an approach may cause artefacts such as unnatural color levels. Improved schemes take also the pixel locality (edge information) into account. However, contrast enhancement typically leads to some reduction of color *saturation*. Improved solutions keep saturation, hue, and intensity of every pixel constant to ensure color accuracy. AMLCD TVs generally have 16:9 aspect ratios. Older video sources that provide the 4:3 formats will be expanded to fill the display area and, as a result, the video edge may become blurred. To avoid such defects, sharpness enhancement is performed by enlarging the intensity between vicinal edge pixels. Improved solutions avoid overshoot effects at the strong edges by dynamically varying the *sharpness* enhancement, enlarging (reducing) the enhancement for low-swinging (high-swinging) regions. Simple Frame Rate Control (FRC) schemes used to increase the color depth adopt fixed pattern to vary the intensity of pixels in temporally consecutive frames. Occasionally the fixed pattern may result in unwanted textures when the upstream video board adopts another fixed FRC pattern. DFRC (*Dynamic FRC*), reduces the risk of conflict of cascaded FRCs [LWH2005].

In this framework, since 1999 Samsung began upgrading its LCD driving architecture in a series of steps code-named Mont Blanc (MB) [BMK2005]. Table 6.1 summarizes these steps. The 12.1-in. WXGA module resulting from step MB5 is illustrated in Fig. 6.9b, a single flexible printed circuit connects the TCON board to the panel, carrying all data, power, and gamma information.

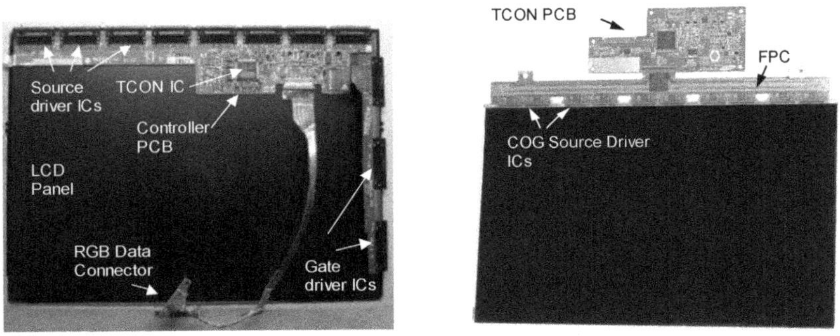

(a) (b)

Fig. 6.9 Photographs of AMLCD panels for notebook computers with driving circuitry: (**a**) 14-in. conventional style; (**b**) 12.1-in. with integrated a-Si Gate Drivers and COG PAD ring of Source Drivers by Samsung [BMK2005]

Table 6.1 Samsung developments for notebook computers LCD drivers.

Mont Blanc Step (year)	**Improvement**
MB1 (1999)	The FPC connecting the source and gate PCBs was removed by routing the gate signals on the glass panel itself.
MB2 (2000)	The gate driver PCB was eliminated by routing the gate signals into and out of film-based row drivers mounted directly to the LCD glass using adhesive coated film (ACF).
MB3 (2000)	Reduced Swing Differential Signalling (RSDS) was adopted as a second-generation interface between the timing controller and column driver ICs.
MB4 (2004)	Gate drivers were integrated onto the glass itself in a-Si.
MB5 (2004)	The source driver PCB was eliminated by realizing Chip on Glass (COG) source driver ICs mounted directly onto the glass, held in place by ACF.

6.2 Video Interfaces

We have already dealt with serial and parallel interfaces adopted by PMLCD drivers in Section 4.2. The data transfer rate required by high-resolution AMLCDs cannot be supported by low speed serial interfaces which, as a rule, are employed up to the QCIF resolution. 8-bit parallel interfaces are employed from 96×128 to WQVGA, while 16-bit parallel from QCIF to WQVGA. RAM-less drivers adopt RGB or YUV (see Appendix D) parallel interfaces employing CMOS signal levels and typically for resolutions higher than QVGA. The data bus, depending on the display architecture and color depth (6 or 8 bit/color) can be a single or dual bus of 18 or 24 conductors each. However, CMOS signalling and the increased frequency of commutations cause and unacceptable increase in power consumption and in electromagnetic interferences (EMI).

To avoid the above drawbacks while achieving higher transfer rates required by video streaming and high color depths, **high-speed serial interfaces** (HSSI) based on **low voltage differential signalling** (LVDS) techniques have been developed (always for RAM-less drivers) [IEEE1996, HG1998, K2000, M2002].

LVDS is adopted in two contexts: (1) as input interface from baseband processor to display driver and (2) as output interface from TCON (either PCB or IC) to the Source Driver ICs (see Figs. 6.5, 6.6, 6.7).

6.2.1 Low Voltage Differential Signalling

LVDS was introduced by National Semiconductor in the early 1990s and was adopted as a standard by Electronic Industries Association, later the Electronic Industries Alliance, as EIA-644. It was developed in order to provide a low-power and low-voltage alternative to other high-speed I/O interfaces for point-to-point transmission, such as emitter-coupled logic (ECL).

In an LVDS communication system, input data are first serialized, distributed to a limited number of wire pairs and then transmitted at a clock rate 7-times higher than the original. The pixel clock is also transmitted via a dedicated differential pair. A single LVDS pair with the associated transmitter, receiver, and transmission line with resistive termination (R_{TR}) is illustrated in Fig. 6.10.

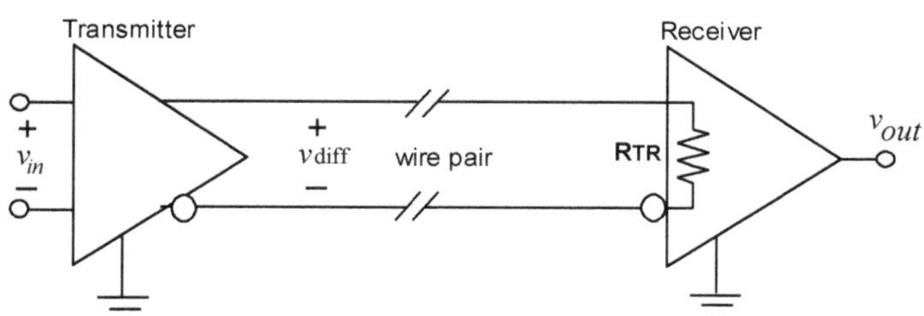

Fig. 6.10 Single transmission line of an LVDS system

The wire pair operates in a differential voltage mode, with a differential swing of ±350 mV. This limited signal excursion (compared to the full swing of, say, 1.8 V of CMOS) improves settling time, thereby reducing propagation delay at the transmission line end. The value of the termination resistance is set to 100 Ω. This value keeps low the RC time constant of the transmission line. The differential-mode signalling causes two equal amplitude currents (150 mV/100 Ω = 3.5 mA), but with opposite signs, to flow into the two wires. As a result, the associated electromagnetic fields are nearly cancelled out if the wires are in close proximity (EMI are strongly reduced up to 90%). Another advantage, useful in handset equipments, is that reducing the number of wires facilitates design of narrow rotating part of rotating or sliding displays, such as those in cellular phones.

As an example, consider the input interfacing between the baseband processor and the LCD driver adopting an 18-bit data format. A parallel data bus would hence require 18 wires. LVDS reduces the number of wires to only 8 through a 7:1 multiplexing scheme. In other words, 3 couples of wires are exploited for RGB data transmission at a clock rate 7 times the original, as illustrated in Fig. 6.11. The remaining additional 3 bit per clock are exploited to convey the horizontal and vertical sync signals (HSYNC and VSYNC) and the display enable (DE). A Phase Locked Loop circuit (PLL) is also used to match the clock frequency to the transmission data rate, while a frequency divider is used in the receiver. A similar solution holds for a 24-bit data format. In this case we have 4 couples of wires for serial transmission of 24-bit RGB data and 4 bit for commands (one is a custom bit).

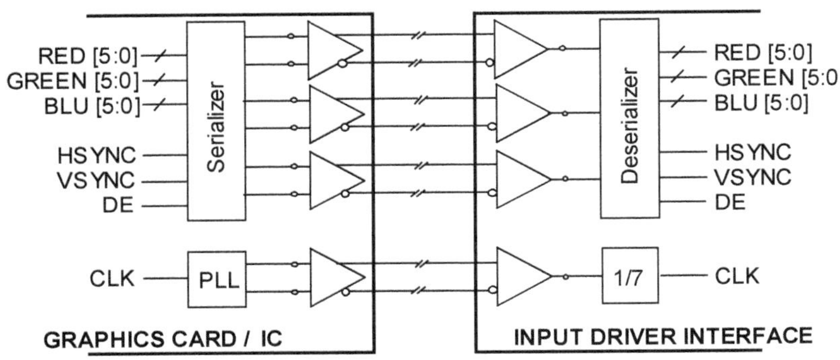

Fig. 6.11 Example of 18-bit LVDS interface

A simplified transistor-level implementation of LVDS transmitter and receiver is depicted in Fig. 6.12a, b. The current-mode transmitter is made up of four MOS switches, M_1-M_4, in bridge configuration. As said, I_{B1} is equal to 3.5 mA. Depending on the sign of the input voltage, $v_{in}^+ - v_{in}^-$, current I_{B1} can be switched in both positive and negative polarity over the transmission line, thereby generating the differential signal voltage, v_{diff}, across R_{TR}. If $v_{in}^+ - v_{in}^- > 0$, then M_2 and M_3 are OFF whereas M_1 and M_4 are ON. Therefore, a negative differential voltage nominally equal to 355 mV is produced. Conversely, a positive differential voltage is set across R_{TR} for $v_{in}^+ - v_{in}^- < 0$.

The receiver (Fig. 6.12b) is a simple differential amplifier acting as a comparator converting the differential-mode input voltage into a single-ended output voltage.

The LVDS standard prescribes the following admissible ranges for the *differential* and *common-mode* levels over the full range of process, supply voltage, and temperature variations:

$$250\text{mV} \le |v_{diff}| \le 450\text{mV} \ \text{ and } \ 1.125\text{mV} \le |v_{cm}| \le 1.375\text{V} .$$

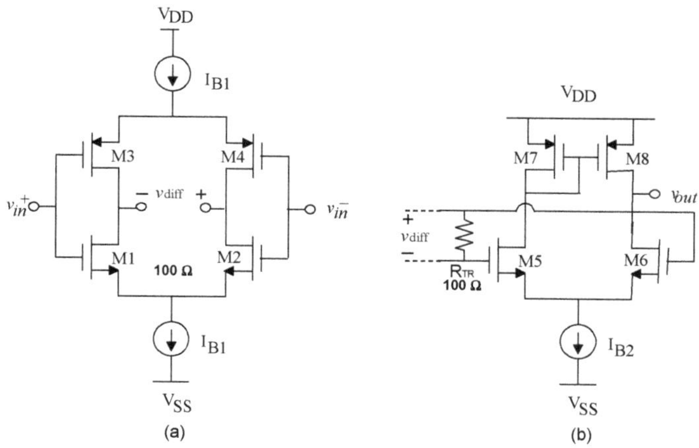

Fig. 6.12 Simplified LVDS transmitter (**a**) and receiver (**b**)

Because of the above wide limitations, the circuits in Fig. 6.12 are oversimplified and not practical. Besides, we have to set properly the output common-mode voltage of the transmitter that otherwise would be free to drift. Finally, the receiver cannot be a simple comparator, but should provide an appropriate hysteresis to avoid comparator multiple commutations for zero crossing input levels due to superimposed disturbances.

The schematic diagram of a practical and simple LVDS transmitter offering transmission speeds in the range of Gbit/s and common-mode control is depicted in Fig. 6.13 [BPV2001]. For maximum speed transfer, a termination resistor at the transmitter end, R_{TT}, is also adopted to provide impedance adaptation ($R_{TT} = 100\ \Omega$). As a drawback, this doubles the current required to generate a 350-mV differential swing ($I_B = 7$ mA is now required for driving the 50-Ω equivalent system resistance). The common-mode output voltage is sensed at the center tap of R_{TT} and compared with a 1.25-V reference via the auxiliary error amplifier implemented by M_5-M_8 and M_{B1}-M_{B1a}-M_{B2}. Current mirrors M_{B1a}-M_{B1} and M_{B2a}-M_{B2} have current gain equal to k (>>1) in order to keep low the current consumption of the error amplifier and increase its loop gain. Of course, the tail current of the error amplifier must be $2\ I_B / k$.

The schematic diagram of a practical and simple LVDS receiver is shown in Fig. 6.14. It implements a conventional CMOS Schmitt trigger which ensures a 25 mV–50 mV hysteresis. Full output swing is by the inverter M_7-M_8 [USS1996]. An improved implementation that better supports the full variation of the common-mode voltage as prescribed by the standard is reported in [BPV2001].

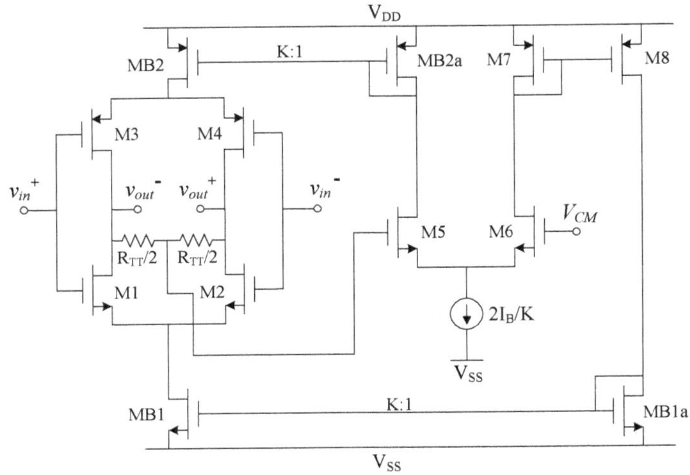

Fig. 6.13 Schematic diagram of LVDS Transmitter with common-mode control [BPV2001]

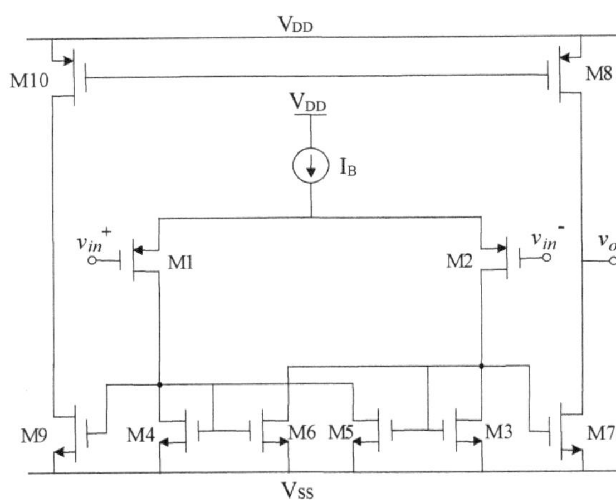

Fig. 6.14 Schematic diagram of LVDS receiver based on a Schmitt trigger (termination resistor R_{TR} is not shown) [USS1996]

LVDS is now adopted as physical layer in different communication protocols developed for usage in LCD drivers.

For example, the Mobile Industry Processor Interface (MIPI) Alliance, which deals with the definition of open standards employed in processors for mobile applications, has defined a high-speed (>100 Mbit/s) communication protocol called MIPI-DSI (Display Serial Interface). It is organized in different layers with a version of LVDS as physical layer. Here, the differential voltage is ±100 mV and the multiplexing ratio can be higher than 7:1 (usually, from 10:1 to 30:1). In order to optimize the power consumption, the high-speed interface is enabled only for video data transmission. The same bus is also used for low-power low-speed communication to transmit the data commands to the display. The transition from high-speed to low-power communication is done by pulling up the common-mode voltage of the differential transmission lines. Vice versa, the interface enters high-speed mode via a proper command sent when the interface is set in low-power mode.

Other proprietary high-speed LVDS-based serial interfaces are the Compact Display Port (CDP) by Nokia, the mini-RGB serial interface by Solomon Systech, Mobile CMADS (Current Mode Advanced Differential Signalling) by NEC Electronics, and the Mobile Display Digital Interface (MDDI) originated by Qualcomm and accepted by the VESA (Video Display Digital Interface) in 2002. All these approaches provide the benefits inherent to LVDS and expedite time-to-market, as the complexity of handset integration is reduced.

As already said, LVDS and its derivations are also adopted in point-to-point communication between TCON and the Source Driver ICs in large-size AMLCDs. Here, the high data rates required for video streaming cannot be supported by the traditional data transmission methods. For instance, NEC adopts the Mini-LVDS technique, while National adopts the Reduced Swing Differential Signalling (RSDS) technique only for RGB data (not commands) in which the differential swing is further lowered to ±200 mV and multiplexing is 2:1 to relax speed requirements [LL2000].

6.3 Source Drivers

Basically, Source Drivers receive the RGB data as inputs, convert them into analog voltages and drive the columns through analog buffers.

The internal architecture of Source Drivers for small- and large-size AMLCDs does not differ substantially. The main differences are in the voltages managed (5 V or below for the former and 16 V for the latter) and in the preemphasis which is absent in small-size AMLCD drivers. Besides, the analog reference voltages of large-size AMLCDs (e.g., 0 V, 8 V and 16 V) are externally distributed (i.e., they are common to all the Source Driver ICs). These voltages cannot be generated within each different IC because they would be subject to unacceptable spreads causing in turn differences in the grey levels of each Source Driver IC.

The block diagram of a Source Driver for large-size AMLCDs is shown in Fig. 6.15. Digital RGB data are stored sequentially into the input registers and latched. These operations are usually performed with high-speed low-voltage CMOS circuits. Level shifters must be then used to boost the digital signals to higher levels compatible with the Digital-to-Analog Converters (DACs). Pre-emphasis can be used to contrast the RC delay of the columns (already described in Section 5.3.1). When required by the highly capacitive data lines of the LCD panel, output analog buffers provide both low driving impedance and high current drive capability. Charge recycler is the final block attached for power saving purposes.

The digital section is quite conventional and level shifters will be discussed in Section 6.4. Therefore, we will focus our attention on the last four blocks. Another function managed by the Source Driver is the inversion method, that will be discussed in Section 6.3.2.

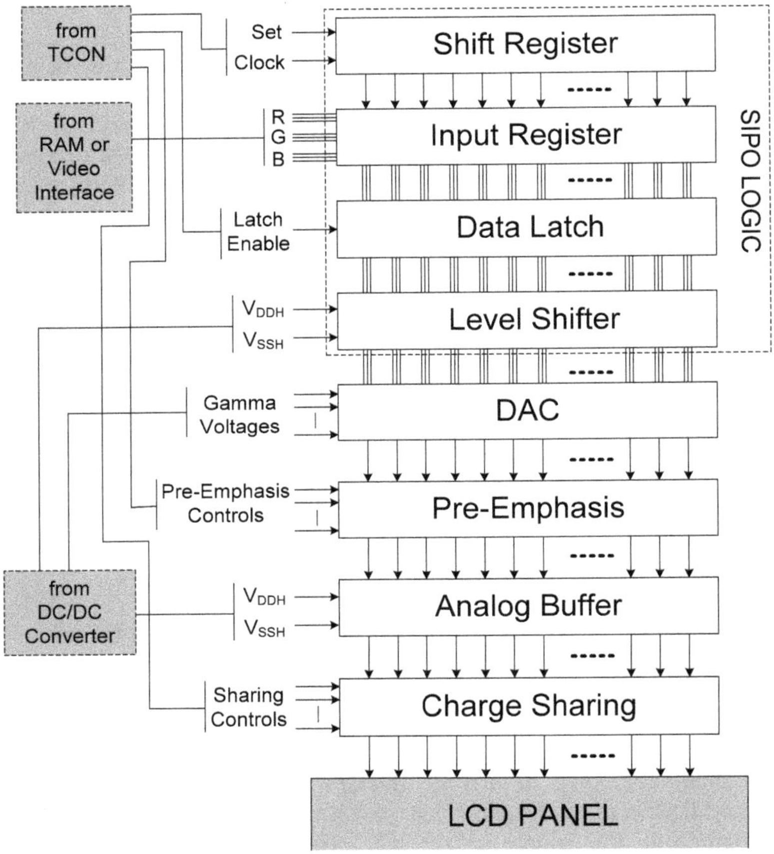

Fig. 6.15 Source Driver architecture of large size AMLCDs

6.3.1 DACs

The overall Source Driver resolution and accuracy mostly relies on the DAC section.

Up to 8-bit resolution, AMLCD drivers adopt **resistor-string DACs (R-DACs)**.

In principle, an R-DAC is made up of one resistor string providing 2^b voltage levels (b is the number of bits) and one multiplexer for each column conveying the analog column voltage associated to the input data word. Actually, this simple scheme must be modified to allow gamma voltages programmability.

As illustrated in Fig. 6.16 (which shows a 6-bit R-DAC plus buffering section), 12 **primary** gamma voltages are generated by a first-level resistor string (10 of these primary voltages can be programmed by the control word associated to the first group of multiplexers) and buffered. A second-level string is then exploited to fit the nonlinear gamma curve by generating the remaining 52 voltage levels. Also these voltages are buffered by a second level of Buffers. Output multiplexers drive the columns. Video data appropriately select the multiplexer output voltage providing digital-to-analog conversion. The 12 primary voltages as well as their allowed programmable range are qualitatively shown in Fig. 6.18.

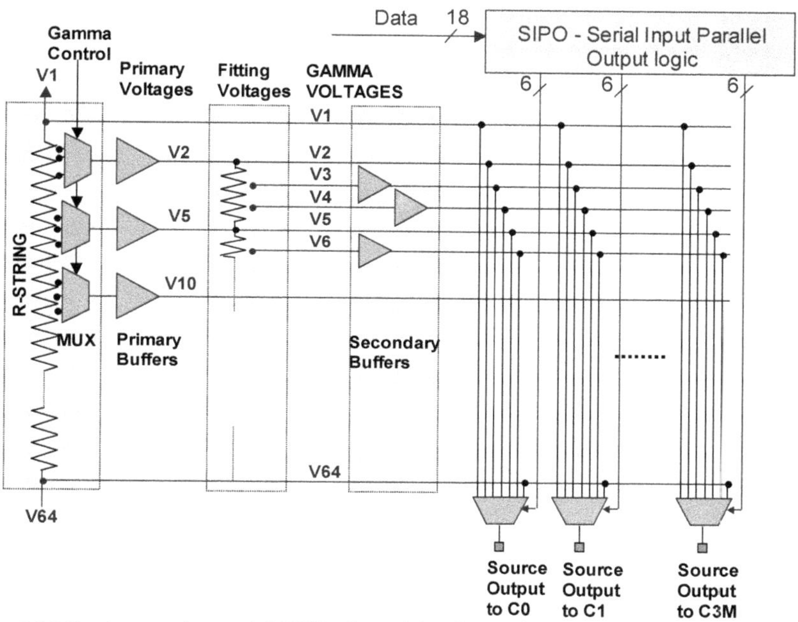

Fig. 6.16 Resistor string and MUX adopted implementing the R-DAC for positive voltages

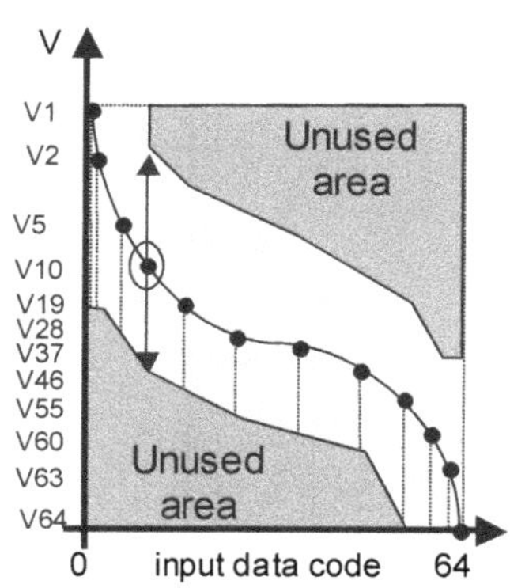

Fig. 6.17 Primary Gamma Voltages distribution, unused and allowed programmable areas

As a result, Fig. 6.16 implements a (nonlinear) R-DAC for the positive gamma voltages. A complementary R-DAC is also needed for the generation of negative polarity gamma voltages.

The MUXs in Fig. 6.17 are easily implemented through switches. A binary tree solution, depicted in Fig. 6.18a is usually adopted for its simplicity. For n bit the MUX requires a number of switches equal to $\sum_{i=0}^{n-1} 2^{n-i} = 2\left(2^n - 1\right)$. The main drawback of this solution is that the output resistance is given by the series of each resistance switch. An alternative scheme requires an n-to-2^n decoder and employs a number of 2^n switches [AH2002], as illustrated in Fig. 6.18b. The output resistance is now that of a single switch.

An intermediate solution employs decoded couple of bits to halve the number of switches in series with respect to the solution in Fig. 6.18a.

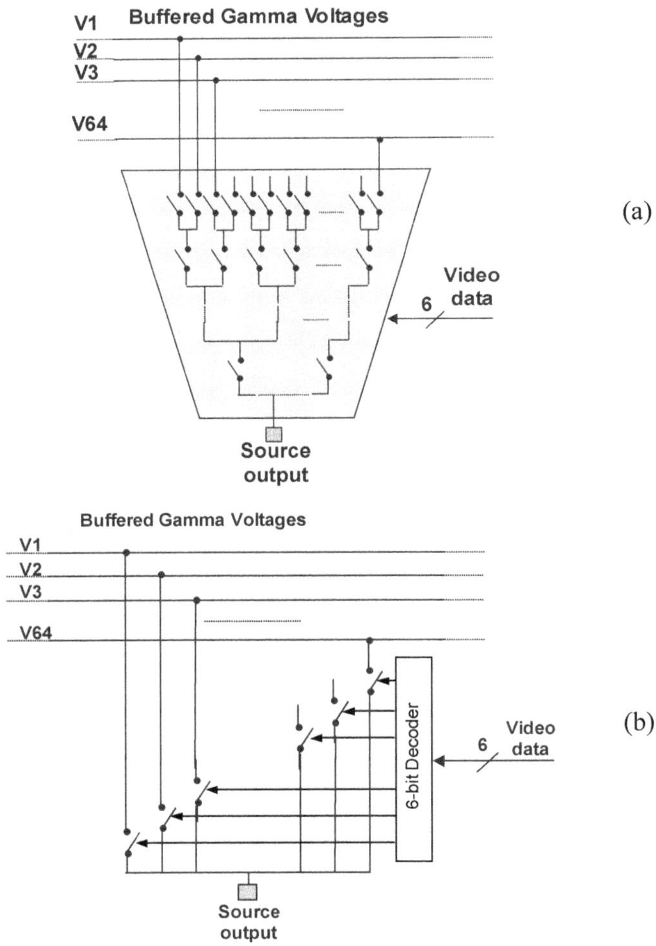

Fig. 6.18 Implementation of the multiplexers with: (**a**) binary tree switches, (**b**) with decoder

Returning to Fig. 6.16, we see also the inclusion of voltage buffers to provide the gamma voltages with current driving capability. Buffers can be placed using two different approaches. The former is in a *centralized* architecture as already described in Section 4.3.3 and illustrated in Fig. 6.17. This approach is convenient when the number of columns times three, 3*M*, is much higher than the gamma levels. The alternative *distributed* buffering approach is illustrated in Fig. 6.19. Due to the large number of buffers (equal to 3*M*) both area and current consumption is considerable. Therefore, attention must be paid in optimizing such performance parameters. The advantage of this solution is in the low driving point output resistance that reduces the series resistance added to that of the source line (in contrast, the output resistance offered by the MUX is dominated by the series of the *n* switches). The buffer performance and implementation strategies will be discussed in Section 6.3.4.

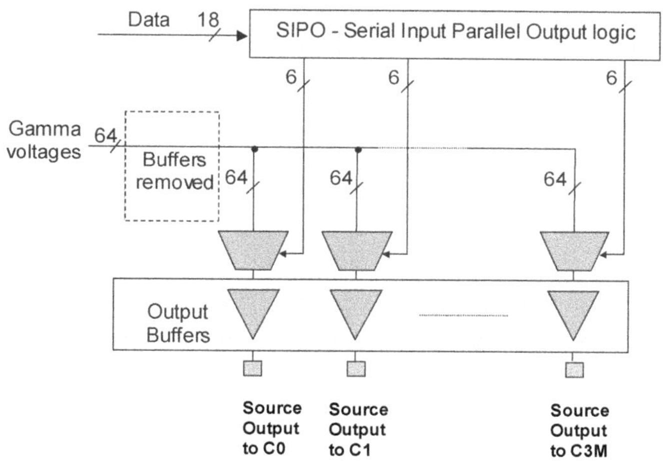

Fig. 6.19 Distributed buffer architecture

In general, the advantages of R-DACs are their monotonicity and immunity against process variations. The main drawback is the area occupation. In a 6-bit/color Source Driver, the R-DAC section occupies 30% of whole driver area, mainly because of the switches and metal routing. It can be shown that each increase in number of bits doubles the area of the R-DAC. Consequently, the area of a 10-bit DAC is 16 times as large as that of a 6-bit counterpart and the area occupation of the simple R-DAC becomes prohibitively large. To increase the color depth, many drivers for mainstream LCD TVs adopt 8-bit R-DACs expanded to 10 bit by Frame Rate Control or Spatial Dithering [HIK1992, MFT1991]. As explained in Section 3.4, these approaches work at the expense of higher frame frequency or reduction in pixel resolution.

To implement true 10-bit DACs while contrasting the area increase, various two-stage DAC architectures for TFT-LCD drivers have been proposed. One (Fig. 6.20a) cascades an 8-bit R-DAC and a 2-bit R-DAC directly [SSK2005]. According to the authors the area increment of this DAC may be only 30% compared to the 8-bit R-DAC, but its performance is impaired by the loading effect of the second resistor string to the first one. To avoid this drawback the two strings are interfaced by two intermediate buffers [G1999] (Fig. 6.20b).

Another 10-bit DAC scheme employs an 8-bit resistor-string DAC and a 2-bit binary weighted capacitor DAC (Fig. 6.20c) [KCK2005]. This RC-DAC has a very simple structure and is area-efficient, as the area increase is only 30% compared to the area of an 8-bit R-DAC, but it is exposed to the disadvantages of a switched-capacitor circuit such as capacitor mismatch, charge injection, and clock feedthrough. Remedies to these problems will be briefly discussed in the following.

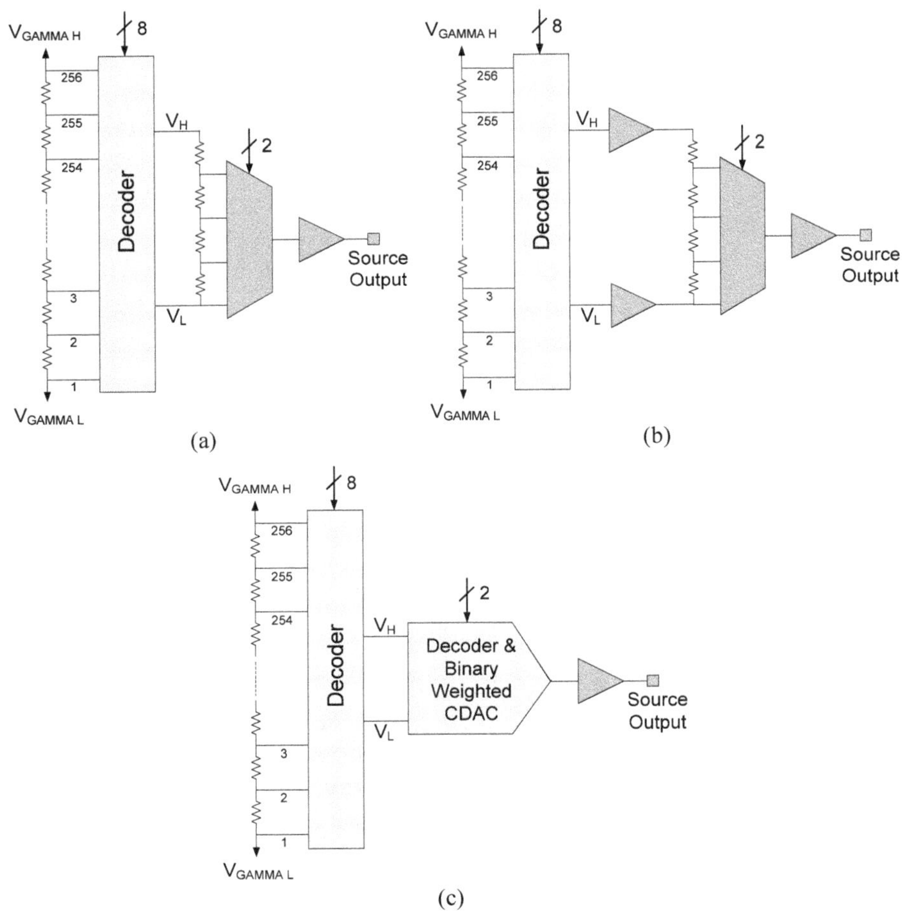

Fig. 6.20 Two-stage DACs: (**a**) 8-bit R-DAC followed by a 2-bit R-DAC; (**b**) 8-bit R-DAC followed by a 2-bit R-DAC with intermediate buffers; (**c**) 8-bit R-DAC followed by a 2-bit binary weighted capacitor DAC

In order to drastically reduce the die area of the source driver for high-resolution displays, the best solution is the serial charge-redistribution (or cyclic) switched-capacitor digital-to-analog converter SC-DAC, first described in [SGH1975] and applied to AMLCDs in [B2005]. Figure 6.21a shows the simplified schematic of the charge-redistribution DAC. It consists of four switches, two equal-valued capacitors, a reference voltage, and an output unity-gain amplifier. The function of the switches is as follows: S_1 is used to precharge capacitor C_S to V_{REF} (if the *i*-th bit, D_i, is a logical 1), while S_2 is used to discharge C_S to 0 V (if the i-th bit is a 0). Switch S_3 enables the charge redistribution between C_S and C_I. Switch S4 is used at the beginning of the conversion cycle to discharge C_I to 0 V. The conversion always begins with the LSB and ends with the MSB. A conversion example for the six-bit code 110100 is also illustrated in Fig. 6.21b.

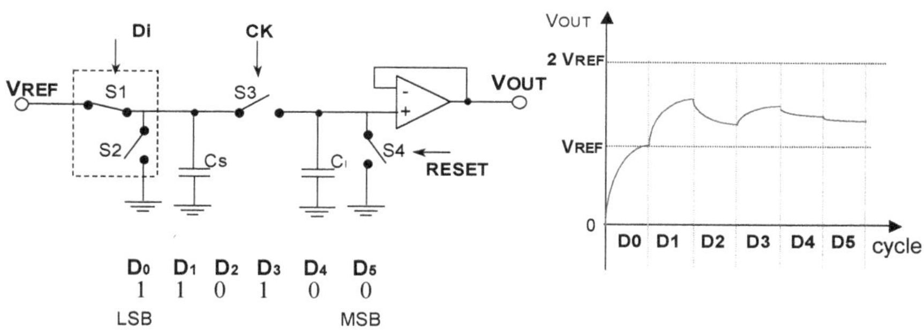

(a) (b)

Fig. 6.21 Charge redistribution DAC (and example with 6-bit conversion)

The output voltage, after 6 clock cycles, is given by

$$V_{OUT} = V_{REF}\left(\frac{D_0}{2^6}+\frac{D_1}{2^5}+\frac{D_2}{2^4}+\frac{D_3}{2^3}+\frac{D_4}{2^2}+\frac{D_5}{2^1}\right) \qquad (6.1)$$

In general for an *N*-bit converter we get

$$V_{OUT} = V_{REF}\sum_{n=0}^{N-1}\frac{D_n}{2^{N-n}} \qquad (6.2)$$

It is apparent that the circuit requires auxiliary digital circuitry to generate the phase signals for the switches.

The amplifier offset and parasitic capacitances of the two capacitors may be the primary causes limiting resolution. An offset-compensated solution insensible to parasitic capacitances should sample the offset in both capacitors thereby compensating it when charge redistribution occurs. Besides, the two capacitors must be placed in such a manner that the charge on parasitic capacitors should not contribute to the output voltage. Figure 6.22 shows the conversion steps of such an offset-compensated charge redistribution DAC. Furthermore, the amplifier is operated in inverting configuration (i.e., in which both the input terminals are grounded as the inverting input terminal is at virtual ground), thus minimizing the required common-mode input range of the amplifier. The last conversion step produces the ideal analog voltage unaffected by offset. In practice, offset specification is relaxed well below ½ LSB.

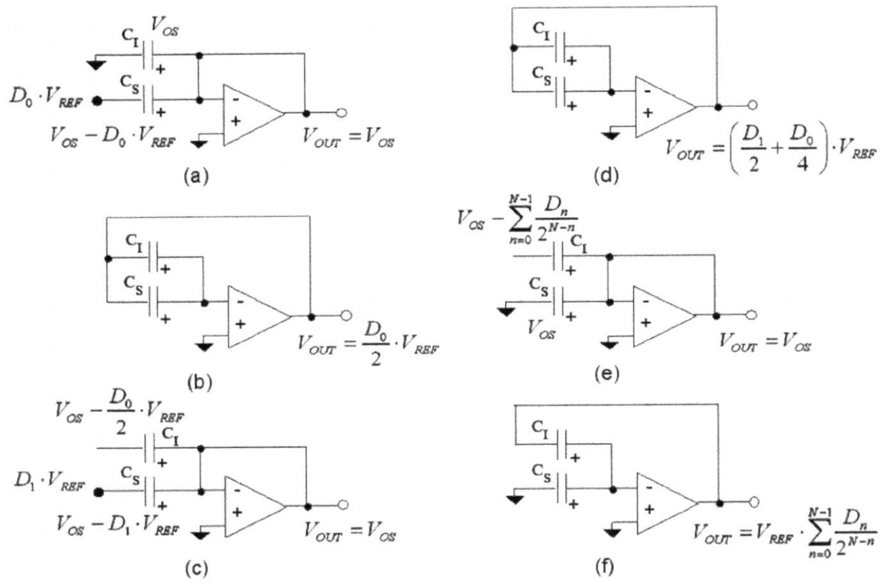

Fig. 6.22 Conversion steps of the offset-compensated charge redistribution DAC. Output and capacitor voltages are indicated

The remaining source of errors, limiting the DAC performance, are due to the capacitor mismatches and switch nonidealities (clock feedthrough and charge injection). The latter are contrasted by adopting improved switch structures (that use complementary and/or dummy transistors). Capacitor mismatches, as known, are minimized by adopting large geometries and careful layout. Two 0.25-pF capacitors were used for a 10/12-bit DAC in [B2005]. A technique to further reduce mismatches and improve linearity is obtained by exchanging the role of C_I and C_S at each clock cycle. In this manner, any mismatch is time averaged [RW1996, RWR1998].

The SC-DAC provides an 80% reduction in area occupation over a 10-bit R-DAC. Actually, the DAC circuit is the same for any number of bits. What changes is the number of charge redistributions, i.e., the number of cycles. Since the row time is constant, this means that the clock frequency must increase for higher resolutions.

Gamma correction is in this case performed digitally by increasing the converter resolution and exploiting only a reduced set of output voltages. For instance a 12-bit SC-DAC is necessary for 10-bit color depth. This is also an advantage of the SC-DAC in order to support three independent gamma curves for each RGB color without severely increasing the die size.

To understand frequency and settling time requirements of an amplifier used in an SC-DAC, we have to consider that the cyclic conversion of an n-bit DAC requires n charge redistribution steps. The conversion occurs with the DAC not connected to the column (and hence to the heavy capacitive load). Therefore, during conversion, the amplifier is loaded only by the DAC capacitance. The minimum value of these capacitances is dictated by the required matching properties which in turn impact resolution. Unless using the above mentioned capacitor swapping techniques, that average capacitor mismatch, the minimum capacitance allowed is in the range of 200–500 fF. If T_{CK} is the DAC clock period, the amplifier must settle in $T_{CK}/2$ under this (light) load. The above performance must be achieved under a very small standby current budget. Besides, to optimize the area occupation, the same amplifier should be used to drive the column. Therefore, the converted analog voltage must be applied to the "heavy" capacitive load (in the range of tens of picofarads for small-size panels). Fortunately, in this case the amplifier is allowed to settle in a larger time interval.

For example assuming T_{CK} equal to 2 μs, a 10-bit SC-DAC needs 20 μs to perform the conversion. During this phase the amplifier is loaded with a 200–500 fF load and for each charge redistribution the settling requirement is below 1μs. Assuming also a 60-Hz frame frequency and 320 display rows, the time left for settling, when the amplifier is loaded by the column, is 32 μs (=16.6 ms/320–20 μs) which is a sufficiently large time.

Design aspects of the analog Buffers will be discussed in Section 6.3.4.

6.3.2 Polarity Inversion

Another function managed by the Source Driver is the inversion method. Either Frame Inversion or n-Line inversion techniques are adopted in small-size LCDs for their good trade-offs between power consumption and crosstalk compensation, whereas the better Dot Inversion is prevalently adopted in large-size LCDs. For large-area panels, inversion techniques require Source Drivers with both positive and negative output polarity capability for the same digital input code.

The additional bit required by the inversion can be implemented through reverse-polarity switches that can follow two approaches. Switching polarity can be accomplished either before or after the Analog Buffers as depicted in Fig. 6.21a, b [SCK2007, MB2004, LH2004]. Switches S_1 connect the positive output voltages to the column lines, while S_{1B} provide polarity reversal. This means that each buffer in Fig. 6.23a should have a rail-to-rail swing both at the input and output because the amplifier should cover all data voltages in the positive and negative reference voltage/gamma ranges [HT1994, KS2006, L2004, CM2007]. In contrast, the solution in Fig. 6.23b does not require rail-to-rail amplifiers because two types of buffers are adopted: B_H dedicated to the positive voltage range and B_L dedicated to the negative voltage range.

The second solution is preferable in terms of design simplicity and current consumption leaving performance unchanged. In fact, current consumption can be halved by staking buffers B_H and B_L buffers that now share the same biasing current. However, the mismatch of different buffers B_H and B_L, especially in their offset voltages, can be an issue that must be taken into account through suitable offset-compensating schemes.

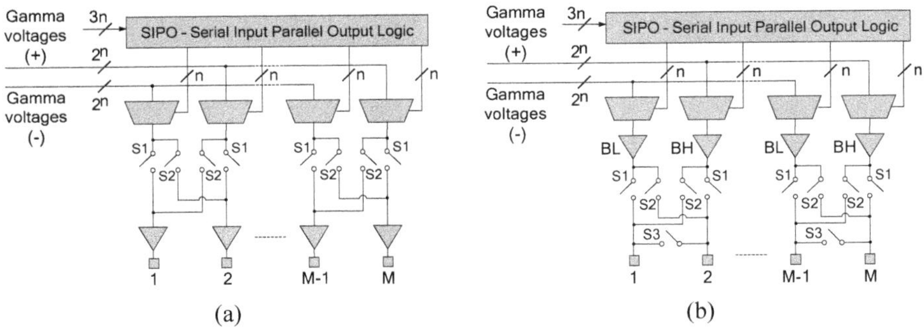

Fig. 6.23 Polarity inversion implementation with polarity reversal switches: (**a**) at the input and (**b**) at the output of the buffer amplifiers

6.3.3 Pre-emphasis

Pre emphasis (also referred to as overdrive) can be easily accomplished in the digital domain [KFT2001, SN2001, M2003, SOY2003, KPP2004] but it requires additional frame memories or look-up tables for data processing. To preserve system simplicity, analog implementations directly within the source driver are preferable [SK2007].

As discussed in (5.3.1), the pre-emphasis technique requires the generation of an adaptive pre-emphasis voltage proportional to the voltage swing of the analog video signal. For example, this can be achieved without using additional reference gamma voltages thanks to a pre-emphasis driving circuit, inserted between the DAC output and the Buffer input. It can be implemented by adding six switches and two capacitors to the conventional unity-gain amplifier as shown in Fig. 6.24, [SK2007]. The switches phases, and near/end voltage waveforms as well as the voltages stored on both capacitors, C_1 and C_2, are shown in Fig. 6.25.

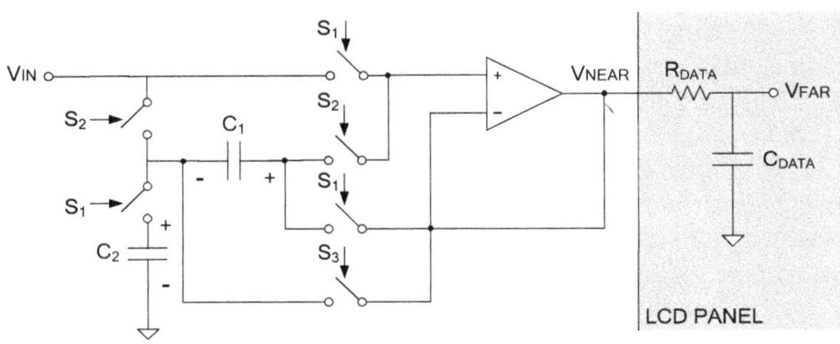

Fig. 6.24 Circuit diagram for the pre-emphasis driving method

The circuit requires three steps at the beginning of each row-line time. The synchronization phases are extracted from the master clock of the source driver.

In the first step, V_{OUT} (that is equal to V_{NEAR}) slews up to the incoming video signal voltage from the digital to analog converter (DAC). As defined in Section 5.3.1, V_{SWING} is the difference between the actual pixel voltage $V_{IN}(i)$ and the previous $V_{IN}(i-1)$ that is stored onto capacitor C_2 at the end of each row time (as shown in Fig. 6.26c). The precharge voltage V_{PRE} is proportional to V_{SWING} as described in (5.18) and reported below for the sake of clarity.

$$V_{PRE}(i) = \frac{e^{-\frac{T_{PRE}}{R_{DATA}C_{DATA}}}}{1-e^{-\frac{T_{PRE}}{R_{DATA}C_{DATA}}}} V_{SWING}(i) \tag{6.3}$$

V_{PRE} is obtained in step 1 (Fig. 6.26a) since C_1 and C_2 are connected in series and the voltage on C_1 is

$$V_{C1}(i) = \frac{C_2}{C_1 + C_2} V_{SWING}(i) \tag{6.4}$$

Then, the voltage stored on C_1 equals $V_{PRE}(i)$ provided that

$$\frac{C_2}{C_1 + C_2} = \frac{e^{-\frac{T_{PRE}}{R_{DATA}C_{DATA}}}}{1-e^{-\frac{T_{PRE}}{R_{DATA}C_{DATA}}}} \tag{6.5}$$

After the detection step, the pre-emphasis voltage stored in capacitor C_1 is added to the original incoming pixel voltage and applied to the unity gain buffer during the preemphasis time (Fig. 6.26b). Therefore, the positive buffer output voltage is

$$V_{OUT}(i) = V_{IN}(i) + V_{C1}(i) \tag{6.6}$$

In step 2, the far-end voltage of the data line rises up to about the original incoming analog video signal voltage from DAC.

Finally, in step 3 (Fig. 6.26c), the current video signal voltage is applied directly to the unity gain OP-AMP, and capacitor C_1 is shorted and reset and VIN(i) is stored onto C_2. The near-end voltage of the data line is driven to the target video signal voltage. The three steps of driving sequences are repeated at every row-line-time.

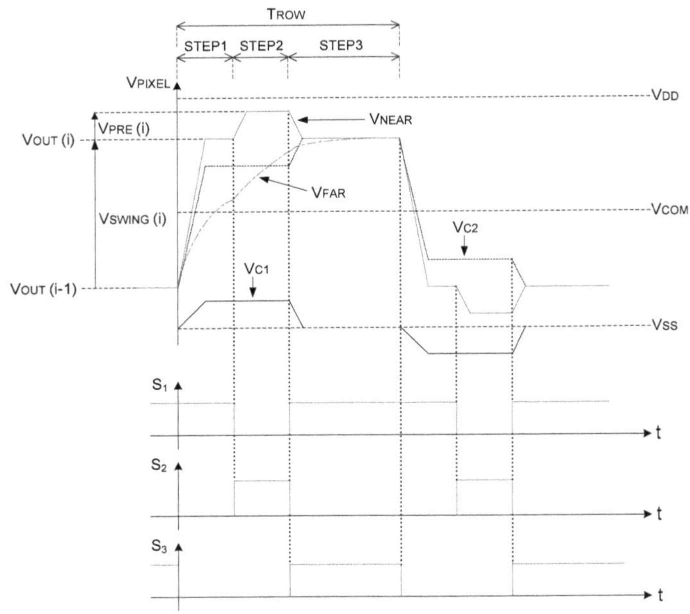

Fig. 6.25 The timing diagram and conceptual waveforms of proposed circuit for pre-emphasis driving method

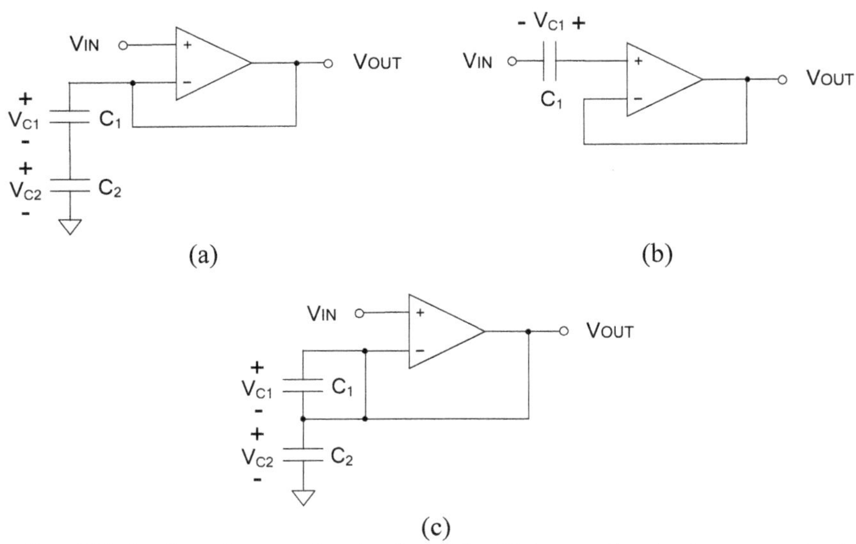

Fig. 6.26 Configurations of the circuit in Fig. 6.24 according to the pre-emphasis steps: (**a**) first step; (**b**) second step; (**c**) third step

6.3.4 Analog Buffers

The voltages coming from the DAC section must be properly buffered to provide low impedance driving point and current drive capability for the relatively high capacitive source load.

R-DACs usually adopt a two-stage operational amplifier in unity gain as voltage buffer. At this purpose, class AB or even B configurations have been exploited to provide sufficient slew rate values with reduced standby currents [IMS2003, L2004, KCK2005]. Describing the detailed design of a feedback amplifier is outside the scopes of this book. Good references, especially for two-stage CMOS amplifiers are found in [PP2002, GT1986, AH1987, LS1994, PPP2001].

A simplified scheme of a possible class-AB two-stage amplifier is shown in Fig. 6.27. Except for transistors M_7-M_8 and M_{11}-M_{14}, the amplifier is a conventional two-stage class A topology. By suitably choosing the aspect ratios of M11 and M7 (M12 and M8) with respect to M5 and M3 so that

$$\frac{(W/L)_7}{(W/L)_3} > 2\frac{(W/L)_{11}}{(W/L)_5} \text{ and } 2\frac{(W/L)_{12}}{(W/L)_5} > \frac{(W/L)_8}{(W/L)_3}$$

transistors M7 and M12 are biased in triode, so that M13 and M14 are normally OFF. They are however able to draw a large amount of current in a push-pull fashion under large-signal operation. It is worth noting that the limited positive common-mode range of the differential input stage requires a higher positive supply voltage. Otherwise, a rail-to-rail input stage, made up of two complementary *p*-channel and *n*-channel differential stages, must be adopted. Though two-stage amplifiers usually adopt Miller frequency compensation, the large capacitive load can be profitable exploited to provide dominant pole compensation and avoid the use the Miller capacitor.

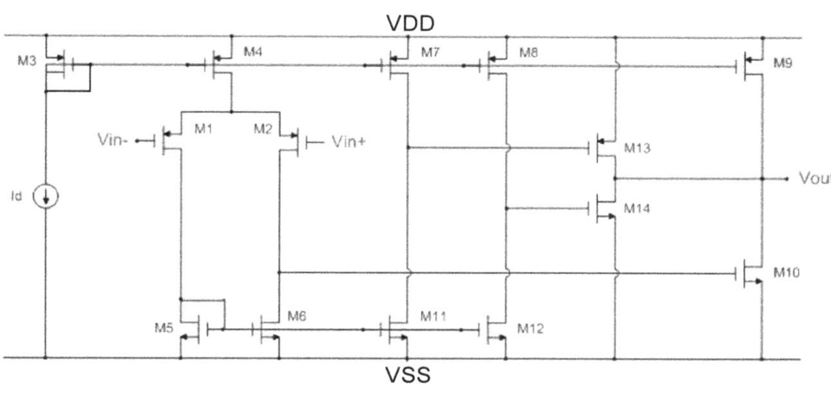

Fig. 6.27 Class-AB two-stage amplifier

The schematic diagram in Fig. 6.28 represents the amplifier adopted in the switched-capacitor DAC in [B2005]. It is made up of a folded cascode first gain stage, M_1-M_9, followed by a rail-to-rail class AB output stage, M_{OP}, M_{ON}. A somewhat complex quiescent current control section is implemented by M_A-M_B (which also causes an undesirable positive feedback path, for small-signal operations). Frequency compensation is obtained through C_{C1} and C_{C2}. Note that the positive zero due to the feedforward path is avoided through the current buffer action of M_{8P} and M_{6N} (thus preventing the use of nulling resistors). All the transistors (except transistors M_{DEG}) work in their *subthreshold* region with a total DC current of about 20 μA. The application was for LCD-TV. The main advantage of the amplifier is its high gain (equivalent to three stages) and rail-to-rail output capability (due to the presence of only two transistors M_{ON} M_{OP} at the output). Disadvantages are due to the class A function of the input stage and to the Miller compensation, causing limited slew rate and gain-bandwidth product values and an increase of the area occupation.

To fit the requirements of battery-operated LCD applications, the standby current of the above circuit must be reduced more than one order of magnitude. Besides, since area occupation is more critical in hand-held devices, degeneration resistors of the differential pair M_1-M_2 included in the original scheme should be replaced by triode-biased MOS transistors (M_{DEG} in Fig. 6.28). However, the class A input stage seems a hard obstacle for substantial current reduction without compromising the slew rate.

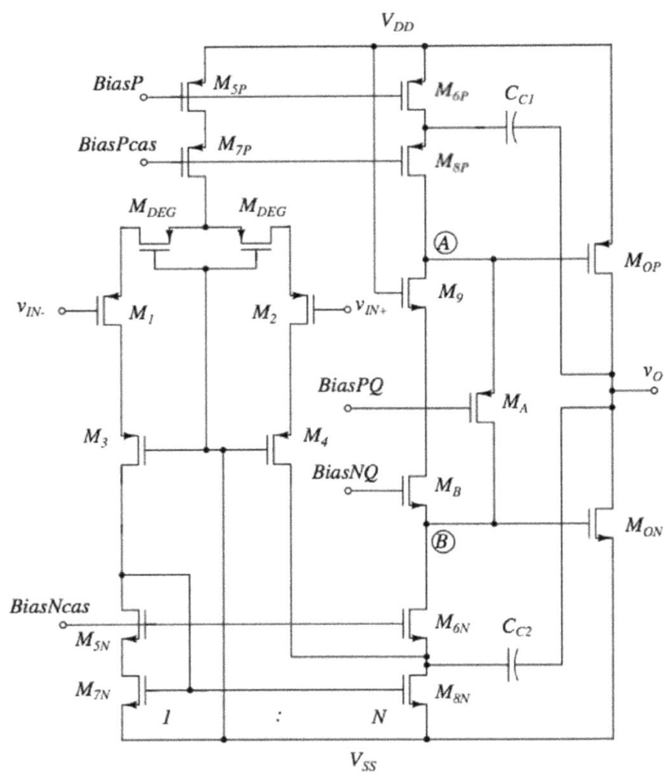

Fig. 6.28 Schematic of the transconductance amplifier in [B2005]

The amplifier, whose schematic is shown in Fig. 6.29, avoids the mentioned drawback. This topology was proposed in [LCG1984] and redesigned to meet the requirements of an SC-DAC for AMLCD drivers in [DPS2007]. It exploits a single-stage cascode topology featuring class-AB operation. Quiescent current is set by M_{B1} and M_{B2}, but under large signal condition M_1 (M_3) and M_2 (M_4) can deliver a large current, solely limited by the overdrive voltage and transistor dimensions. This current is mirrored to the output multiplied by N. The main advantage of this solution is its high achievable slew rate (owing to the full class AB operation), superior unity gain bandwidth and the absence of an explicit compensation capacitor. This reduces the area of the circuit notwithstanding separate wells for M_1-M_8 are required to avoid the increase in threshold voltages caused by the body effect. The main drawbacks are the lower gain (it is equivalent to two gain stages, if required this value could be increased by adopting well-known gain-boosting techniques [SG1990, BG1990, A2006]) and reduced output swing, due to cascoding.

It should be also observed that the limited common-mode input range is not a real limitation as the amplifier will operate in inverting configuration (i.e., with both inputs grounded). Finally, the transfer accuracy of current mirrors that may be impaired by the subthreshold operation results to be adequate thanks to the adopted cascoded topologies that strongly reduce the channel-length modulation effect.

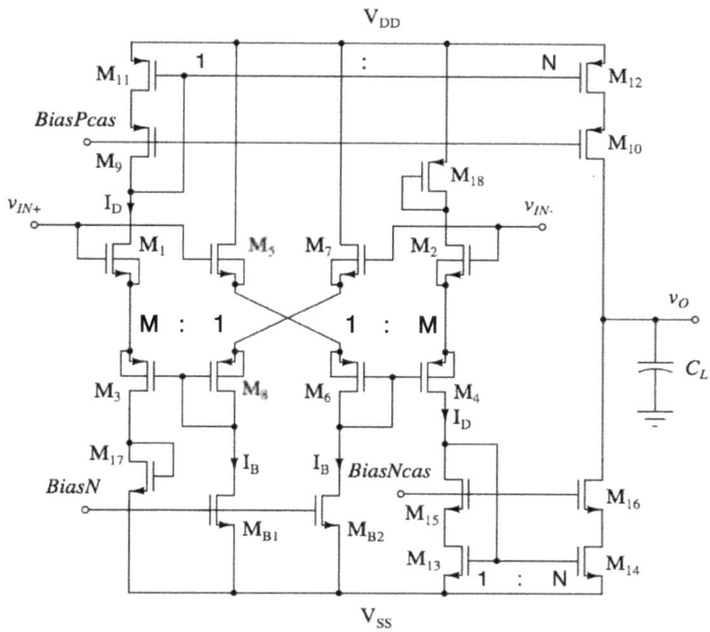

Fig. 6.29 Schematic of the transconductance amplifier in [DPS2007]

As reported in [DPS2007], the amplifier was implemented in the STMicroelectronics HVG8 process with 0.18-μm CMOS technology. Using a total standby current as low as 670 nA, it exhibits a DC open-loop gain of about 80 dB, and a gain-bandwidth product and phase margin around 2 MHz and 70°, with a load capacitance of 500 fF (the one used in the SC-DAC). Besides, the solution provides a slew rate equal to 27 V/μs and a settling time at 0.01% of the final value equal to 800 ns. Area was smaller compared to alternative solutions and was 17.6 μm×40 μm.

The amplifier can be used in unity gain also to drive the column load. Assuming a 10-pF load, the settling time at 0.01% of the final value was about 9.4 μs.

6.3.5 Charge Sharing

Power dissipation in a conventional source driver has three contributions:

- *Digital part.* It represents usually less than 1/10 of the total power dissipation. It is hence negligible.
- *Static power.* It is due to DC bias currents in resistor strings and buffer amplifiers.
- *Dynamic power.* It is consumed in charging capacitive outputs. It is the relevant part.

Charge sharing is a technique for recovering some of the energy that is stored into the display, reducing the dynamic power consumption of source drivers. In a dot-inversion driving method, half of the pixels in a column is driven positively and the other half is driven negatively. Charge sharing is performed by shorting the column terminals together at the beginning of each row addressing before the source buffers are connected to the columns. In such a way, the source lines voltage is averaged so that when the buffers are let to drive the columns they do not start from the most extreme voltage, but from an intermediate value, saving power. If properly designed, this technique may even fasten the rise and fall times because the short provides a very low impedance high-speed path. Figure 6.30 shows a possible implementation of the charge sharing technique. Two series of switches controlled by signal CR are used to connect the columns either to the buffers or together. The associated main voltage waveforms are exemplified in Fig. 6.31. The positively (negatively) driven pixel voltage is assumed equal to V_H (V_L). During the small time interval in which CR is high the charge sharing process occurs and the pixel voltages are averaged to V_{COM}.

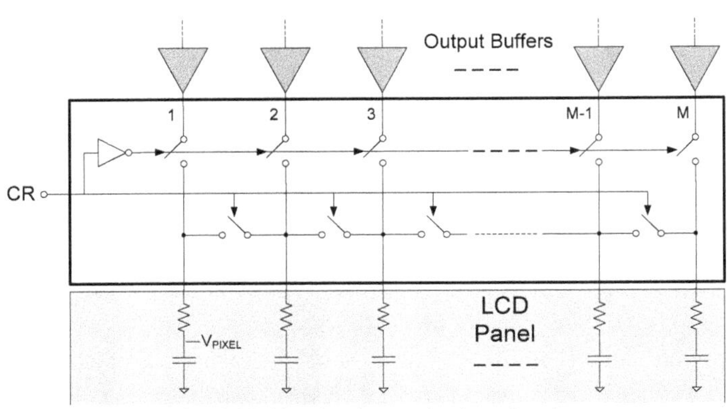

Fig. 6.30 Simple charge sharing circuit

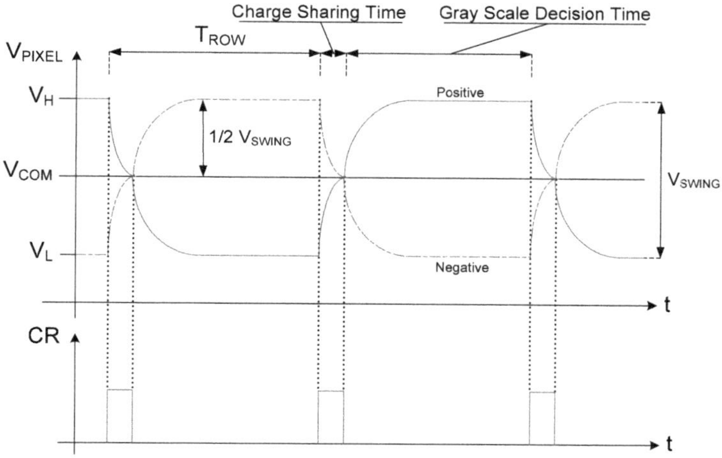

Fig. 6.31 Timing diagram and output waveforms of the charge sharing solution in Fig. 6.30

The simple charge-sharing technique has a power efficiency limited to only 50%. An efficiency of 66.6% is obtainable by the so called *triple charge-sharing method* [SCK1999]. Here, about two-thirds of the total voltage swing of data line is provided by charge sharing between source line and an additional large external capacitor, C_{EXT}, with a negligible area penalty and without increase in the circuit complexity.

C_{EXT} acts as an additional power supply provided that $C_{EXT} >> M\,C_L$, where M is the number of the outputs and C_L is the capacitance of a single source line.

The source driver architecture that uses triple charge sharing is shown in Fig. 6.32. Compared to the simple approach, other than C_{EXT}, it requires four control signals, AMP, SEL_1 SEL_2, SEL_3, and another set of M switches.

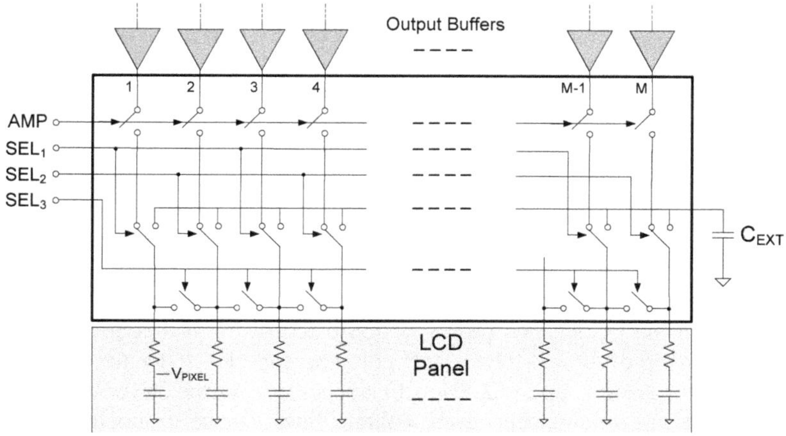

Fig. 6.32 Triple charge sharing circuit

The operation of the triple charge sharing can be understood through the aid of Fig. 6.33. Here it should be noted that the swing of the output buffers to charge the source line capacitance is reduced to a third compared to that of the conventional source driver. The remaining two-thirds of the swing are accomplished through triple charge sharing: (1) among odd lines, (2) among all lines, and (3) among even lines and the external capacitor.

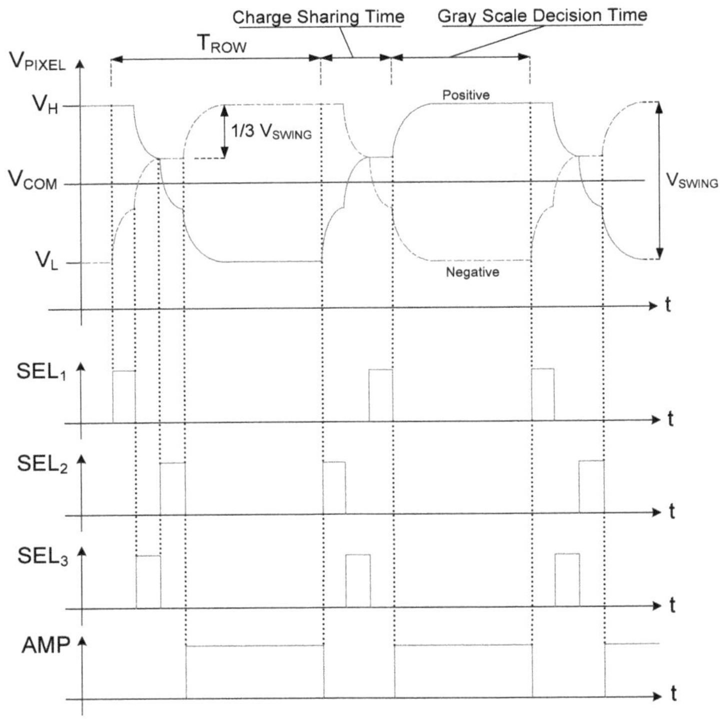

Fig. 6.33 Timing diagram and output waveforms of the triple charge sharing solution in Fig. 6.32

Up to 75% saving efficiency in AC power consumption of the TFT-LCD column driver can be obtained by multi-phase charge-recycling technique (75% is theoretically obtainable with infinite phases) [YGC2003]. This technique is easy to be implemented because it does not require any external capacitor in dot inversion driving. Figure 6.34 exemplifies the approach in which four columns are grouped into one set with switches controlled by signals S_1-S_3. Related waveforms are illustrated in Fig. 6.35, where column voltages for the four columns O_1, O_2, O_3, and O_4 are shown in particular. AC power is dissipated in the last step with the worst case voltage swing of $(V_H–V_L)/2$ for half columns and $(V_H–V_L)/4$ for the remaining half. The power saving is up to 62.5% with respect to a source driver without charge sharing. Expanding this concept, more column lines can be grouped as one set to further reduce power consumption.

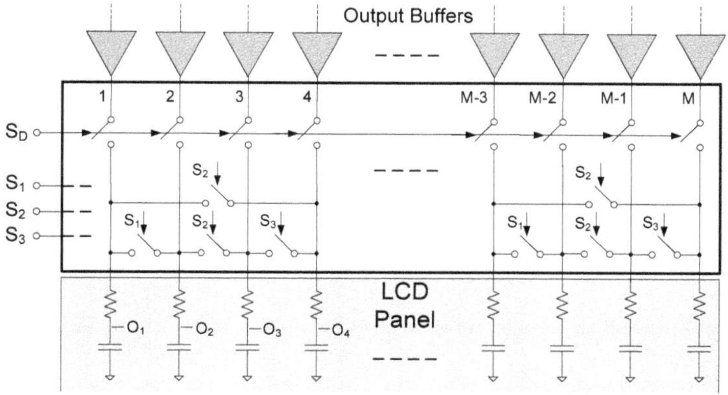

Fig. 6.34 Multiphase charge sharing circuit

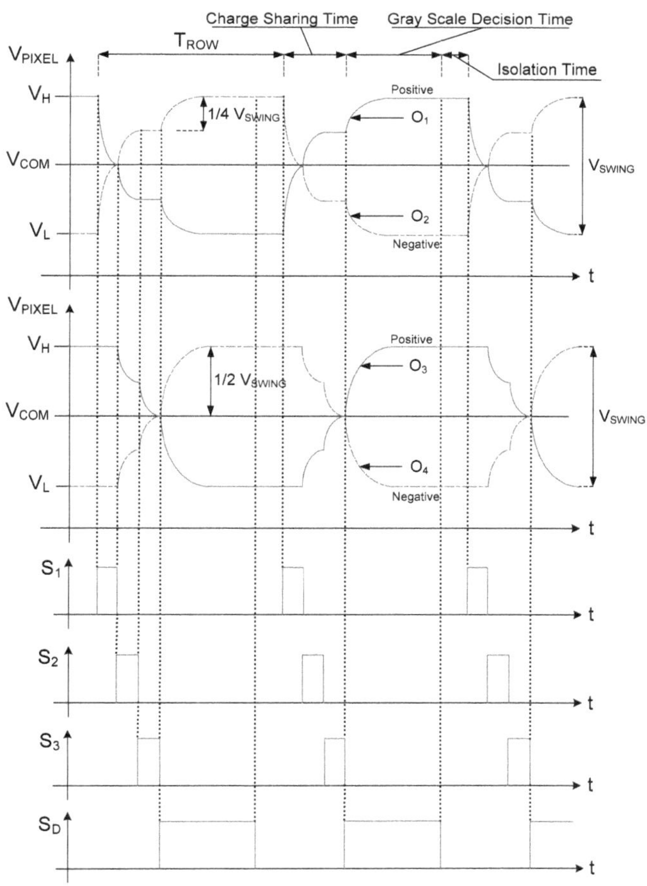

Fig. 6.35 Timing diagram and output waveform of the multiphase charge sharing solution in Fig. 6.34

6.4 Gate Drivers

The Gate Driver has to turn on the TFT switches of each row in a row-by-row succession. A gate driver generally includes a shift register, a level shifter and output buffers, as shown by the block diagram in Fig. 6.36. The scan impulse voltage is provided by a shift register operating at low supply voltage to contain power consumption. However, as discussed in Section 5.1, TFTs must be driven by large gate-source voltages (from 15 V to 20 V in the ON state and from −5 V to −8 V in the OFF state), a dual level shifter is therefore necessary. The output digital buffers, required in large-area and/or high-resolution displays, are used to enhance the driving capability for adequate switching speed.

The level shifter *must precede* the shift register if the gate driver is integrated in the panel.

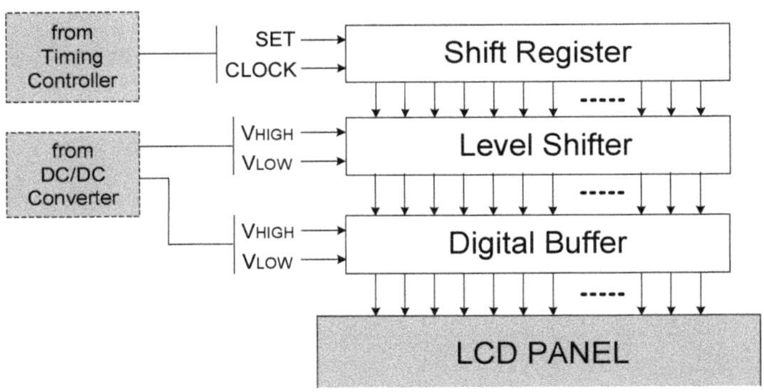

Fig. 6.36 Block diagram of the gate driver

6.4.1 Shift Registers

The simplest gate driver includes a shift register with a walking ‘1’ data running at the row scan rate. The well-known diagram of a shift register is illustrated in Fig. 6.37a. It consists of a chain of flip flops (or latches). A set pulse is presented at the input and is shifted right one stage for each high-to-low (or low-to-high) transition of the clock signal as shown in Fig. 6.37b. If the shift register is formed by latches the input is transferred in correspondence of the clock high (or low) level.

Driver ICs fabricated in monocrystalline silicon typically adopt the solution in Fig. 6.37a, implemented with standard CMOS logic which minimizes the static consumption.

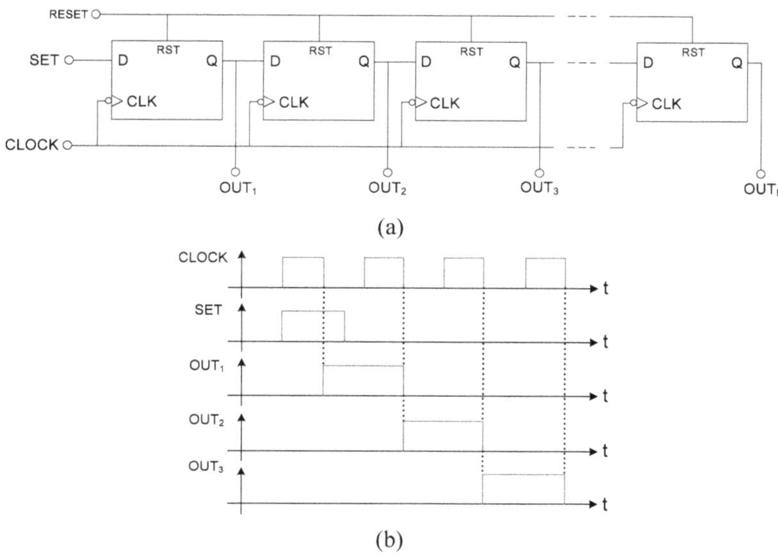

Fig. 6.37 Flip flop shift register: (**a**) block diagram and (**b**) waveforms

Figure 6.38a shows a shift register suitable for integration in the a-Si backplane, as it exploits only nMOS transistors. Diode-connected TFTs are used as load of the inverter stages [KSS2004]. Two complementary nonoverlapping high-voltage phases, ϕ_1 and ϕ_2, drive respectively the pass transistor M_1 and M_4. The input data (OUT_{i-1}) is therefore shifted at the output (OUT_i) after one clock period. Due to the diode-connected active load topology, this solution is however characterized by high power consumption, which can be tolerated in applications such as large area printers, facsimile machines, and scanners.

In high-resolution and/or large-area panels, the outputs of the shift register cannot be directly connected to the gate line because of the unacceptable delay caused by the RC load. Indeed, if the delayed *i*-th output pulse is transferred to the next stage, the delay can accumulate as the number of shift register unitary cells is increased. This aspect is especially critical at low temperature operation, since the ON-current of a-Si TFTs is reduced. The shift register cell in Fig. 6.38b adds a buffer at the output of each stage with the purpose of decoupling the gate line from the subsequent shift register stage [DM1994]. Four non overlapping clock phases are required to transfer the input to the gate line. When ϕ_1 is high, the gate capacitance of M_4 is precharged through M_3 and when subsequently ϕ_2 goes high M_2 is switched on and the charge on M_4 is either removed or held depending on the input OUT_{i-1}. Transistors M_7-M_{10} implement the inverting buffer that drives the gate line clocked by ϕ_2. Since the pulse at the buffer output is an inverted version of the input, a walking '0' approach must be adopted.

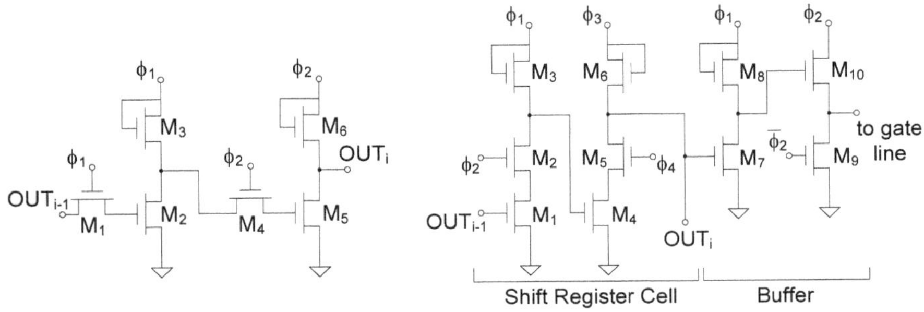

(a) (b)

Fig. 6.38 Schematics of the shift register cell using a-Si TFTs: (**a**) simple version [KSS2004] and (**b**) buffered version [DM1994]

The shift register in Fig. 6.37 implemented with CMOS-logic is also suitable for LTPS AMLCDs. An alternative solution using pMOS transistor only, and therefore suitable for low-cost processes with minimum number of masks is illustrated in Fig. 6.39 [L2006]. This circuit performs a latch operation: when ϕ goes high, transistors M_9 and M_{10} are turned off whereas M5 and M6 are turned on. At this time, the input signal, OUT_{i-1}, is transferred to the register output by the inverters M_3-M_4 and M_1-M_2; when ϕ goes low, the shift register keeps the previous state because M_5 and M_6 are turned off (the input signal is not transferred to the output) and the inverters M_1-M_2 and M_7-M_8 are connected in a positive feedback path to form a static memory element. Of course, compared to a CMOS implementation, also this solution is characterized by higher power consumption.

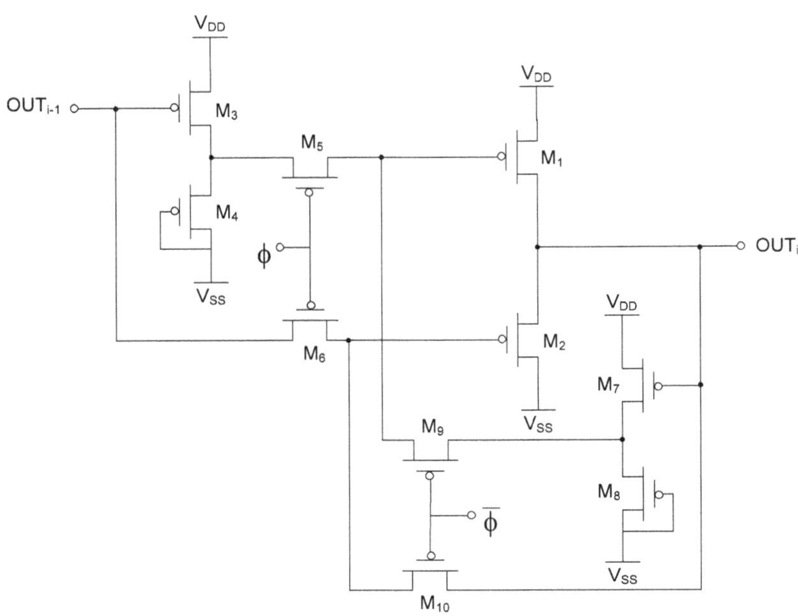

Fig. 6.39 Schematic of the LTPS shifter register cell

Another shift register employing a bootstrap technique and that is suitable for large-area AMLCDs is shown in Fig. 6.40 [MLL2007]. **Bootstrapping** is a dynamic circuit technique that obtains a full rail–to–rail output swing and reduces propagation delays. Such technique uses a capacitor to couple charge from a source. The charge boosts either an internal node (indirect bootstrapping) or output node (direct bootstrapping) above the power supply voltage, achieving a full output swing and a high switching speed. In the circuit of Fig. 6.39, the input pulse, OUT_{i-1}, turns on M_1 through M_4. M_1 pulls up the gate line when ϕ is high, while M_2 and M_3 reset both source and gate voltages of M_1 to V_{SS} when the next stage output, OUT_{i+1}, goes high. However, for stable operation during the OFF-period, clock-feedthorugh from ϕ to node X and caused by the gate-drain parasitic capacitor of M_1 must be avoided. Voltage V_X should be less than the threshold voltage of M_1 to prevent abnormal operation. By enlarging C, V_X can be effectively reduced, however, it is difficult to keep V_X below V_{TH} of M_1 because V_{TH} varies of about ±2 V at room temperature. Improved solutions that compensate clock-feedthrough have been proposed in [HTL2005, YJK2005, MLL2007].

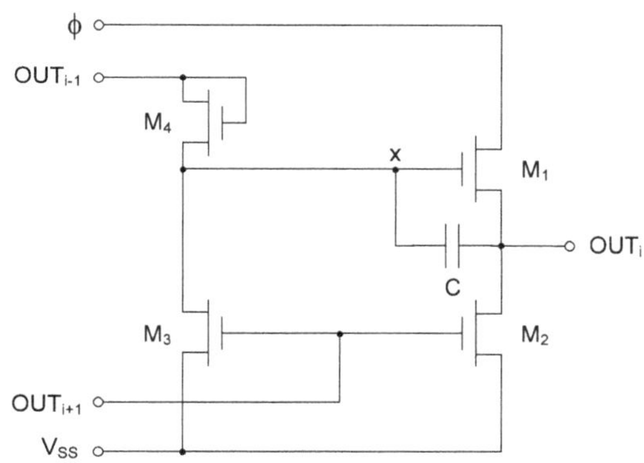

Fig. 6.40 Schematic of the a-Si shift register cell with bootstrap

In mobile AMLCDs one of the key specifications is power consumption. The architecture of the gate driver also has a significant effect on power consumption. The power dissipated by an integrated shift register is proportional to the scan frequency f and to the total capacitance offered by the shift register C_{SR}. A reduction of power consumption is then obtained by adopting a multi-phase drive clock strategy. For instance, each phase of a six-phase clock drives an equivalent capacitance of $C_{SR}/6$ at a $f/6$ clock frequency. The total power consumption is hence theoretically reduced by a factor of 1/6. Up to 12 different clock signals have been employed [YID2007].

Further power saving can be obtained if the clock transitions are modified by creating a third selected level at ground, making three level waveforms at each transient from V_{DDH} to V_{SSH} and vice versa, as shown in Fig. 6.41.

By using this 3-level approach, the power in a clock line can be halved, because when the clock line is connected to ground, the existing charge on clock line is discharged to ground without consuming power.

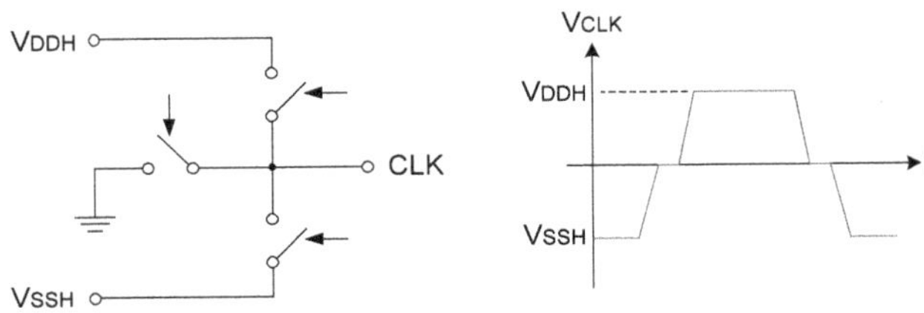

Fig. 6.41 Driving signal waveform detail for a "3-level" drive

6.4.2 Level Shifters

A digital level shifter converts an input signal with a (low) voltage swing ($V_{DDL}-V_{SSL}$) to an output signal with a higher voltage swing ($V_{DDH}-V_{SSH}$). The two couples of voltage levels are qualitatively shown in Fig. 6.42. Of course, the new voltage levels must be able to switch ON/OFF the TFTs of the LCD panel.

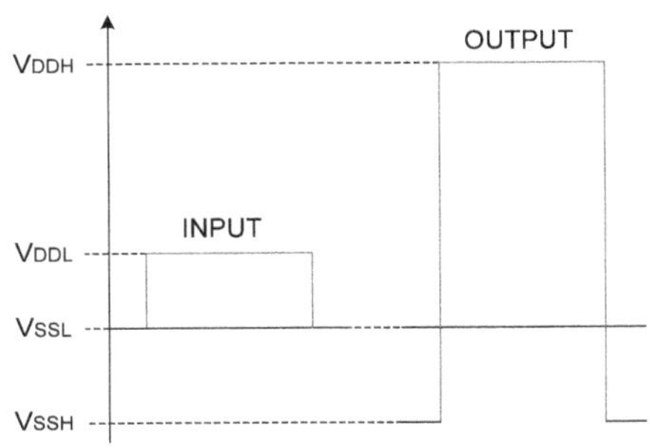

Fig. 6.42 Input and output waveforms of Level Shifter

A conventional level shifter, suitable for integration in monocrystalline silicon, is shown in Fig. 6.43a. The circuit comprises a CMOS inverter, M_1-M_2, supplied from $V_{DDL}-V_{SSL}$ and two latches. The first one M_3-M_6, supplied by $V_{DDL}-V_{SSH}$, is used for negative supply shifting. The second latch, M_7-M_{10}, is powered from $V_{DDH}-V_{SSH}$. By exploiting the two complementary outputs of the first latch, the second latch can be directly driven by the first one without any additional inverter. An undesired limitation of this solution is the **cross-conduction** current (flowing from the positive supply rail towards the negative one) that occurs when a pull-up and a pull-down transistor changing their logic state are momentarily both ON due to a delay in their respective controlling signals. Although this takes place for a small fraction of the clock cycle, the amount of cross-conduction current can be quite large because only the ON-resistance of the transistors limits its magnitude.

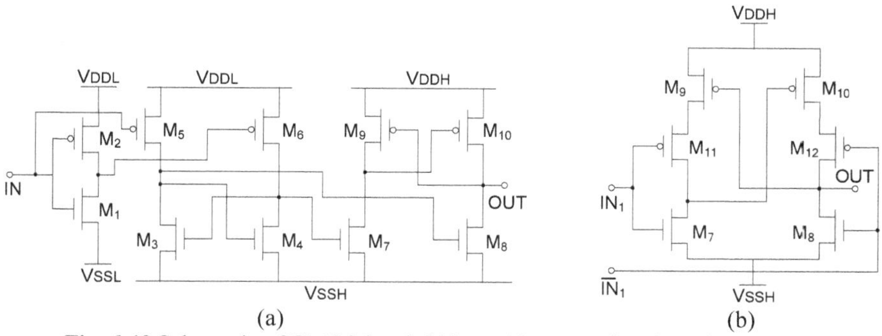

Fig. 6.43 Schematic of CMOS level shifters: (**a**) conventional, (**b**) improved latch

A solution suitable for integration with a-Si, but for applications with no power dissipation constraints, is shown in Fig. 6.44. It exploits two cascaded (ratioed-logic) inverters, with bootstrap capacitors C_1 and C_2 [BCO2006]. Because of the high threshold voltage (3 V–5 V) of a-Si TFTs, two stages are required to efficiently shift up the low voltage range input signal (0 V–5 V) to the high voltage output swing (0 V–30 V). The use of two cascaded inverters also avoids input-output phase inversion.

Consider the first inverter; if the input voltage becomes a logic 0, M_2 turns off and then the first output voltage rises. During this rise, the gate voltage of M_1 is bootstrapped by capacitor C_1, as the charge at this node cannot change due to reverse biased diode M_3.

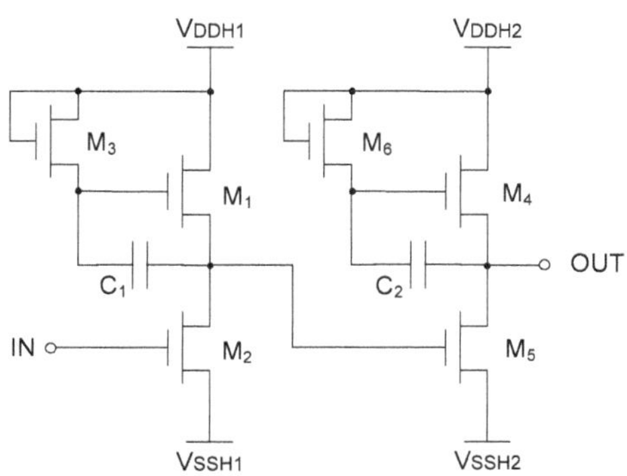

Fig. 6.44 Schematic of level shifter made up of cascaded bootstrapped inverters

Level shifter architectures have been also optimized for LTPS TFTs [FO2004, SKC2003, T1996, MKO2004, LJ2006]. The last solution is shown in Fig. 6.45. The circuit operates in two steps. First, if the input IN is high, M_3 is switched on and the X node is pulled down to V_{SSH}. This switches M_5 ON, so that Y node is charged to V_{DDH}. After the completion of feedback, the potential across the capacitor C becomes V_{DDH}–V_{IN}. Since voltage across a capacitor cannot be changed instantaneously, the gate of M_4 swings between V_{DDH}–V_{IN} and V_{DDH} when input is logic 0 and logic 1, respectively. Therefore, the voltage at node X would be V_{DDH}

when the input signal is logic 0 while it would be V_{SSH} when the input is logic 1. During the second step, M_1 and M_2 participate to maintain the potential of Y node to the desired level. This part of the circuit is especially useful when input stays in one logic state for a relatively long time. In this case, the potential across a capacitor can change significantly and results in circuit failure.

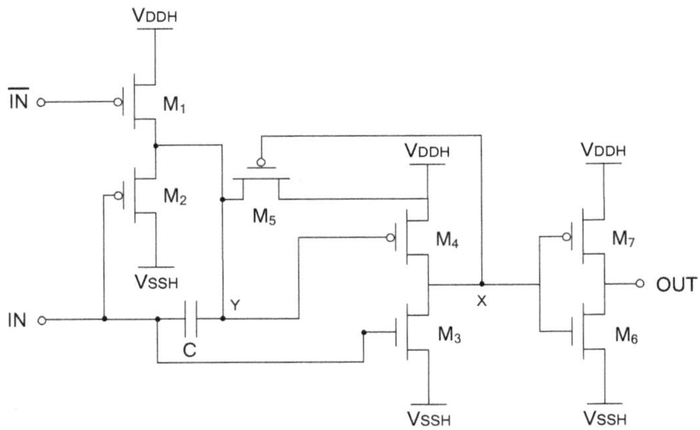

Fig. 6.45 Schematic of an LTPS level shifter [LJ2006]

6.4.3 Digital Buffers

A possible architecture of Digital Buffer implemented in monocrystalline silicon with high voltage gate-oxide CMOS transistors is shown in Fig. 6.46a. It consists of three main sections. The half bridge stage, M_1-M_2, which drives the gate line high and low with proper current capability. The transfer-gate section, M_3-M_6, which has the function to precharge (discharge) the gate line at the intermediate voltage V_{SSL} during rising (falling) output transitions, with the purpose to save power consumption. Finally, two level shifters are necessary to properly drive the half bridge as well as two transistors of the transfer gate. The complete schematics of these level shifters are illustrated in Fig. 6.46b, c.

Assuming IN_1 and IN_2 both high, transistor M_1 is ON and the buffer output is at V_{SSH}. A full commutation from V_{SSH} to V_{DDH} is performed in two steps as shown in Fig. 6.47. In the former step IN_1 is forced low, thus M_1 and M_3 are respectively switched OFF and ON, pulling the buffer output up to V_{SSL}. Actually, as M_4 tends to turn OFF when the voltage is close to V_{SSL}, the output tends asymptotically to $V_{SSL}-V_D$, where V_D is the body-diode voltage drop. In the latter step IN_2 is forced low, thus M_2 is switched ON, pulling the buffer output up to V_{DDH}.

Analogously, a full commutation from V_{DDH} to V_{SSH} is performed as follows. In the former step IN_2 is forced high, thus M_2 and M_5 are respectively switched OFF and ON, pulling the buffer output down to V_{SSL}. As above, since M_6 tends to turn OFF when the voltage is close to V_{SSL}, the output tends asymptotically to $V_{SSL}+V_D$. In the latter step IN_1 is forced high, thus M_1 is switched ON, pulling the buffer output down to V_{SSH}. As above mentioned, this double transition through V_{SSL} allows up to a 50% maximum power saving as long as V_{SSL} is in the middle between V_{SSH} and V_{DDH}.

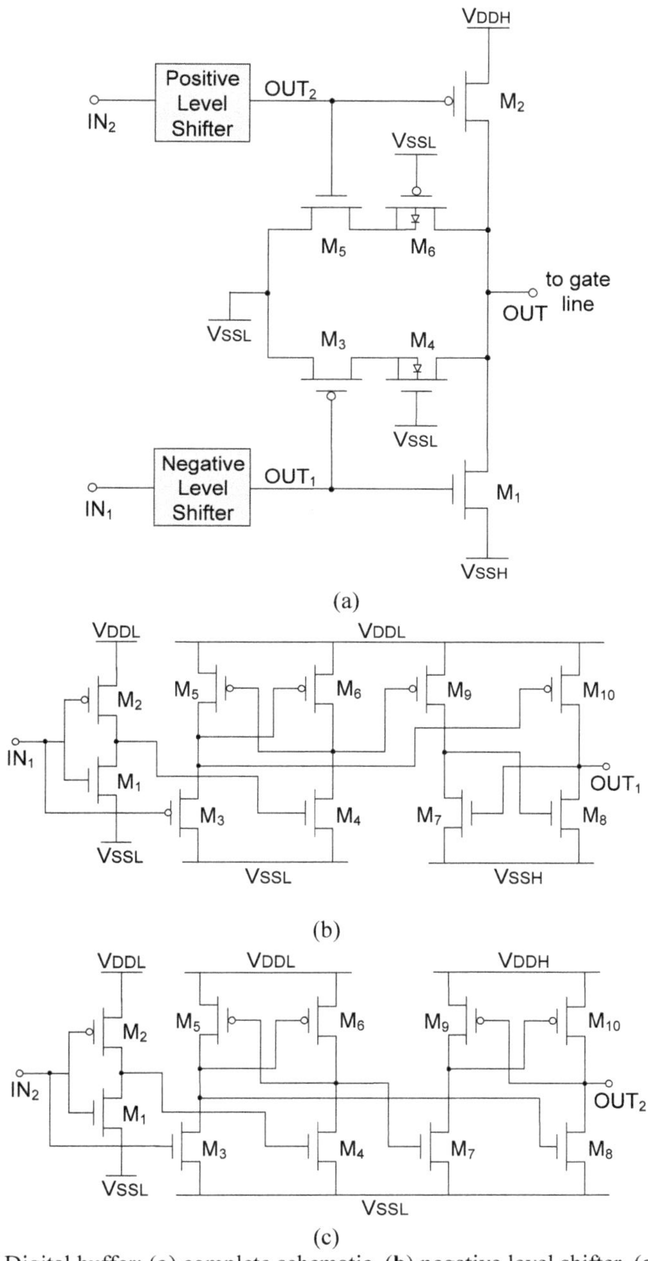

Fig. 6.46 Digital buffer: (**a**) complete schematic, (**b**) negative level shifter, (**c**) positive level shifter

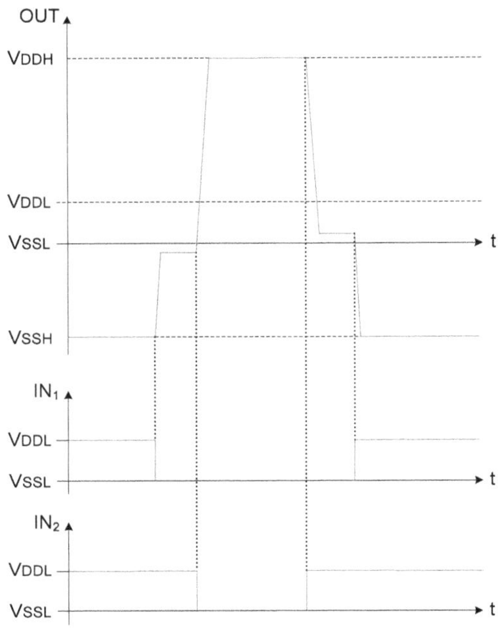

Fig. 6.47 Timing diagram of the circuit shown in Fig. 6.46a

The bootstrap technique has also been widely used in the implementation of the buffers in low-voltage driver circuits to enhance the speed performance. Indirect bootstrap (a node of the bootstrap capacitor is connected to the gate of the output device) [LK1997, LYS1999, YMD2000], may not be as effective as the direct bootstrap (a node of the bootstrap capacitor is the output port of the driver) [GMN2005, GMS2004, CK2002]. However, direct bootstrapping usually requires two capacitors; one for bootstrapping during the pull-up switching transition and one for the pull-down transition. Area and power dissipation can be reduced using the single capacitor bootstrap technique [CLL2007]. A complementary dual-bootstrap technique is also reported in [LL2007] to increase further the speed of the TFT-LCD scan-line signals for driving heavy load. This is achieved by driving the gate of the CMOS devices and the output node at the same time.

References

[A2006] M.M. Ahmadi, "A New Modeling and Optimization of Gain Boosted Cascode Amplifier for High-Speed and Low-Voltage Applications", *IEEE Transactions on Circuits and Systems - Part II*, Vol. 53, No. 3, pp. 169–173, 2006.

[AH1987] P. Allen and D. Holberg, *CMOS Analog Circuit Design*, Oxford University Press, 1987.

[AH2002] P.E. Allen and D.R. Holberg, *CMOS Analog Circuit Design*, second edition, Oxford University Press, 2002.

[B2005] M.J. Bell, "An LCD Column Driver Using a Switch Capacitor DAC," *IEEE Journal of Solid-State Circuits*, Vol. 40, No. 12, pp. 2756–2765, 2005.

[BCO2006] B.S. Bae, J.W. Choi, J.H. Oh, and J. Jang, "Level Shifter Embedded in Drive Circuit with Amorphous Silicon TFTs," *IEEE Transactions on Electron Devices*, Vol. 3, No. 3, pp. 494–498, 2006.

[BG1990] K. Bult and G. J. G. M. Geelen, "A Fast-Settling CMOS Op Amp for SC Circuits with 90-dB DC Gain," *IEEE Journal of Solid-State Circuits*, Vol. 25, No. 6, pp. 1379–1383, 1990.

[BMK2005] B.H. Berkeley, R. I. McCartney, T. Kim, and C. Zajac, "World's First ABSI LCD Module," *SID Digest*, Vol. 36, No. 1, pp. 1684–1687, 2005.

[BPV2001] A. Boni, A. Pierazzi, and D. Vecchi, "LVDS I/O Interface for Gb/s-per-Pin Operation in 0.35-µm CMOS," *IEEE Journal of Solid-State Circuits*, Vol. 36, No. 4, pp. 706–711, 2001.

[CK2002] P. C. Chen and J. B. Kuo, "Sub-1V CMOS Large Capacitive-Load Driver Circuit Using Direct Bootstrap Technique for Low-Voltage CMOS VLSI," *Electronics Letters*, Vol. 38, No. 6, pp. 265–266, 2002.

[CLH2001] H.-J. Chung, S.-W. Lee, and C.-H. Han, "Poly-Si TFT Push-Pull Analogue Buffer for Integrated Data Drivers of Poly-Si TFT-LCDs," *Electronics Letters*, Vol. 37, No. 17, pp. 1093–1095, 2001.

[CLL2007] Y.Z. Cao, Z.M. Lin, and H.C. Lai, "Single Capacitor Bootstrapped Driver for TFT-LCD Gate Driver Circuit," *Proceedings of IEEE Conference on Electron Devices and Solid-State Circuits*, pp. 885–888, 2007.

[CM2007] J. Choi, K. Min and C. Yoo, "High-Speed and Low-Power Analog Source Driver for TFT-LCD using Dynamic Biased Operational Amplifier," *SID Digest*, Vol. 38, No. 1, pp. 1647–1650, 2007.

[DM1994] V.M. Da Costa and R.A. Martin, "Amorphous Silicon Shift Register for Addressing Output Drivers," *IEEE Journal of Solid-state Circuits*, Vol. 29, No. 5, pp. 596–600, 1994.

[DPS2007] S. Di Fazio, F. Pulvirenti, T. Signorelli, C. Lao, S. Pennisi, "Low Quiescent Current High Speed Amplifier for LCD Column Driver", *Proceedings of European Conference on Circuit Theory and Devices*, pp. 523–526, 2007.

[ET1996] C.C. Enz and G.C. Temes, "Circuit Techniques for Reducing the Effects of Op-Amp Imperfection: Autozeroing, Correlated Double Sampling, and Chopper Stablilization," *Proceedings of IEEE*, Vol. 84, No. 11, pp. 1584–1614, 1996.

[EWK2006] S. Edo, M. Wakagi, and S. Komura, "A 2.2" QVGA a-Si TFT LCD with High Reliability Integrated Gate Driver," *SID Digest*, Vol. 37, No. 1, pp. 1551–1554, 2006.

[FO2004] S. Fugikawa and T. Ozawa, "Level Shifter and Electro-Optical Apparatus Incorporating The Same," *U.S. Patent #6 714 046 B2*, issued 2004.

[G1999] R. Gregorian, *Introduction to CMOS OP-AMPs and Comparators*, Wiley, pp. 218–231, 1999.

[GMN2005] J.C. Garcia, J.A. Montiel-Nelson, and S. Nooshabadi, "A Single Capacitor Bootstrapped Power-Efficient CMOS Driver," *Proceedings of IEEE Midwest Symposium on Circuits and Systems*, Vol. 1, pp. 287–290, 2005.

[GMS2004] J.C. Garcia, J.A. Montiel-Nelson, J. Sosa, and H. Navarro, "A Direct Bootstrapped CMOS Large Capacitive-Load Driver Circuit," *Proceedings of IEEE Design, Automation and Test in Europe Conference and Exhibition*, Vol. 1, pp. 680–681, 2004.

[GT1986] R. Gregorian and G.C. Temes, *Analog MOS Integrated Circuits for Signals Processing*, Wiley, 1986.

[HG1998] S.B. Huq, J. Goldie, "An Overview of LVDS Technology," *National Semiconductor Application Note 971*, pp. 1–6, 1998.

[HIK1992] M. Hirose, F. Inoue, N. Kabuto, H. Kataoka, and K. Masuda, "5-bit TV Pictures on 3-bit Color TFT-LCDs Using Alternating Dither," *SID Digest*, pp. 597–600, 1992.

[HT1994] R. Hogervorst, J.P. Tero, G.H. Eschaizier and J.H. Huisjing, "A Compact Power-Efficient 3V CMOS Rail-to-Rail Input/Output Operational Amplifier for VLSI Cell Libraries," *IEEE Journal of Solid-State Circuits*, Vol. 29, No. 12, pp. 1505–1513, 1994.

[HTL2005] H.-R. Han, J.-F. Tsai, W.-T. Liao, and W.-C. Wang, "Reliable Integrated s-Si Select Line Driver for 2.2-in. QVGA TFT-LCD," *SID Digest*, Vol. 36, No. 1, pp. 946–949, 2005.

[IEEE1996] IEEE Standard for Low-Voltage Differential Signalling (LVDS) for Scalable Coherent Interface (SCI), *IEEE Standard 1596.3-1996*, 1996.

[IMS2003] T. Itaku, H. Minamizaki, T. Satio, and T. Kuroda, "A 402-Output TFT-LCD Driver IC with Power Control Based on Number of Colors Selected," *IEEE Journal of Solid-State Circuits*, Vol. 38, No. 3, pp. 503–510, 2003.

[JCL2004] J. Jeon, K.-S. Choo, W.-K. Lee, J.-H. Song, and H.-G. Kim, "Integrated a-Si Gate Driver for TFT-LCD Panel," *SID Digest*, Vol. 35, No. 1, pp. 10–13, 2004.

[JPH2004] S.H. Jung, J. H. Park, C.W. Han, and M.K. Han, "New Source Follower Type Analog Buffers Using Poly-Si TFTs for Active Matrix Displays," *SID Digest*, Vol. 35, No. 1, pp. 1452–1455, 2004.

[K2000] S. Kempainen, "Low-Voltage differential Signalling (LVDS)," *Insight*, Vol. 5, No. 2, pp. 4–9, 2000.

[K2001] F.J. Kahn, "Recent Trends in Electronic Projection," *Information Display*, Vol. 17, No. 2, pp. 24–27, 2001.

[K2005] S.S. Kim, "The World's Largest (82-in.) TFT-LCD," *SID Digest*, Vol. 36, No. 1, pp. 1842–1847, 2005.

[KCK2005] J.-H. Kim, B.-D. Choi, and O.-K. Kwon, "1-Billion-Color TFT-LCD TV with Full HD Format," *IEEE Transactions on Consumer Electronics*, Vol. 51, No. 4, pp. 1042–1050, 2005.

[KDH2006] M.-D. Ker, C.-K. Deng, and J.-L. Huang, "On-Panel Output Buffer With Offset Compensation Technique for Data Driver in LTPS Technology," *IEEE Journal of Display Technology*, Vol. 2, No. 2, pp. 153–159, 2006.

[KFT2001] K. Kawabe and T. Furuhashi, and Y. Tanaka, "New TFT-LCD Driving Method for Improved Moving Picture Quality," *SID Digest*, Vol. 32, No. 1, pp. 998–1001, 2001.

[KPP2004] T.-S. Kim, B.-I. Park, J.-H. Park, B. Berkeley, and S.-S. Kim, "An Optimized Boost Table Measurement Method for Response Time Acceleration in LCD," *SID Digest*, Vol. 35, No. 1, pp. 372–375, 2004.

[KS2006] S. Kim, Y. Son and G. Cho, "Low-Power High-Slew-Rate CMOS Buffer Amplifier for Flat Panel Display Drivers," *Electronics Letters*, Vol. 42, No. 4, pp. 214–215, 2006.

[KSS2004] A. Kumar, S. Sambandan, K. Sakariya, P. Servati, and A. Nathan, "Amorphous Silicon Shift Registers for Display Drivers," *Journal of Vacuum Science and Technology A*, Vol. 22, No. 3, pp. 981–986, 2004.

[L2004] C.W. Lu, "High-Speed Driving Scheme and Compact High-Speed Low- Power Rail-to-Rail Class-B Buffer Amplifier for LCD Applications," *IEEE Journal of Solid-State Circuits*, Vol. 39, No. 11, pp. 1938–1947, 2004.

[L2006] H.-Y. Lin, "Shift Register Unit and Signal Driving Using the Same," *U.S. Patent #7,027,550 B2*, issued 2006.

[LCG1984] P.W. Li, M.J. Chin, P.R. Gray, and R. Castello, "A Ratio-Independent Algorithmic Analog-to-Digital Conversion Technique," *IEEE Journal of Solid-State Circuits*, Vol. 19, No. 6, pp. 828–836, 1984.

[LH2004] C. Lu and K. Hsu, "A High Speed Low-Power Rail-to-Rail Column Driver for AMLCD Application," *IEEE Journal of Solid-State Circuits*, Vol. 39, No. 8, pp. 1313–1320, 2004.

[LJ2006] J.C. Lee and J.Y. Jeong, "High Speed, Small Area, Reliable, LTPS TFT-based Level Shifter for System-On-Panel Technology," *Proceedings of IEEE International Conference on Integrated Circuit Design and Technology*, pp. 1–4, 2006.

[LK1997] J.H. Lou and J.B. Kuo, "A 1.5V Full-Swing Bootstrapped CMOS Large Capacitive-Load Driver Circuit Suitable for Low-Voltage CMOS VLSI," *IEEE Journal of Solid-State Circuits*, Vol. 32, No. 1, pp. 119–121, 1997.

[LL2000] A. Lee and D.W. Lee, "Integrated TFT-LCD Timing Controllers with RSDS Column Driver Interface," *SID Digest*, Vol. 31, No. 1, pp. 43–45, 2000.

[LL2007] H.C. Lai and Z.M. Lin, "A High-Speed Full Swing CMOS Driver for TFT-LCD Scan-Line Circuit," *Proceedings of TENCON 2007 - IEEE Region 10 International Conference*, pp. 1–3, 2007.

[LLY2005] K.M. Lim, K.E. Lee, J.S. Yoo, J.-M. Yoon, M.K. Baek, J.-S. Yoo, Y.-S. Jung, J.K.Park, S.-W. Lee, H.C. Kang, C.-D. Kim, and I.-J Chung, "A 3.5 in. QVGA Poly-Si TFT-LCD with Integrated Driver Including New 6-bit DAC," *Solid-State Electronics*, Vol. 49, No. 7, pp. 1107–1111, 2005.

[LS1994] K. Laker, and W. Sansen, *Design of Analog Integrated Circuits and Systems*, Mc Graw-Hill, 1994.

[LWH2005] C.T. Liu, A. Wang, H.-J. Hong, Y.-J. Hsieh, M.-S. Lai, A. Tsai, T.M. Wang, M.-J. Jou, W.-C. Chang, S.-L. Sui, J.H. Liao, and M.-F. Tien, "Color and Image Enhancement for Large-Size TFT-LCD TVs," *SID Digest*, Vol. 36, No. 1, pp. 1730–1733, 2005.

[LYS1999] L.C. Fatt, Y.K. Seng, and S.S. Rofail, "Sub-1V Bootstrapped CMOS Driver for Giga-Scale-Integration Era," *Electronics Letters*, Vol. 35, No. 5, pp. 392–393, 1999.

[M2003] R. McCartney, "A Liquid Crystal Display Response Time Compensation Feature Integrated into an LCD Panel Timing Controller," *SID Digest*, Vol. 34, No. 1, pp. 1350–1353, 2003.

[MB2004] A. Milanesi and P. Buchschacher, "A Novel Offset Cancellation Circuit for TFT-LCD Driver," *SID Digest*, Vol. 35, No. 1, pp. 1568–1571, 2004.

[MFT1991] H. Mano, T. Furuhashi, and T. Tanaka, "Multicolor Display Control Method for TFT-LCD," *SID Digest*, pp. 547–550, 1991.

[MKO2004] S. Matsumoto, N. Komiya, M. Okuyama, and K. Hirosawa, "Level Shifter for Use in Active Matrix Display Apparatus," *U.S. Patent #6,801,181 B2*, issued 2004.

[MLL2007] S.-H. Moon, Y.-S. Lee, M.-C. Lee, B.H. Berkeley, N.-D. Kim, and S.-S. Kim, "Integrated a-Si:H TFT Gate-Driver Circuits on Large-Area TFT-LCDs," *SID Digest*, Vol. 38, No. 1, pp. 1478–1481, 2007.

[NKM2004] Y. Nakajima, Y. Kida, M. Murase, Y. Toyoshima, and Y. Maki, "Latest Development of "System-on-Glass" with Low Temperature Poly-Si TFT Display," *SID Digest*, Vol. 35, No. 1, pp. 864–867, 2004.

[PLY2005] Y.I. Park, D.Y. Lee, J.S. Yoo, H.C. Kang, K.M. Lim, C.-D- Kim, and I.-J. Chung, "LTPS PMOS Four-Mask Process for AMLCDs," *SID Digest*, Vol. 36, No. 1, pp. 341–343, 2005.

[PP2002] G. Palumbo and S. Pennisi, *Feedback Ampilfiers: Theory and Design*, Kluwer Academic Publishers, 2002.

[PPP2001] G. Palmisano, G. Palumbo, and S. Pennisi, "Design Procedures for Two–Stage CMOS OTAs: A Tutorial," *Analog Integrated Circuits and Signal Processing*, Vol. 27, No. 3, pp. 179–189, 2001.

[RW1996] P. Rombouts and L. Weyten, "Linearity Improvement for the Switched-Capacitor DAC," *Electronics Letters*, Vol. 32, No. 4, pp. 293–294, 1996.

[RWR1998] P. Rombouts, L. Weyten, J. Raman, and S. Audenaert, "Capacitor Mismatch Compensation for the Quasi-Passive Switched-Capacitor DAC," *IEEE Transactions on Circuits and Systems - part I*, Vol. 45, No. 1, pp. 68–71, 1998.

[SCK1999] Y. Sung, B. Choi and O. Kwon, "Low-Power TFT-LCD Source Driver Using Triple Charge Sharing Method," *Proceedings of International Conference on VLSI and CAD*, pp. 317–320, 1999.

[SCK2007] M.-S. Shin, J.-W. Choi, Y.-J. Kim, K.-R. Kim, and I. Lee, "Design of Low-Cost 2.2-inch qVGA LTPS TFT-LCD Panel," *SID Digest*, Vol. 38, No. 1, pp. 260–263, 2007.

[SG1990] E. Sackinger and W. Guggenbuhl, "A High-Swing High-Impedance MOS Cascode Circuit," *IEEE Journal of Solid-State Circuits*, Vol. 25, No. 1, pp. 289–298, 1990.

[SGH1975] R.E. Suarez, P.R. Gray, and D.A. Hodges, "All-MOS Charge Redistribution Analog-to-Digital Conversion Techniques - Part II," *IEEE Journal of Solid-State Circuits*, Vol. 10, No. 6, pp. 379–385, 1975.

[SK2007] Y.-C- Sung and O.-K. Kwon, "Low-Cost TFT-LCDs with Pre-Emphasis Driving Method for Large-Size and High-Definition TVs," *IEEE Transactions on Consumer Electronics*, Vol. 53, No. 4, pp. 1674–1681, 2007.

[SKC2003] H.-S. Shim, J.-H. Kim, B.-C. Cho, and O.-K. Kwon, "A New Level Shifter Using Low Temperature Poly-Si TFTs," *International Meeting on Information Display Digest*, pp. 117–120, 2003.

[SMH2000] H. Sakamoto, N. Makita, and M. Hijikigawa, "2.6-in. HDTV Panel Using CG Silicon," *SID Digest*, Vol. 31, No.1, pp. 1190–1193, 2000.

[SN2001] K. Sekiya and H. Nakamura, "Overdrive Method for TN-mode LCDs–Recursive System with Capacitance Prediction," *SID Digest*, Vol. 32, No. 1, pp. 114–117, 2001.

[SOY2003] J. Someya, N. Okuda, H. Yoshii, and M. Yamakawa, "A New LCD-Controller for Improvement of Response Time by Compression FFD," *SID Digest*, Vol. 32, No. 1, pp. 1346–1349, 2003.

[SSK2005] Y.–C. Sung, S.–M. So, J.–K. Kim, and O.–K. Kwon, "10bit Source Driver with Resistor-Resistor Digital to Analog Converter," *SID Digest*, Vol. 36, No. 1, pp. 1099–1101, 2005.

[T1996] S. Tanoi, "Level Shifter Circuit," *U.S. Patent #5,498,991*, issued 1996.

[TSL2005] C.S. Tan, W.T. Sun, S.H. Lu, C.H. Kuo, I.T. Chang, S.H. Yeh, C.C. Chen, L. Liu, Y.C. Lin, and C.S. Yang, "A Simple Architecture for Fully Integrated 2.4" Poly-Si TFT-LCD," *SID Digest*, Vol. 36, No. 1, pp. 336–339, 2005.

[USS1996] Y. Unekawa, K. Seki-Fukuda, K.Sakaue, T. Nakao, S. Yoshioka, T. Nagamatsu, H. Nakakita, Y. Kaneko, M. Motoyama, Y. Ohba, K. Ise, M. Ono, K. Fujiwara, Y. Miyazawa, T. Kuroda, Y. Kamatani, T. Sakurai, and A. Kanuma, "A 5-Gb/s 8x8 ATM Switch Element CMOS LSI Supporting Five Quality-of-Service Classes with 200-MHz LVDS Interface," *Proceedings of IEEE Solid-State Circuits Conference*, pp. 118–119, 1996.

[YCL2007] S.-H. Yang, W.-C. Chen, H.-Y. Lin, C.-H. Lee, V. Lai, C.-M. Chiu, H.-C. Wang, and C.-C. Lee, "High Integrated 10.2-inch WVGA LTPS LCD Manufactured by PMOS Process," *SID Digest*, Vol. 38, No. 1, pp. 245–248, 2007.

[YCS2004] Y.S. Yoo , J.Y. Choi, H. S. Shim, and O. K. Kwon, C "A High Accurate Analog Buffer Circuit Using Low Temperature Poly-Si TFT," *SID Digest*, Vol. 35, No. 1, pp. 1460–1463, 2004.

[YGC2003] S. Yang, P. Guo, T. Chang, and J. Hong, "A Multi-Phase Charge-Sharing Technique without External Capacitor for Low-Power TFT-LCD Column Drivers," *Proceedings of IEEE International Symposium on Circuits and Systems*, Vol. 5, pp. 365–368, 2003.

[YID2007] K. Yamashita, T. Inada, S. Deane, P. Collins, S. Kawata, Y. Onda, M. Yoshiga, J. Hector, and M. Inoue, "a-Si Gate Driver Design for Low Power AMLCD Operation", *Information and Media Technologies*, Vol. 2, pp. 1–5, 2007.

[YJK2005] S.Y. Yoon, Y.H. Jang, B. Kim, M.D. Chun, H.N. Cho, N.W. Cho, C.Y. Sohn, S.H. Jo, C.-D. Kim and I.-J. Chung, "Highly Stable Integrated Gate Driver Circuit Using a-Si TFT with Dual Pull-Down Structure," *SID Digest*, Vol. 36, No. 1, pp. 348–351, 2005.

[YKL2005] C. Yoo, D.-J. Kim, and K.-L. Lee, "Threshold Voltage and Mobility Mismatch Compensated Analogue Buffer for Driver-Integrated Poly-Si TFT LCDs," *Electronics Letters*, Vol. 41, No. 2, pp. 65–66, 2005.

[YMD2000] K.-S. Yeo, J.-G. Ma, and M.-A Do, "Ultra Low Voltage Bootstrapped CMOS Driver for High Performance Applications," *Electronics Letters*, Vol. 36, No. 8, pp. 706–708, 2000.

Chapter 7

Charge Pumps for LCD Drivers

by Gaetano Palumbo and Domenico Pappalardo

It was shown in Chapters 4 and 6 that LCD drivers, especially those for small-size panels, make extensive use of charge pumps. High-voltage boosting circuits are indeed required to generate the row and column voltages if the external supply is kept low to contain power consumption and area of the digital section [YKC2003], [SK2008]. Owing to the consequent relevance of an optimized CP design and also because a comprehensive treatment of the subject is missing in electronic textbooks, we decided to include this chapter as a conclusion of the book.

A Charge Pump (CP) is an electronic circuit that converts the supply voltage V_{DD} to a DC output voltage V_{Out} that is several times higher than V_{DD} (i.e., it is a DC-DC converter whose input voltage is lower than the output one) [D1976], [WGM1989]. Unlike the other traditional DC-DC converters, which employ inductors, CPs are made up of capacitors and switches (or diodes) only, thereby allowing integration on silicon. CPs were originally used in smart power ICs [H1990, WVK1992, GP1994, GP1996] and nonvolatile memories [JNH1992, UKA1992, AKU1994, CGO1999, ZLD2001, IOS2001, PPG2006] and, given the continuous scaling down of IC power supplies, they have also been employed in a vast variety of integrated systems [CT1991, DD1998, BN1998, MS1998, NGN2001, MC2001].

In practical applications, CPs may be either loaded by a simple capacitor (or equivalently the gate of a MOS transistor) or by more complex networks. In the last case the CP load can be simply modeled by means of an equivalent capacitance and an equivalent current generator whose value is given by the load current, which for LCD drivers is proportional to the frame frequency and the driven capacitance. A current generator can also be used to model a pure resistive load. The load type determines a change in the CP behavior, and consequently on the design strategy to be followed. Although for LCD drivers only the case with the current load has to be considered, in this chapter both the situations are considered for the sake of completeness.

D.J.R. Cristaldi et al., *Liquid Crystal Display Drivers*, DOI 10.1007/978-90-481-2255-4_7,

7.1 Analysis of the Charge Pump with a Pure Capacitive Load

7.1.1 One-Stage Charge Pump

To show the behavior of an ideal CP, let us consider the one-stage topology (Fig. 7.1) which comprises a single pumping capacitor C, two switches, S_1 and S_2 (driven by two complementary phases), a clock signal whose amplitude is equal to the power supply V_{DD} (Fig. 7.2) and a pure capacitive load C_L (also referred to as the bulk capacitor).

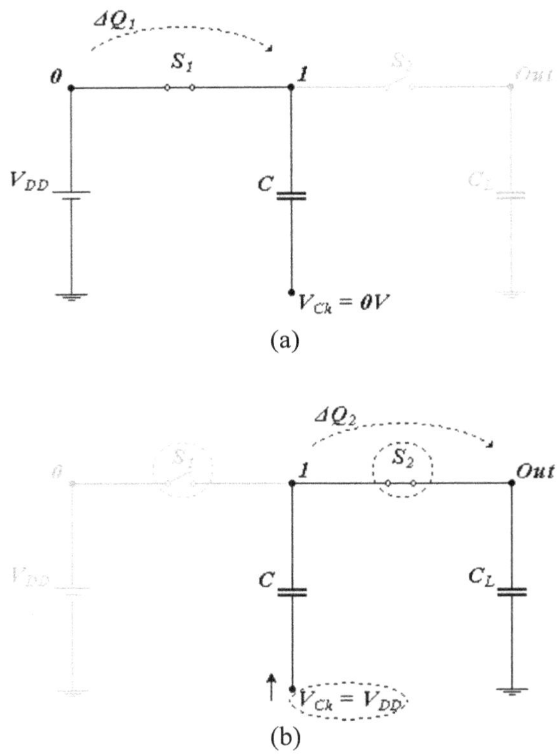

Fig. 7.1 Ideal one-stage CP: (**a**) first half period; (**b**) second half period

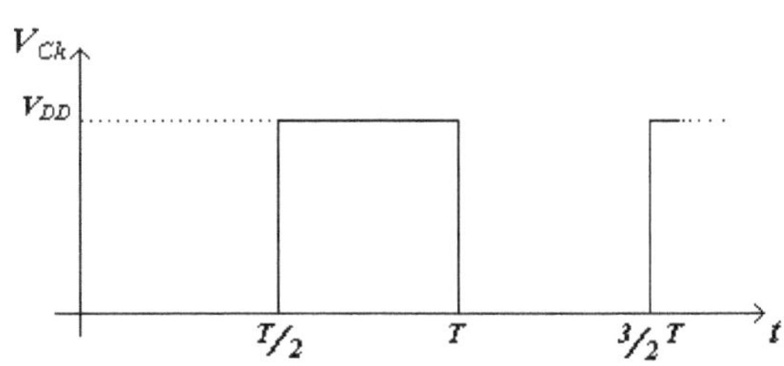

Fig. 7.2 Clock signal, V_{Ck}, of the CP in Fig. 7.1

During the first half period (0 to $T/2$), S_1 and S_2 are respectively closed and open and C, being connected to the power supply, is charged to V_{DD} (Fig. 7.1a). In the second half period ($T/2$, T) the switches change their state, the clock signal now equals V_{DD}, and part of the charge stored in C is transferred to C_L. Hence, at each cycle the output voltage will increase up to the final asymptotic value, $2V_{DD}$. In a generic period j, considering that C is charged to V_{DD} before redistribution and that the charge into C_L is the one stored in the previous period being equal to $C_L V_{Out}(j-1)$, we get the recursive equation

$$V_{Out}(j) = \frac{C \cdot 2V_{DD} + C_L \cdot V_{Out}(j-1)}{C + C_L} \tag{7.1}$$

The above equation is plotted in Fig. 7.3. It is apparent that several cycles are needed to approach $2V_{DD}$ and that the step increment of the output voltage in each successive clock period becomes smaller. Indeed, the output voltage will steeply increase in the first part of the transient and will slowly tend to its final value, where no charge redistribution will ideally happen.

In conclusion, according to its name, a CP takes charges from the power supply, via the capacitor C, and pumps these charges into the output capacitor C_L, thus increasing the output voltage up to an ideal value that is twice the power supply.

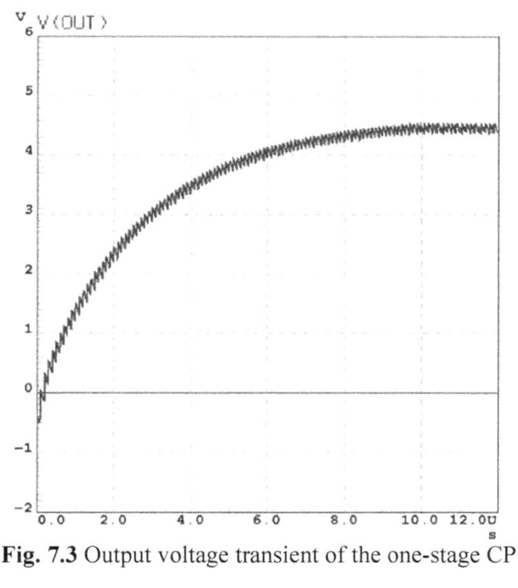

Fig. 7.3 Output voltage transient of the one-stage CP

7.1.2 *N*-stage Charge Pump

The one-stage CP topology can be generalized by including additional cascaded stages as shown in Fig. 7.4, where a generic N-stage CP is depicted. Each stage is made up of a pumping capacitor C and a switch S_i; the CP needs a two-phase clock and to properly connect the output load to the final stage switch S_{Out} is also required.

The behavior of the N-stage CP is similar to that of a one-stage CP. During the first half clock period, V_{Ck} is low and only the odd switches are closed (see Fig. 7.5a). The first pumping capacitor is thus charged to V_{DD} and all the other pumping capacitors in the odd stages receive the charge from the capacitor of the previous stage (this happens also for the load capacitor if the number of stages is even).

During the subsequent half clock period the signal V_{Ck} is equal to V_{DD} and only the even switches are closed (see Fig. 7.5b). Now all the capacitors in the odd stages give the charge to the capacitor in the subsequent stage (the load capacitor is maintained separated from the CP if the number of stages is even, since the switch S_{Out} is open).

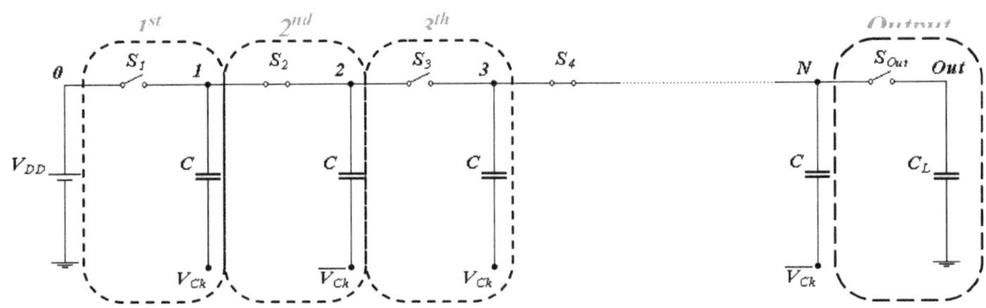

Fig. 7.4 N-stage charge pump

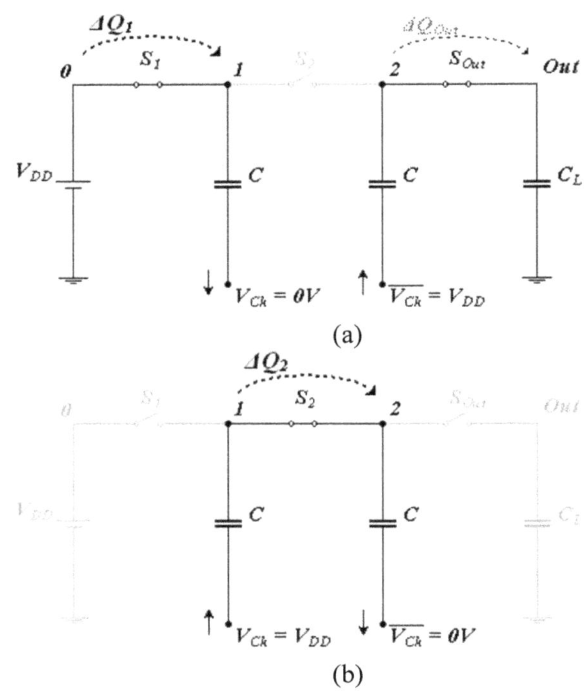

Fig. 7.5 N-stage CP: (**a**) first half period; (**b**) second half period

In summary, in a complete clock period, each CP capacitor receives an amount of charge from the capacitor at its left side and gives a part of this charge to the capacitor at its right side. Thus, in each period there is a charge transfer from the power supply to the output load.

7.1.3 Charge Pump Parameters

The most important design parameters of a CP are the number of stages, the silicon area occupation, the rise time and the charge consumption.

The **number of stages,** N, is strictly related to the required output voltage, V_{Out}. Indeed, the output voltage of a CP with a pure capacitive load in the steady state (i.e., when the charge transfer from the power supply to the output is ideally zero) is

$$V_{Out}\Big|_{Steady\ State} = (N+1)\cdot V_{DD} \tag{7.2}$$

To physically implement a CP, the required **total silicon area**, A_{Tot}, may be non negligible. Since this area is mainly due to the capacitors, we can approximate A_{Tot} with only the area required to implement the capacitors

$$A_{Tot} = k\cdot N\cdot C = k\cdot C_{Tot} \tag{7.3}$$

In the above equation, the parameter k depends on the processes used to realize the capacitors and C_{Tot} is the sum of the pumping capacitors (i.e., $N\cdot C$).

The **rise time**, t_r, is defined as the time required to achieve a defined output value $V_{Out}(t_r)$. From the dynamic models developed in [TT1997, PBB2000, PP2006], the CP rise time can be approximated with the relationship

$$t_r = T\cdot\left(N\frac{C_L}{C}+0.3N+0.6\right)\cdot\ln\left[\frac{(N+1)V_{DD}-V_{Out}(0)}{(N+1)V_{DD}-V_{Out}(t_r)}\right] \tag{7.4}$$

where T is the clock period. Moreover, for N much higher than 1, and normalizing the output voltage to the power supply

$$v_x = \frac{V_{Out}(t_r)}{V_{DD}} \tag{7.5}$$

$$v_{x0} = \frac{V_{Out}(0)}{V_{DD}} \tag{7.6}$$

we get

$$t_r = T \cdot N^2 \cdot \frac{C_L + C_{Eq}}{C_{Tot}} \cdot \ln\left(\frac{N+1-v_{x0}}{N+1-v_x}\right) \tag{7.7}$$

where capacitance C_{Eq} is equal to $C_{Tot}/3$.

As expected, to reach the output steady state voltage, $(N+1)V_{DD}$, (7.7) anticipates an infinite rise time. Moreover, it is apparent that t_r is affected by the load capacitance and the total CP capacitance, C_{Tot}.

Since from (7.7) we can write

$$V_{Out}(t_r) = (N+1)V_{DD} - \left[(N+1)V_{DD} - V_{Out}(0)\right] \cdot e^{-\frac{t_r}{\frac{N}{Cf}(C_L + C_{Eq})}} \tag{7.8}$$

the dynamic behavior of CP with a pure capacitive load is equivalent to that of a simple RC circuit [PBB2000]. More specifically, the CP model is the RC circuit in Fig. 7.6 where

$$R_{Eq} = \frac{N}{C \cdot f} \tag{7.9}$$

and $V_{Out}(+\infty)$ is the output voltage in the steady state, given by (7.2).

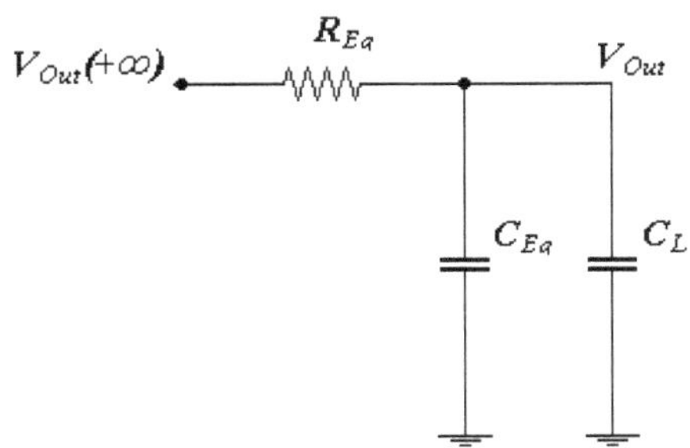

Fig. 7.6 CP equivalent model

To evaluate the **charge consumption** of a CP, we must compute the charge delivered by the power supply to the CP during its rise time. The charge consumption can be divided into three main contributes: the charge given to the load, Q_L, the charge required during the transient by the pumping capacitances, Q_{Pump}, and the contribution wasted in the parasitic elements shown in Fig. 7.7, Q_{Par}.

Note that, referring to Fig. 7.7, the parasitic capacitance at the bottom plate, C_P, of the pumping capacitors is generally more than one order of magnitude higher than that of the other plate. Thus, the charge consumption can be written as

$$Q_T = Q_L + Q_{Pump} + Q_{Par} \tag{7.10}$$

Taking into account the equivalent RC circuit in Fig. 7.6, we get

$$Q_L(t_r)+Q_{Pump}(t_r)=(N+1)\cdot(C_{Eq}+C_L)\cdot[V_{Out}(t_r)-V_{Out}(0)] \tag{7.11}$$

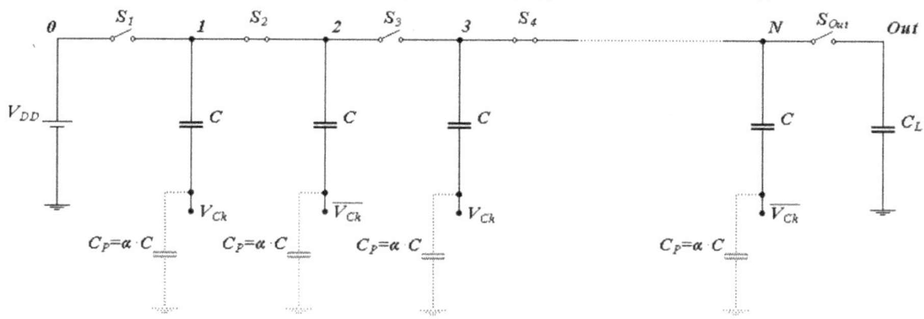

Fig. 7.7 *N*-stage CP with parasitic effects

During each clock period, the charge loss due to parasitic effects (required to charge and discharge the parasitic capacitances on the bottom plate of a pumping capacitor) can be modeled by NC_PV_{DD}, and the contribute during the rise time results to be

$$Q_{Par}=NC_PV_{DD}\frac{t_r}{T}=\alpha C_{Tot}V_{DD}\frac{t_r}{T} \tag{7.12}$$

where the bottom plate parasitic capacitance, C_P, is assumed proportional to the pumping capacitance C through a factor α.

Substituting each contribute into (7.10), we get the expression of the total charge sunk by the power supply up to the rise time

$$Q_T(t_r)=\left[(N+1)(v_x-v_{x0})+\alpha N^2\ln\frac{N+1-v_{x0}}{N+1-v_x}\right](C_{Eq}+C_L)\cdot V_{DD} \tag{7.13}$$

7.2 Optimized Design of the Charge Pump with a Pure Capacitive Load

The design parameters involved in the design of a CP are summarized in Table 7.1 and can be obtained from relationships (7.3), (7.7) and (7.13). The first parameter to be set is the number of stages. However, even if (7.2) relates N and the steady state output voltage in a simple way, it cannot be really used since this value is ideally reached after and infinite time.

From a practical viewpoint, it is useful to analyze the plot in Fig. 7.8, where Q_{Tot} and C_{Tot}, normalized to their minimum values, are plotted versus N. By inspection, it is apparent that two different minima exist. Hence, before starting the design procedure, we have to clearly identify the main design target, in order to use the design approach which allows the chosen performance to be optimized. Two different design strategies can be developed: one minimizes the area (and rise time

[DP1996]), the other minimizes the charge consumption (i.e. the power consumption) during the transient behavior.

Table 7.1 CP design parameters

Parameter	**Comment**
V_{DD}	Technology dependent
Q_{Tot}	**Not known a priori**
$V_{Out}(t_r)$	Design constraint
t_r	Design constraint
$F=1/T$	Technology dependent
$A_{Tot} \propto C_{Tot}$	**Not known a priori**
N	**Not known a priori**
$\alpha = C_P/C$	Technology dependent

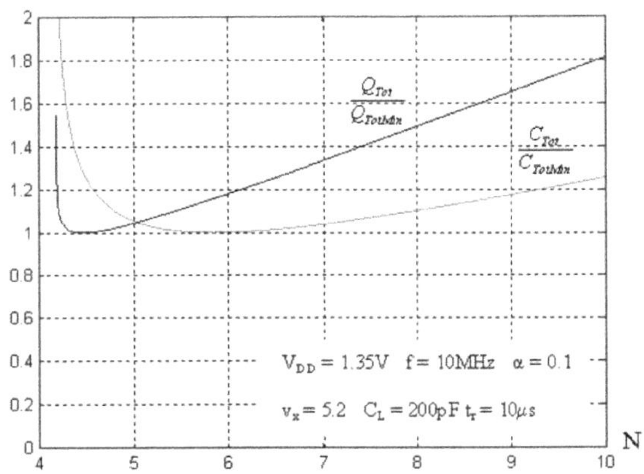

Fig. 7.8 Q_{Tot} and C_{Tot} normalized to their minimum values versus N

7.2.1 Minimizing Area Occupation and Rise Time

To establish the optimum number of stages that minimizes the silicon area we must first evaluate C_{Tot} through (7.7), and set to zero its derivative with respect to N. Similarly, the optimum N that minimizes the rise time is found by directly setting to zero the derivative of (7.7). In both cases we get

$$2\ln\frac{(N+1)-v_{x0}}{(N+1)-v_x}+N\left[\frac{1}{(N+1)-v_{x0}}-\frac{1}{(N+1)-v_x}\right]=0 \tag{7.14}$$

The above expression can be simplified using the empirical approximation (very accurate for x ranging from 0.3 to 1 [PP2006])

$$\ln(x)\approx\frac{2x^2-x-1}{3x} \tag{7.15}$$

And, after some algebraic simplifications (7.14) becomes

$$\frac{(v_x - v_{x0})\left[4v_x + 2v_{x0} - 3(N+2)\right]}{\left[(N+1) - v_x\right]\left[(N+1) - v_{x0}\right]} = 0 \tag{7.16}$$

from which, the optimum N minimizing both the total capacitance and rise time is

$$N_{Aop} = \frac{4}{3}v_x + \frac{2}{3}v_{x0} - 2 \tag{7.17a}$$

and since v_{x0} is often equal to 1, we get

$$N_{Aop} = \frac{4}{3}(v_x - 1) \tag{7.17b}$$

Once defined N_{Aop}, we can use the rise time constraint, given by (7.7), to evaluate the required total pumping capacitance and, hence, the value of C.

7.2.2 Minimizing Charge Consumption

To find the optimum number of stages that minimizes charge consumption, we take the derivative of (7.13) and set it to zero

$$(v_x - v_{x0}) + 2\alpha N \ln\frac{(N+1) - v_{x0}}{(N+1) - v_x} - \alpha N^2 \frac{v_x - v_{x0}}{\left[(N+1) - v_x\right]\left[(N+1) - v_{x0}\right]} = 0 \tag{7.18}$$

To solve (7.18) we again approximate the logarithmic term. We now use the linear approximation which minimizes the error in the range 0–1 [PP2006]

$$\ln(x) \approx 2(x-1) \tag{7.19}$$

Thus (7.18) becomes

$$1 + \frac{4\alpha N}{(N+1) - v_{x0}} - \frac{\alpha N^2}{\left[(N+1) - v_x\right]\left[(N+1) - v_{x0}\right]} = 0 \tag{7.20}$$

or equivalently

$$(1+3\alpha)N^2 - \left[v_x + v_{x0} + 4\alpha(v_x - 1) - 2\right]N + (v_x - 1)(v_{x0} - 1) = 0 \tag{7.21}$$

Since v_{x0} is often equal to 1, the optimum number of stages minimizing charge consumption is expressed by

$$N_{Qop} = \frac{1+4\alpha}{1+3\alpha}(v_x - 1) \tag{7.22}$$

Again, once calculated N_{Qop}, we find the pumping capacitor, C, for the required rise time by using relationship (7.7).

7.2.3 Comparison Between the Optimized Design Strategies

To perform a comparison of the design strategies discussed before, we evaluate the area overhead caused by the minimum power consumption design and the charge consumption overhead caused by the minimum silicon area design [PP2006].

As already stated, the silicon area can be directly derived from the CP total capacitance, hence evaluating (7.7) for each design strategy, we get

$$t_{r,Aop} = T\frac{C_L + C_{Eq}}{C_{T,Aop}}\left[\frac{4}{3}(v_x - 1)\right]^2 \ln 4 \tag{7.23}$$

$$t_{r,Qop} = T\frac{C_L + C_{Eq}}{C_{T,Qop}}\left[\frac{1+4\alpha}{1+3\alpha}(v_x - 1)\right]^2 \ln\left(\frac{1}{\alpha}+4\right) \tag{7.24}$$

where $C_{T,Aop}$ and $C_{T,Qop}$ are the pumping total capacitances in the cases of area and charge consumption minimization, respectively. Assuming a given rise time to be achieved by both design strategies, we can evaluate, by equating (7.23) and (7.24), the increased total CP capacitance for the optimized power consumption design with respect to the minimum area design. It results to be

$$\frac{C_{T\,Qop} - C_{T\,Aop}}{C_{T\,Aop}} = \left(\frac{3}{4}\frac{1+4\alpha}{1+3\alpha}\right)^2 \frac{\ln\left(\frac{1}{\alpha}+4\right)}{\ln 4} - 1 \tag{7.25}$$

Relationship (7.25) is a decreasing function of α as can be seen by its plot in Fig. 7.9. It is apparent the area increase of 25% to 5% for α ranging from 0.1 to 0.5.

To compare the charge consumption of the two design strategies, we substitute N from (7.17) and (7.22) into relationship (7.13) (assuming as usual v_{x0} equal to 1), we get

$$Q_{T\,Aop} = \left[\frac{4}{3}(v_x - 1)^2 + \alpha\left(\frac{4}{3}\right)^2 (v_x - 1)^2 \ln 4\right]\left(C_L + C_{Eq}\right)V_{DD} \tag{7.26}$$

for the area minimization design, and

$$Q_{T\,Qop} = \left[\frac{1+4\alpha}{1+3\alpha} (v_x - 1)^2 + \alpha \left(\frac{1+4\alpha}{1+3\alpha} \right)^2 (v_x - 1)^2 \ln\left(\frac{1}{\alpha} + 4 \right) \right] \left(C_L + C_{Eq} \right) \cdot V_{DD} \tag{7.27}$$

for the charge consumption minimization design. Hence, the increase of the charge consumption incurred by the minimum area design is given by

$$\frac{Q_{T\,Aop} - Q_{T\,Qop}}{Q_{T\,Qop}} = \frac{\frac{4}{3}\left(1 + \alpha \frac{4}{3} \ln 4 \right)}{\frac{1+4\alpha}{1+3\alpha} \left[1 + \alpha \frac{1+4\alpha}{1+3\alpha} \ln\left(\frac{1}{\alpha} + 4 \right) \right]} - 1 \tag{7.28}$$

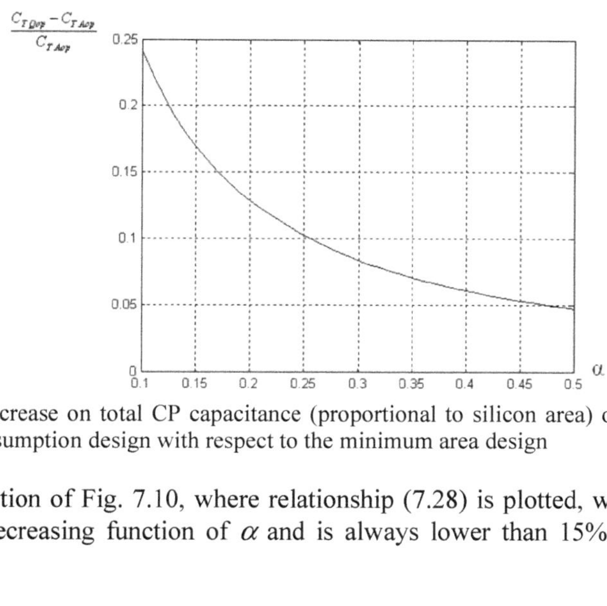

Fig. 7.9 Increase on total CP capacitance (proportional to silicon area) of minimum power consumption design with respect to the minimum area design

By inspection of Fig. 7.10, where relationship (7.28) is plotted, we can see that (7.28) is a decreasing function of α and is always lower than 15% for typical α values.

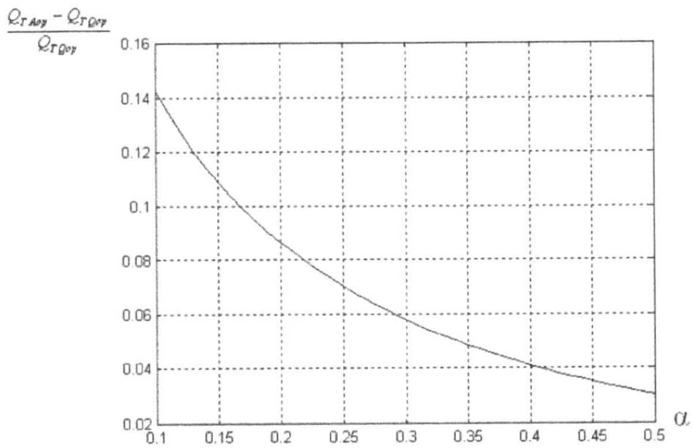

Fig. 7.10 Increase on charge consumption (i.e. power consumption) of minimum area design with respect to minimum consumption design

7.3 Analysis of the Charge Pump with a Current Load

A CP with a current load is shown in Fig. 7.11. In this case, once reached the steady state, the CP has to provide in each time period, T, an amount of charge equal to

$$\Delta Q = I_L \cdot T \tag{7.29}$$

which is constant, assuming constant the current load, I_L. This is also the amount of charge exchanged in each half period (in the steady state) between two adjacent capacitors. More specifically, ΔQ is transferred from the capacitor on the left to the capacitor (or the load) on the right connected in each half period (see Fig. 7.12).

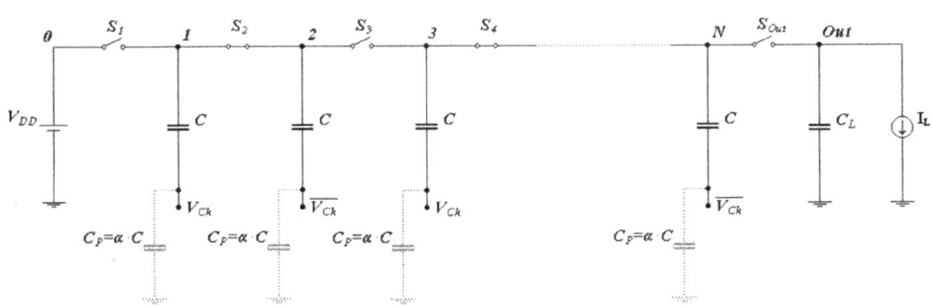

Fig. 7.11 N-stage CP with a current load

The amount of charge ΔQ affects the steady state output voltage, whose average value in each period is

$$V_{Out}\Big|_{Steady\ State} = (N+1) \cdot V_{DD} - N \cdot \frac{I_L \cdot T}{C} \tag{7.30}$$

Consider for example a two-stage CP with a current load, as shown in Fig. 7.12. During the first half period (in the steady state), the first capacitor is connected to the power supply, and is recharged by ΔQ, the same amount provided by the CP to the output. In the next half period, when the first and the second pumping capacitors are connected together, since switch S_2 is closed, they exchange the same amount of charge, ΔQ, and the highest node voltage results to be

$$V_1\Big|_{V_{Ck}=V_{DD}} = V_2\Big|_{\overline{V_{Ck}}=0V} = V_1\Big|_{V_{Ck}=0V} + V_{Ck} - \Delta V = 2\cdot V_{DD} - \Delta V \tag{7.31}$$

where

$$\Delta V = \frac{\Delta Q}{C} \tag{7.32}$$

Finally, in the subsequent first half period, when the switches S_1 and S_{Out} are closed, the output voltage is

$$V_{Out} = V_2\Big|_{\overline{V_{Ck}}=V_{DD}} = V_2\Big|_{\overline{V_{Ck}}=0V} + V_{Ck} - \Delta V = 3\cdot V_{DD} - 2\Delta V \tag{7.33}$$

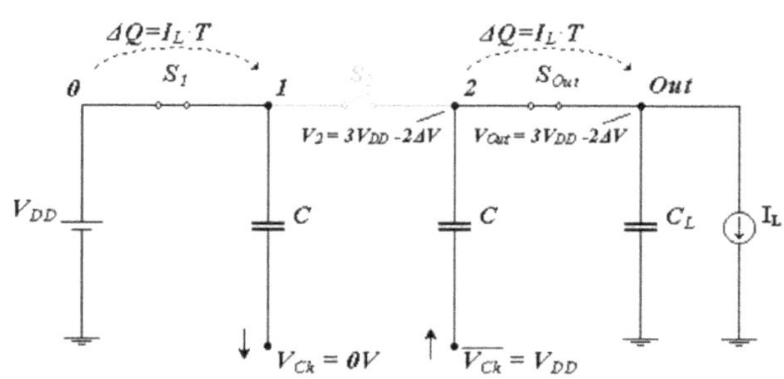

Fig. 7.12 A two-stage CP with a current load

In conclusion, generalizing the analysis for an N-stage CP, we write the output voltage

$$V_{Out}\Big|_{Steady\ State} = (N+1)\cdot V_{DD} - N\cdot\Delta V \tag{7.34}$$

and substituting (7.32) and (7.29) we get relationship (7.30).

Carrying out a more detailed analysis by considering the continuous charge transfer from the CP to the output load, we observe that when the output load is not connected to the CP (i.e., when switch S_{Out} is open), the current load discharges the load capacitance by an amount equal to $\Delta Q/2$. An output voltage ripple, V_r, arises, whose amplitude is

$$V_r = \frac{\Delta Q}{C_L} = \frac{I_L \cdot T}{C_L} \tag{7.35}$$

It decreases with C_L and with the clock frequency.

The area of CP with a load current is of course still given by (7.3), but using (7.30) it can also be rewritten as

$$A_{Tot} = k \cdot \frac{N^2}{(N+1) \cdot V_{DD} - V_{Out}} \frac{I_L}{f} \tag{7.36}$$

The current consumption, I_{VDD}, can be divided into two main terms [PPG2002]

$$I_{VDD} = I_{Id} + I_{Par} \tag{7.37}$$

The former, I_{Id}, is related to the ideal CP behavior and the latter, I_{Par}, accounts for the parasitic effects. I_{Id} can be evaluated by considering that the charge amount delivered to the load, ΔQ, and provided by the power supply, is transferred in each period from one capacitor to another, thus it results to be

$$I_{Id} = (N+1) \cdot \frac{\Delta Q}{T} = (N+1) \cdot I_L \tag{7.38}$$

The same result can be achieved by using the transformation factor of the CP. We amplify the power supply of an N+1 factor, and in an ideal case without wasting energy (i.e., transferring all the power taken from the power supply into the output load), the current sunk by the CP is N+1 times higher than the load current.

Neglecting the contribute of the cross conduction currents, which arises when two adjacent switches provide a conductive path during commutation,[1] current I_{Par} is mainly due to the charging and discharging in each time period, T, of the total parasitic capacitance, C_P,

$$I_{Par} = N \frac{C_P V_{DD}}{T} = \alpha C_{Tot} f V_{DD} \tag{7.39}$$

[1] Cross conduction depends on the CP topology and increases with the switching frequency. It sets a limit to the maximum frequency adoptable.

Thus, from (7.38), (7.39) and using (7.30) to express the total pumping capacitance, the current consumption in the steady state is given by

$$I_{VDD} = \left[(N+1) + \alpha \cdot \frac{N^2}{(N+1) \cdot V_{DD} - V_{Out}} \cdot V_{DD} \right] \cdot I_L \qquad (7.40)$$

It is worth noting that the current consumption, and hence the power consumption, depends neither on the clock frequency, nor on total CP capacitance, but it is linearly related to the current load, I_L.

7.4 Optimized Design of the Charge Pump with a Current Load

The design parameters involved in the design of a CP with a current load are summarized in Table 7.2. Among the unknown entries (evidenced in bold characters) two can be evaluated by using (7.36) and (7.40).

We can start the CP design by finding parameter N, and as for the design of a CP with a pure capacitive load, we can follow two possible strategies minimizing either the area or power consumption.

As shown in Fig. 7.13 (which exemplifies the case in whch V_{DD}=1.35 V, V_{Out}=5 V, f=10 MHz, α=0.1, and I_L=300 μA) area and current consumption are minimized for different values of N.

Table 7.2 CP design parameters

Parameter	Comment
V_{DD}	Technology dependent
I_{VDD}	**Not known a priori**
V_{Out}	Design constraint
I_{Out}	Design constraint
$f=1/T$	Technology dependent
$A_{Tot} \propto C_{Tot}$	**Not known a priori**
N	**Not known a priori**
$\alpha = C_P/C$	Technology dependent

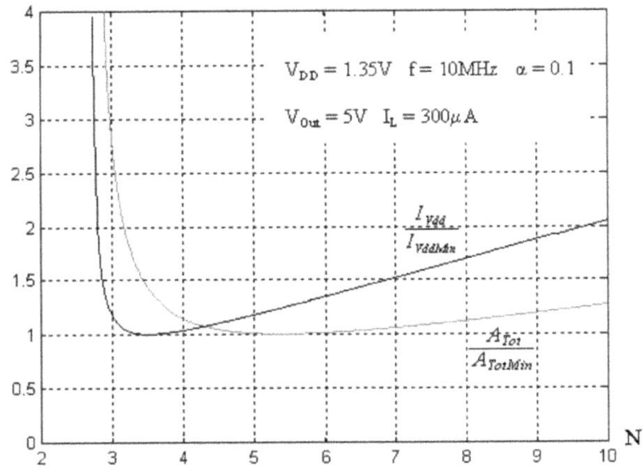

Fig. 7.13 Current consumption and area normalized to their minimum values versus N

7.4.1 Minimizing Area Occupation

To find the optimum N minimizing the silicon area, we must set to zero the derivative of (7.36) with respect to N. On the other hand, if we want to maximize the current provided to the load, we have to set to zero the derivative of current I_L evaluated from (7.30). In both cases, we get

$$\left\{2N\cdot\left[(N+1)\cdot V_{DD}-V_{Out}\right]-N^2\cdot V_{DD}\right\}\cdot\frac{I_L}{f}=0 \tag{7.41}$$

which, solved for N, gives [TA1999]

$$N_{Aop}=2\cdot\left(\frac{V_{Out}}{V_{DD}}-1\right) \tag{7.42}$$

After solving (7.30) for C,

$$C=N\cdot\frac{I_L\cdot T}{(N+1)\cdot V_{DD}-V_{Out}} \tag{7.43}$$

and substituting N_{Aopt} in (7.43), we find the stage capacitance C.

7.4.2 Minimizing Current (Power) Consumption

The optimum N that minimizes current consumption is obtained by setting to zero the derivative of (7.40) with respect to N.

$$1+\alpha\frac{N\cdot\left[(N+2)\cdot V_{DD}-2\cdot V_{Out}\right]}{\left[(N+1)\cdot V_{DD}-V_{Out}\right]^2}\cdot V_{DD}=0 \tag{7.44}$$

Hence, solving for N, we get [26]

$$N_{Iop}=\left(1+\sqrt{\frac{\alpha}{1+\alpha}}\right)\cdot\left(\frac{V_{Out}}{V_{DD}}-1\right) \tag{7.45}$$

Finally, substituting the optimum value of (7.45) in (7.42) we get the required value of C for the optimized design.

7.4.3 Comparison Between the Optimized Design Strategies

In order to compare the two considered design strategies, let us start evaluating the increase in area of the minimum power consumption design, compared to the minimum area design [PPG2002].

From (7.43) and using the values N_{Aop} and N_{Iop}, we obtain the CP total capacitance for minimization of area and power consumption, respectively

$$C_{T,Aop}=4\left(\frac{V_{Out}}{V_{DD}}-1\right)\frac{I_L\cdot T}{V_{DD}} \tag{7.46}$$

$$C_{T,Iop}=\frac{\left(1+\sqrt{\frac{\alpha}{1+\alpha}}\right)^2}{\sqrt{\frac{\alpha}{1+\alpha}}}\left(\frac{V_{Out}}{V_{DD}}-1\right)\frac{I_L\cdot T}{V_{DD}} \tag{7.47}$$

From which the increase of area results to be

$$\frac{C_{T\ Iop}-C_{T\ Aop}}{C_{T\ Aop}}=\frac{\left(1+\sqrt{\frac{\alpha}{1+\alpha}}\right)^2}{4\sqrt{\frac{\alpha}{1+\alpha}}}-1 \tag{7.48}$$

Relationship (7.48), plotted in Fig. 7.14, is a decreasing function of α, and it is about equal to 0.4 and 0.2 (i.e. an increase of 40% and 20% in area) for α equal to 0.1 and 0.2, respectively. For α much lower than 0.1 (not plotted), the minimum power consumption strategy requires a huge amount of silicon area.

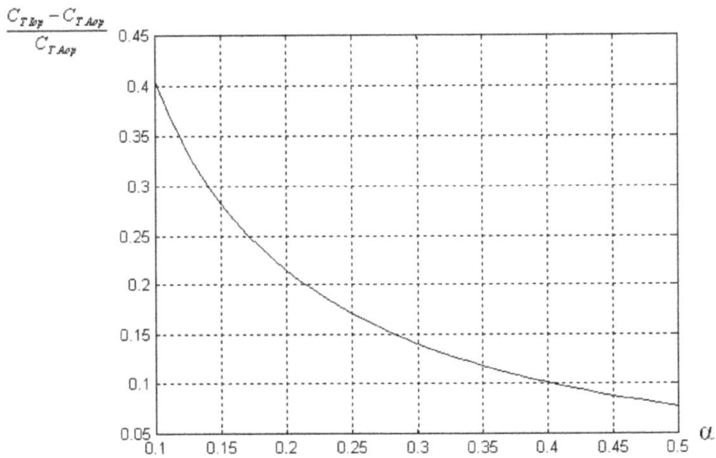

Fig. 7.14 Increment of capacitance (i.e., area) in the case of power consumption optimization with respect the case of area minimization

Let us now evaluate the increase in current consumption incurred by the minimum area design. Substituting (7.42) or (7.45) into (7.40), we find the expression of I_{VDD} for the minimum area and minimum current design strategies

$$I_{VDD,Aop} = \left[2(1+2\alpha)\left(\frac{V_{Out}}{V_{DD}} - 1 \right) + 1 \right] \cdot I_L \tag{7.49}$$

$$I_{VDD,Iop} = \left\{ \left[(1+2\alpha) + 2\sqrt{\alpha+\alpha^2} \right]\left(\frac{V_{Out}}{V_{DD}} - 1 \right) + 1 \right\} \cdot I_L \tag{7.50}$$

The current increase is given by

$$\frac{I_{VDD\ Aop} - I_{VDD\ Iop}}{I_{VDD\ Iop}} = \frac{\left[2(1+2\alpha)\left(\frac{V_{Out}}{V_{DD}} - 1 \right) + 1 \right]}{\left[(1+2\alpha) + 2\sqrt{\alpha+\alpha^2} \right]\left(\frac{V_{Out}}{V_{DD}} - 1 \right) + 1} - 1 \tag{7.51}$$

By inspection of Fig. 7.15, in which (7.51) is plotted, we see that the increment in I_{VDD} is a decreasing function of α and an increasing function of the ratio V_{Out}/V_{DD}. In particular, for an ideal one-stage CP, the current consumption increase is, for the optimum area design, not higher than 20%. On the other hand, for V_{Out}/V_{DD} higher than 4, the current consumption increase is higher than 20% provided that α is lower than 0.15. Similar results that take into consideration CP power efficiency instead of current consumption are found in [PPG2002].

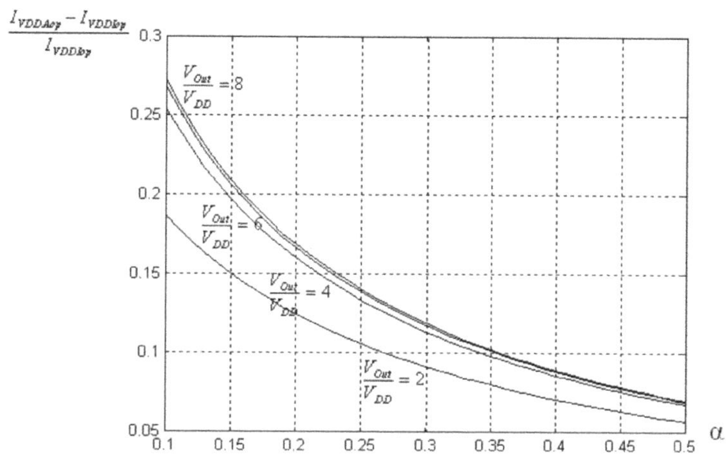

Fig. 7.15 Increment of current consumption (i.e., power consumption) in the case of area minimization with respect the case of minimum current consumption

7.5 Voltages in a CP with a Current Load

7.5.1 Evaluation of Voltages in the Inner Nodes

Since the voltages in the CP nodes are higher than the power supply, to evaluate the voltage stress on each CP component, it is useful to compute the voltage on each CP node and across each switch. At this purpose, let us consider the voltage at the generic node *j*, and analyze its value at the beginning and the end of each half period. Without loss of generality, we can assume that in the first half period the capacitor of the stage *j* (with the clock signal low) is connected together the previous stage through the closed switch S_j,. Hence, during this half period, it shares its charge with the capacitor in the stage (*j*-1).

The voltage at the beginning of this half period is

$$V_{i,start-left-sharing} = j\left(V_{DD} - \Delta V\right) \tag{7.52}$$

and at the end of the half period it is

$$V_{i,end-left-sharing} = j\left(V_{DD} - \Delta V\right) + \Delta V \tag{7.53}$$

During the subsequent half period, when switch S_j is open and switch S_{j+1} is closed, the clock signal driving the stage *j* is high (i.e., there is charge sharing with the capacitor in the stage *j*+1), and the voltage at the beginning of the half period is

$$V_{i,start-right-sharing} = j\left(V_{DD} - \Delta V\right) + \Delta V + V_{DD} \tag{7.54}$$

and at the end of the half period it is

$$V_{i,end-right-sharing} = j\left(V_{DD} - \Delta V\right) + V_{DD} \quad (7.55)$$

7.5.2 Behavior of the Voltages Across the Switches

Let us consider the switch S_j, which interconnects capacitors at nodes j-1 and j, and evaluate the voltage across it, $V_{j,j\text{-}1}=V_j-V_{j\text{-}1}$. This voltage behavior is plotted in Fig. 7.16.

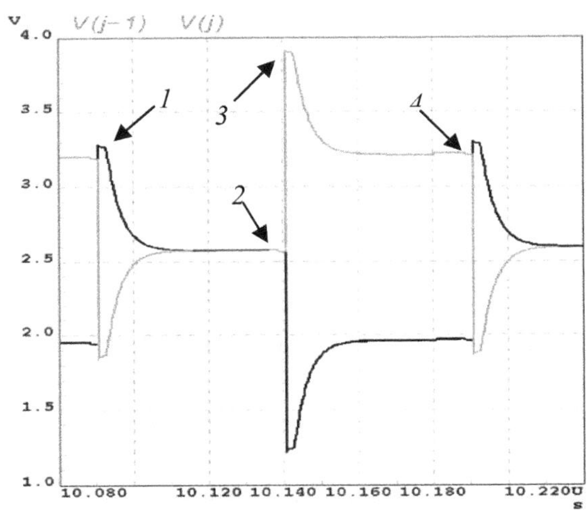

Fig. 7.16 Voltage across switch S_j

At the first half period end S_j is closed and consequently the voltage across the switch is zero (point 2 in Fig. 7.16). At the beginning of the subsequent half period (S_j open) the clock signal changes, and the voltages at nodes j and j-1 fall and rise, respectively, by the clock amplitude. Hence, the voltage across the switch results to be $V_{j,j\text{-}1}=2V_{DD}$ (point 3 in Fig. 7.16). At the end of this half period, the capacitor at node j ($j-1$) gives (receives) charge ΔQ. Thus, as seen from point 4 in Fig. 7.16, the voltage at node j ($j-1$) reduces (increases) of ΔV, and $V_{j,j\text{-}1}=2V_{DD}-2\Delta V$. Finally, when S_j is closed again and the charge transfer between capacitors at nodes j-1 and j begins, the clock signal changes, and the voltage across the switch is $V_{j,j\text{-}1}=-2\Delta V$ (point 1 in Fig. 7.16).

These voltage values are equal for all the CP switches, and are summarized in Table 7.3.

It is worth noting that the voltage across the switches depends on the clock amplitude (assumed, as usual, equal to the power supply) and on the voltage step, ΔV, which is given by

$$\Delta V = \frac{(N+1)V_{DD} - V_{Out}}{N} \tag{7.56}$$

After substituting the appropriate N defined for each optimized design strategy in (7.56), we find the ΔV expressions summarized in Table 7.4, that combined with the results in Table 7.3, allow to accurately define the voltage across the switches.

Table 7.3 Switch voltages versus time

Time	Switch voltages $V_{j,j-1}$
Beginning S_j closed	$-2\Delta V$
End S_j closed	0 V
Beginning S_j open	$2V_{DD}$
End S_j open	$2V_{DD}-2\Delta V$

Table 7.4 Expression of ΔV in Table 7.3

CP load	Optimized design	ΔV
Capacitive	*Minimum Area*	$V_{DD}/4$
	Minimum Charge Consumption	$\frac{1+4\alpha}{2+7\alpha} \cdot V_{DD} \cong \frac{V_{DD}}{2}$
Capacitive and Current	*Minimum Area*	$V_{DD}/2$
	Minimum Current Consumption	$\left[\sqrt{\alpha(1+\alpha)} - \alpha\right] \cdot V_{DD}$

7.6 Charge Pump Topologies

In the previous sections we have considered CP with ideal switches. However, real CPs differ substantially for the way in which switches are implemented.

7.6.1 The Dickson Charge Pump

The IC realization of a CP was demonstrated for the first time by Dickson in 1976 [D1976]. Like previous CPs adopted in discrete implementation, such as the Crockcroft and Walton topology proposed in 1932 [CW1932], the **Dickson CP** makes use of diodes instead of switches, as illustrated in Fig. 7.17. A CP topologically similar to the Dickson but implemented with MOS diodes, as shown in Fig. 7.18, was also implemented on silicon [WGM1989].

The main advantage provided by diodes is the absence of the switch control signals. The main drawback is the reduction of the CP output voltage. Indeed, when a diode is forward biased (i.e., when the corresponding switch must be closed), it causes a voltage loss equal to the diode threshold voltage, V_γ, which reduces the output voltage of a factor $(N+1)V_\gamma$.

This reduction is particularly critical under low power supplies, and determines also a loss in the CP power efficiency. For this reason, the Dickson CP is not a practical topology for present applications.

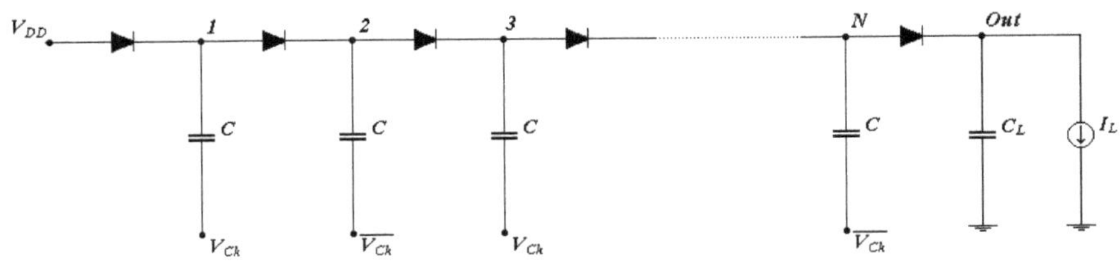

Fig. 7.17 Dickson CP

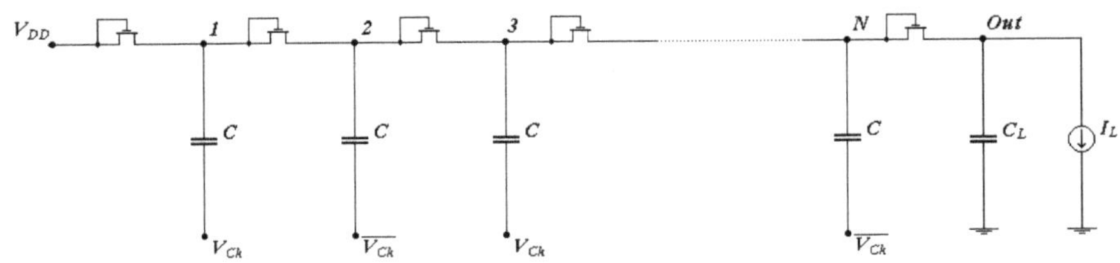

Fig. 7.18 CP with MOS diodes

7.6.2 The Bootstrap Charge Pump

An attractive and widely adopted implementation of the CP switches was presented in [UKA1992], [AKU1994] and is called **bootstrap CP**. The associated CP topology is shown in Fig. 7.19, and the clock signals are plotted in Fig. 7.20.

Even if the implementation of a switch is conceptually simple, for example through a simple MOS transistor (or a transmission gate), in a CP the voltage at the switch terminals are higher than the power supply. As a consequence, the MOS transistors used to implement the CP switches, have to be switched ON by applying suitable gate voltages higher than their source terminal voltages. Specifically, considering the CP in Fig. 7.19, the required high gate voltages are obtained for each stage thanks to a bootstrap circuit, which is realized by adding (for each stage) another capacitor and MOS transistor.

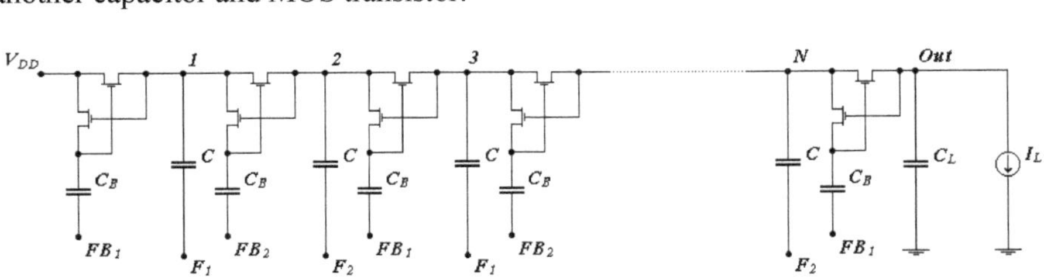

Fig. 7.19 Bootstrap CP

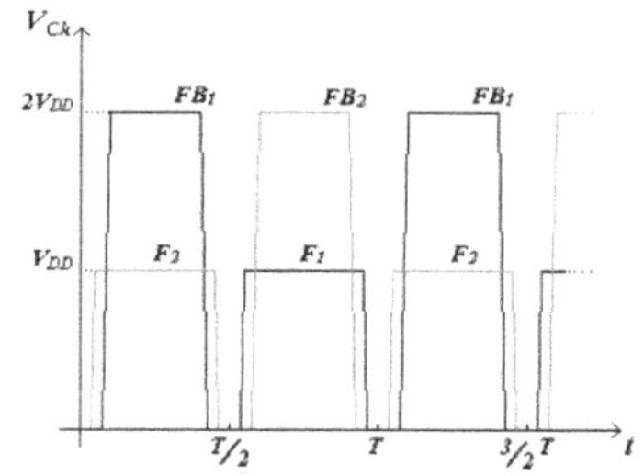

Fig. 7.20 Bootstrap CP clock (F_i) and control (F_{Bi}) signals

To understand the working principle of a bootstrap CP, let us analyze the generic stage (plus a capacitor) depicted in Fig. 7.21.

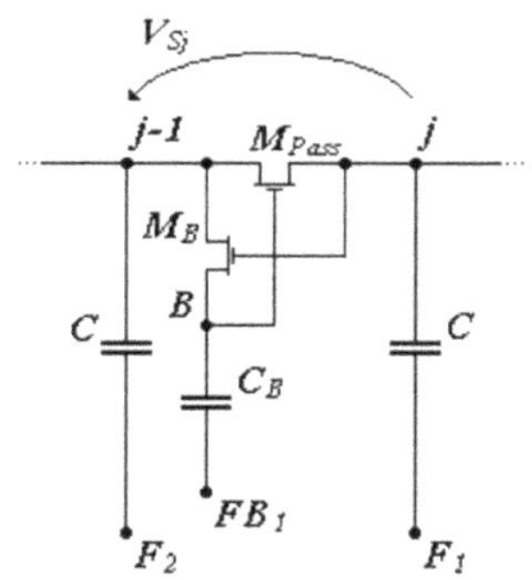

Fig. 7.21 Generic stage (plus a capacitor) of a bootstrap CP

During the half period in which there is no charge transfer (i.e., M_{PASS} is open), signal F_{B1} is low and the added transistor M_B is closed, since the voltage at its gate is higher than a threshold voltage ($V_{GS,MB} = 2V_{DD}$) than the other two nodes. The capacitor C_B is then charged up to a voltage equal to $V_{j\text{-}1}=(j-1)V_{DD}-(j-2)\Delta V$, see (7.51).

During the subsequent half period, the clock signals F_1 and F_2 change their value, and after another small time slot (see Fig. 7.20), F_{B1} goes high to $2V_{DD}$. Now the transistor M_B is open meanwhile the pass transistor M_{PASS} is closed. Indeed, the gate voltage of M_{PASS} is equal to $(j+1)V_{DD}-(j-2)\Delta V$, which is higher than the voltage at of its source (i.e., the node j) by $V_{DD}+2\Delta V$. Besides, during this half period the gate-source of M_{PASS} is sufficiently high to keep this transistor closed (the gate-source voltage is always not lower than $V_{DD}+\Delta V$).

It is apparent that the ingenious behavior of this topology is obtained at the price of a more complex clocking and control section (requiring four phases and $2V_{DD}$ amplitude).

7.6.3 Double Charge Pumps

The **double CP** has been conceived to reduce the output ripple by using the same total CP capacitance, C_T. As depicted in Fig. 7.22, each half part, which has a total capacitance $C_T/2$, feeds the load in a different half period [KKJ1996]. Hence, the charge ΔQ pumped at the output is divided into two equal parts, each for half period. The output voltage (7.30) is the same as the simple CP, but the ripple is now

$$V_r = \frac{1}{2}\frac{\Delta Q}{C_L} = \frac{I_L \cdot T}{2 \cdot C_L} \tag{7.57}$$

Of course, by properly implementing the switches we realize a **double Dickson CP** or a **double bootstrap CP**. The latter is shown in Fig. 7.23.

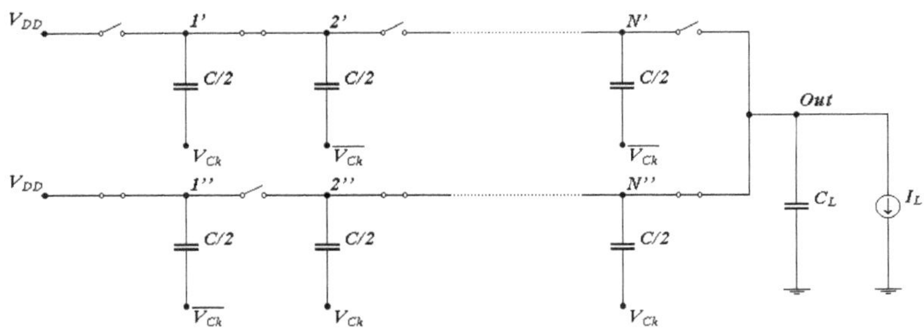

Fig. 7.22 Simplified schematic of a double CP

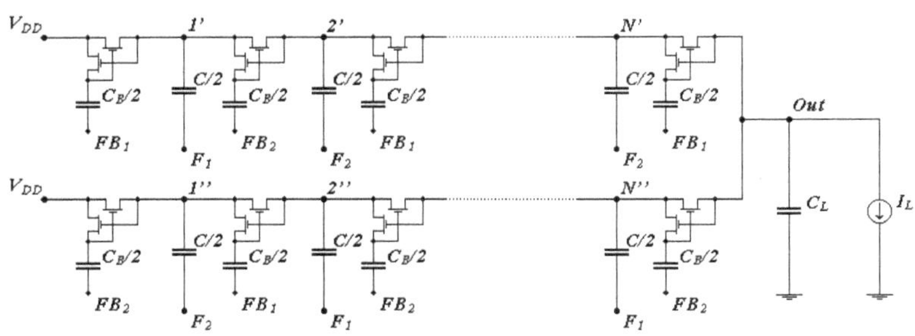

Fig. 7.23 Double bootstrap CP

A very interesting double CP that only recently is gaining popularity, was originally proposed in [GP1994, GP1996], and is shown in Fig. 7.24. This topology, often referred to as **latched CP**, because it includes a latch in each stage, is suited for very high clock frequencies. Indeed, unlike the bootstrap CP, the latched CP needs only a two-phase clock. This topology is suitable for full integration in PMLCD drivers with only C_L as external discrete capacitor.

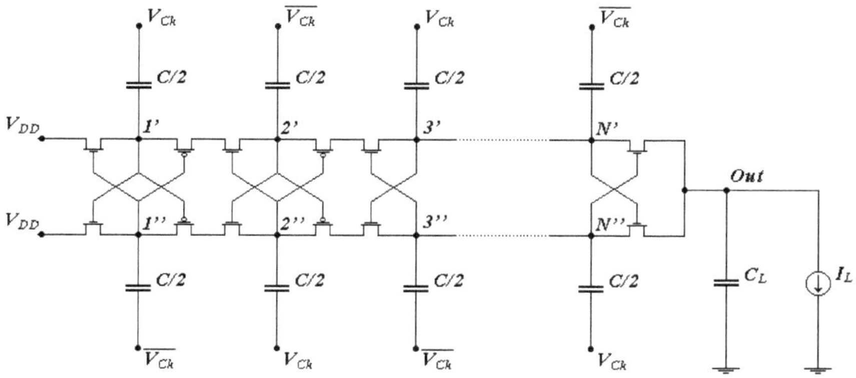

Fig. 7.24 Latched CP

7.6.4 Series-Parallel Charge Pumps

Another topology that reminds those originally adopted for discrete implementations [CW1932], is the **series-parallel CP**. It has been seldom used in an IC implementation [MS1998], since it was believed as inefficient for this use. However, only recently an in-depth analysis to evaluate its suitability for ICs has been carried out [SJO2001, CGT2007].

A two stage series-parallel CP is shown in Fig. 7.25. The peculiarity of this CP is the parallel charging of all the capacitors to the power supply, V_{DD}, during the first half period (i.e., when switches P_i and P_i' are closed and switches S_i are open), and the series connection of all capacitors in the other half period (i.e., when switches P_i and P_i' are open and switches S_i are closed).

The main drawback of this topology is caused by the parasitic capacitances which affect the behavior and performance more than the other topologies. Moreover, another critical aspect concerns the switches implementation.

AMLCDs require higher voltages and higher energy than PMLCDs, therefore the pumping capacitors (other than the C_L) are usually all external. Since this topology has the lowest number of capacitors it is suitable for AMLCD drivers in order to minimize the number of pins.

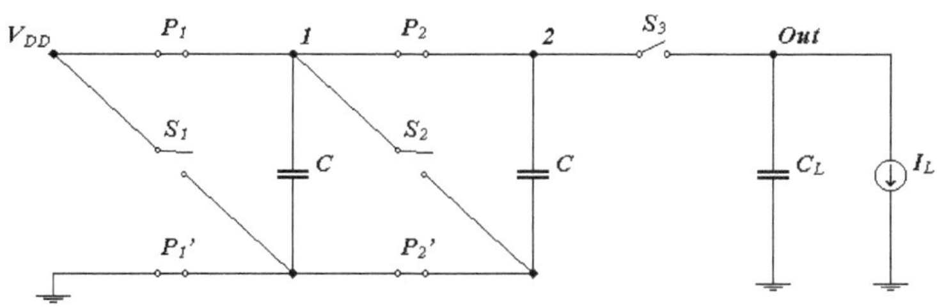

Fig. 7.25 Series-parallel CP

7.6.5 Charge Pumps with Adaptive Number of Stages

In many IC applications more than one CP with different number of stages is required. Thus, when the CPs are used in non overlapping time periods, it can be useful to implement only one CP which dynamically adapts its number of stages. Examples of CPs with adaptive stages are shown in [TTT2002, PPG2006].

More specifically, a solution that switches from a two-stage to a four-stage topology is proposed in [TTT2002]. In [PPG2006], the number of stages is dynamically chosen by rearranging the whole set of capacitors so that the required output voltage could be reached by maximizing the CP's efficiency. Hence, for a defined number of stages, the whole CP capacitance (i.e., silicon area used by the CP) always remains equal.

A version of the topology proposed in [PPG2006], which adapts its number of stages from 1 to 3, is shown in Fig. 7.26. Without loss generality, the topology makes use diodes as switches, but changes can be simply implemented if switches instead of diodes are used.

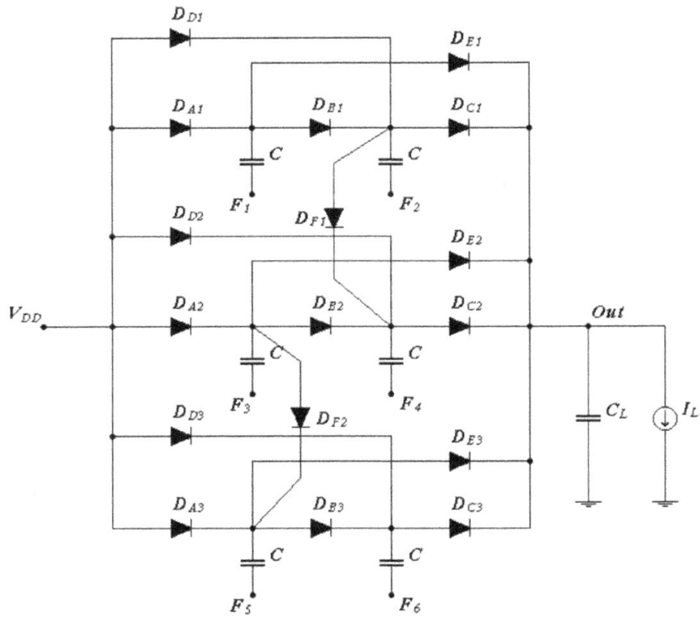

Fig. 7.26 CP with adaptive stages

It is worth noting that to provide adaptability, each capacitor is connected through a diode both to the power supply (D_{Ai} and D_{Di}) and to the output node (D_{Ci} and D_{Ei}). Moreover, there are diodes that connect couple of capacitors (D_{Bi} and D_{Fi}).

To understand the behavior of the CP, let us consider each single case starting from the single-stage one. In this configuration, all the capacitors have to be connected in parallel to form a single capacitor. In other words, this means that all the capacitors have to be driven by the same phase signal and all the diodes connecting couple of capacitors (D_{Bi} and D_{Fi}) are always reverse biased.

In the case in which a two-stage topology is needed, the CP in Fig. 7.26 works like three parallel two-stage topologies. Hence, it can be simply inferred that the set of phases F_1, F_3 and F_5 must be complementary to the other set F_2, F_4 and F_6. Under this condition, the diodes D_{Fi} are always reverse biased. Moreover, since at steady state $V_{2i}>V_{DD}$ and $V_{2i+1}<V_{Out}$, the diodes D_{Di} and D_{Ei} are also reverse biased.

Finally, a three-stage CP can be obtained by configuring the CP like two parallel three-stage CPs. This is obtained by driving the input phases F_1, F_3, F_4 and F_6 together, and the input phases F_2 and F_5 by the complementary phase. Thus, the two three-stage CPs have the two main paths constituted by internal nodes 1, 2 and 4 with the set of phases F_1, F_2 and F_4, and internal nodes 3, 5 and 6 with the set of phases F_3, F_5 and F_6. It can be simply verified that at the steady state the diodes D_{Di}, D_{Ei}, D_{C1}, D_{B2} and D_{A3} are reverse biased.

The phase arrangement for the three cases is summarized in Table 7.5.

Table 7.5 Clock phases for the CP in Fig. 7.26

	One-stage CP	**Two-stage CP**	**Three-stage CP**
FX	F_1, F_2, F_3, F_4, F_5, F_6	F_1, F_3, F_5	F_1, F_3, F_4, F_6
FN	-	F_2, F_4, F_6	F_2, F_5

7.7 Concluding Remarks

Following the suggestions of the authors of the book, derived from their own working practice, we report some additional guidelines particularly oriented to the design of charge pumps for LCD drivers.

For those cases in which the charge pump capacitors cannot be fully integrated, the main design target is to minimize the number of external pins required, and hence the number of stages N. Therefore, we cannot follow any of the optimized strategies previously presented. Instead, once fixed the minimum N, we find the capacitors value from (7.30) by using an adequate switching frequency as low as possible to minimize the losses due to cross conduction but avoiding acoustic noise.

On the other hand, when the integration of the charge pump capacitors is possible, as in small size LCD drivers, we could minimize the silicon area as shown in Section 7.4.1. However, in order to reduce also the current consumption (fundamental objective in portable applications), a trade-off between energy consumption and area must be met. This leads to a number of stages value comprised between that deriving from (7.42) (minimum silicon area) and that obtained by minimizing the current consumption. It is worth noting that to evaluate the current consumption, the cross conduction contribute must be included into (7.40) because in general the switching frequency adopted in fully integrated charge pumps is as high as possible. Since cross conduction currents increase with the switching frequency, a good compromise from a practical point of view is to set that particular clock frequency for which the losses due to cross conduction are equal to those due to parasitic capacitances.

Finally, it must be observed that the chapter deals with the charge pump core only. In practical applications this core is embedded within a negative feedback loop (as shown for instance in Fig. 4.12), which allows the regulation of the output voltage by acting on the switching frequency, thus changing it according to the supply voltage and load values.

References

[AKU1994] S. Atsumi, M. Kuriyama, A. Umezawa, H. Banba, K. Naruke, S. Yamada, Y. Ohshima, M. Oshikiri, Y. Hiura, T. Yamane, and K. Yoshikawa, "A 16-Mb Flash EEPROM with a New Self-Data-Refresh Scheme for a Sector Erase Operation," *IEEE Journal of Solid-State Circuits*, Vol. 29, No. 4, pp. 461–468, 1994.

[BN1998] G. den Besten and B. Nauta, "Embedded 5 V-to-3.3 V Voltage Regulator for Supplying Digital IC's in 3.3 V CMOS Technology," *IEEE Journal of Solid-State Circuits*, Vol. 33, No. 7, pp. 956–962, 1998.

[CGO1999] P. Cappelletti, C. Golla, P. Olivo, and E. Zanoni, *Flash Memories*, Kluwer Academic Publishers, 1999.

[CGT2007] A. Cabrini, L. Gobbi, and G. Torelli, "Voltage Gain Analysis of Integrated Fibonacci-Like Charge Pumps for Low Power Applications," *IEEE Transactions on Circuits and Systems - Part II*, Vol. 54, No. 11, pp. 929–933, 2007.

[CT1991] R. Castello and L. Tomasini, "1.5-V High-Performance SC Filters in BiCMOS Technology," *IEEE Journal of Solid-State Circuits*, Vol. 26, No.7, pp. 930–936, 1991.

[CW1932] J. Crockcroft and E. Walton, "Production of High Velocity Positive Ions," *Proceedings of the Royal Society, A,* Vol. 136, pp. 619–630, 1932.

[D1976] J. Dickson, "On-Chip High-Voltage Generation MNOS Integrated Circuits Using an Improved Voltage Multiplier Technique," *IEEE Journal of Solid-State Circuits*, Vol. 11, No. 3, pp. 374–378, 1976.

[DD1998] T. Duisters and E. Dijkmans, "A -90-dB THD Rail-to-Rail Input Opamp Using a New Local Charge Pump in CMOS," *IEEE Journal of Solid-State Circuits*, Vol. 33, No. 7, pp. 947–955, 1998.

[DP1996] G. Di Cataldo and G. Palumbo, "Optimized Design of an N-th Order Dickson Voltage Multiplier," *IEEE Transactions on Circuits and Systems - Part I*, Vol. 43, No. 5, pp. 414–418, 1996.

[GP1994] R. Gariboldi and F. Pulvirenti, "A Monolithic Quad Line Driver for Industrial Application," *IEEE Journal of Solid-State Circuits*, Vol. 29, No. 9, pp. 957–962, 1994.

[GP1996] R. Gariboldi and F. Pulvirenti, "A 70 mΩ Intelligent High Side Switch with Full Diagnostics," *IEEE Journal of Solid-State Circuits*, Vol. 31, No. 7, pp. 915–923, 1996.

[H1990] S. Hobrecht, "An Intelligent BiCMOS/DMOS Quad 1-A High-Side Switch," *IEEE Journal of Solid-State Circuits*, Vol. 25, No. 6, pp. 1395–1402, 1990.

[IOS2001] T. Ishii, K. Oshima, H. Sato, T. Yoshitake, A. Sato, S. Kubono, K. Manita, T. Nakayama, and A. Hosogane, "A 126.6-mm^2 AND-Type 512-Mb Flash Memory with 1.8-V Power Supply," *IEEE Journal of Solid-State Circuits*, Vol. 36, No. 11, pp. 1707–1711, 2001.

[JNH1992] T. Jinbo, H. Nakata, K. Hashimoto, T. Watanabe, K. Ninomiya, T. Urai, M. Koike, T. Sato, N. Kodama, K. Oyama, and T. Okazawa, "A 5-V-Only 16-

Mb Flash Memory with Sector Erase Mode," *IEEE Journal of Solid-State Circuits*, Vol. 27, No. 11, pp. 1547–1553, 1992.

[KKJ1996] G T. Kawahara, T. Kobayashi, Y. Jyouno, S.-I. Saeki, N. Miyamoto, T. Adachi, M. Kato, A. Sato, J. Yugami, H. Kume, and K. Kimura, "Bit-Line Clamped Sensing Multiplex and Accurate High Voltage Generator for Quarter-Micron Flash Memories," *IEEE Journal of Solid-State Circuits*, Vol. 31, No. 11, pp. 1590–1599, 1996.

[MC2001] K.-S. Min and J.-Y. Chung, "A Fast Pump-Down V_{BB} Generator for Sub-1.5-V DRAMs," *IEEE Journal of Solid-State Circuits*, Vol. 36, No. 7, pp. 1154–1157, 2001.

[MS1998] H. Morimura and N. Shibata, "A Step-Down Boosted-Wordline Scheme for 1-V Battery-Operated Fast SRAM's," *IEEE Journal of Solid-State Circuits*, Vol. 33, No. 8, pp. 1220–1227, 1998.

[NGN2001] A. Novo, A. Gerosa, and A. Neviani, "A Sub-Micron CMOS Programmable CP for Implantable Pacemaker," *Analog Integrated Circuits and Signal Processing*, Vol. 27, No. 3, pp. 209–215, 2001.

[PBB2000] G. Palumbo, N. Barniol, and M. Bethaoui, "Improved Behavioral and Design Model of an N-th Order Charge Pump," *IEEE Transactions on Circuits and Systems - Part I*, Vol. 47, No. 2, pp. 264–268, 2000.

[PP2006] G. Palumbo and D. Pappalardo, "Charge Pump Circuits with only Capacitive Loads: Optimized Design," *IEEE Transaction on Circuits and Systems - Part II*, Vol. 53, No. 2, pp. 128–132, February 2006.

[PPG2002] G. Palumbo, D. Pappalardo, M. Gaibotti, "Charge Pump Circuits: Power Consumption Optimization," *IEEE Transactions on Circuits and Systems - Part I*, Vol. 49, No. 11, pp. 1535–1542, 2002.

[PPG2006] G. Palumbo, D. Pappalardo, and M. Gaibotti, "Charge Pump with Adaptive Stages for Non-Volatile Memories," *IEEE Proceedings of Circuits, Devices and Systems*, Vol. 153, No. 2, pp. 136–142, 2006.

[SJO2001] J. Starzyk, Y. Jan, and F. Oiu, "A DC-DC Charge Pump Design Based on Voltage Doublers," *IEEE Transactions on Circuits and Systems - Part I*, Vol. 48, No. 3, pp. 350–359, 2001.

[SK2008] F. Su and W.-H. Ki, "Component-Efficient Multiphase Switched-Capacitor DC-DC Converter with Configurable Conversion Ratios for LCD Driver Applications," *IEEE Transaction On Circuits and Systems - Part II*, Vol. 55, No. 8, pp. 753–757, 2008.

[TA1999] T. Tanzawa and S. Atsumi, "Optimization of Word-Line Booster Circuits for Low-Voltage Flash Memories," *IEEE Journal of Solid-State Circuits*, Vol. 34, No. 8, pp. 1091–1098, 1999.

[TT1997] T. Tanzawa and T. Tanaka, "A Dynamic Analysis of the Dickson Charge Pump Circuit," *IEEE Journal of Solid-State Circuits*, Vol. 32, No. 8, pp. 1231–1240, 1997.

[TTT2002] T. Tanzawa, T. Tanaka, K. Takeuchi, and H. Nakamura, "Circuit Techniques for a 1.8-V-Only NAND Flash Memory," *IEEE Journal of Solid-State Circuits*, Vol. 37, No. 1, pp. 84–89, 2002.

[UKA1992] A. Umezawa, M. Kuriyama, S. Atsumi, H. Banba, K. Imamiya, K. Naruke, S. Yamada, E. Obi, M. Oshikiri, T. Suzuki, M. Wada, and S. Tanaka, "A 5-V-Only Operation 0.6-μm Flash EEPROM with Row Decoder Scheme in Triple-Well Structure," *IEEE Journal of Solid-State Circuits*, Vol. 27, No. 11, pp. 1540–1545, 1992.

[WGM1989] J. Witters, G. Groeseneken, and H. Maes, "Analysis and Modeling of On-Chip High-Voltage Generator Circuits for Use in EEPROM Circuits," *IEEE Journal of Solid-State Circuits*, Vol. 24, No. 5, pp. 1372–1380, 1989.

[WVK1992] S.L. Wong, S. Venkitasubrahmanian, M.J. Kim, and J.C. Young, "Design of a 60-V 10-A Intelligent Power Switch Using Standard Cells," *IEEE Journal of Solid-State Circuits*, Vol. SC-27, No. 3, pp. 429–432, 1992.

[YKC2003] T.-R. Ying, W.-H. Ki, and M. Chan, "Area-efficient CMOS Charge Pumps for LCD drivers," *IEEE Journal of Solid-State Circuits*, Vol. 38, No. 10, pp. 1721–1725, 2003.

[ZLD2001] M. Zhang, N. Llaser, and F. Devos, "Multi-Value Voltage-to-Voltage Converter Using a Multi-Stage Symmetrical Charge Pump for On-Chip EEPROM Programming," *Analog Integrated Circuits and Signal Processing*, Vol. 27, No. 1–2, pp. 85–95, 2001.

Appendix A

Flat Panel Displays

In the last years, the technological advances enabled flat panel displays (FPDs) to be introduced in a wide range of applications and devices and have enabled applications otherwise impossible, which are limited only by the fantasy of the designer.

A.1 Technologies and Classification

Other than the LCD, the subject of this book, many other types of flat display technologies have been devised such as: PDP (Plasma Display Panel), LED (Light Emitting Diode) OLED (Organic LED), VFD (Vacuum Fluorescent Display), TFELD (Thin Film Electroluminescent Display), FED (Field Emission Display), DMD (Digital Micro-mirror Machine), PALC (Plasma Addressed LCD), Bistable Displays (including the Electrophoretic and the Nematic Bistable).

According to the application and price-to-performance ratio, the most suitable technology has to be chosen. In the following we shall briefly discuss the display technologies that have reached a sufficient development and commercial application.

In such a multitude of applications and display types, it is useful to operate a first classification in: **Direct view** displays and **Projection** displays.

In the first case the name is self explaining, in the second one the image can be projected on a screen either for home/office application or car/airplane application. The image can be also projected on the human eye retina by microdisplays.

Then, a second classification, related to the display technologies, must be performed between: **Emissive** displays and **Nonemissive** displays

In the first case, the device emits light from each pixel. In the second case, the device modulates light (from a light source) by means of absorption, reflection, refraction, and/or scattering mechanisms.

Examples of emissive displays are: PDPs, LEDs, OLEDs, FEDs, VFDs (and CRTs) . As light modulator display we mention: LCDs and the Bistable Displays.

Generally speaking the choice between light modulator and emissive display depends upon the ambient light. The emissive displays have both wide viewing angle and high color saturation but, despite the high brightness, in case of portable equipments for outdoor application they could not compete against the 120k cd/m^2 of the sun light, originating the so called "wash out" phenomena.

A.2 Emissive Displays

Emissive displays do not require backlighting, thereby allowing a substantial reduction of power consumption. The process of luminescence, responsible of the light emission, is due to the electron/hole recombination. Different mechanisms generating the electron/hole couple can be found in emissive displays. If the cause is an UV wavelength irradiating a phosphor layer, the phenomenon is called "photoluminescence" or "phosphor-luminescence". If the electron/hole couple is generated by a p-n junction, it is called "electroluminescence". Finally, it is called "cathode luminescence" if the generation is caused by an electron beam.

A.2.1 LEDs

Light-emitting diodes are made up of crystalline semiconductors. Since the dimension of a single LED pixel is quite large (several millimeters), LEDs are used for large displays for outdoor signaling (e.g., traffic signals) with up to 100-in. and millions of LED pixels. Other than for displays, LEDs are used to replace the cold cathode fluorescent lamps (CCFL) as backlight modules for LCDs and for general lighting.

A.2.2 OLEDs

Among the emissive displays, the OLED technology has been the most studied initially at university level, then in company research laboratories, and is finding now more and more applications. The electroluminescence of organic materials was discovered by Ching Tang in 1979, a researcher of Eastman Kodak, who was experimenting new low cost solar cells, and, irony of the risen ones, he got the opposite: his organic material was able to convert electricity in light emission with a quantum efficiency as high as never seen for an organic material. Important development steps have been: the invention of small molecules heterostructure OLED at Eastman Kodak [TV1987] and in 1990 the conjugated polymers OLED by J.H. Burroughes of CDT, Cambridge Display Technologies [BBB1990].

At the beginning many issues had to be solved. Among them, life of organic layers was very short (indeed the emission was drastically reduced as a function of time), high current consumption, not enough quantum efficiency.

After many years of studies and starting from low-end applications (e.g. displays for electric shavers), the strong potentiality of this technology, mainly in terms of color saturation and viewing angle, is giving the expected results.

Passive-matrix OLED displays were the first to be marketed. In 2005, Samsung demonstrated a 40-in. prototype OLED TV. Sony put in commerce the world's first OLED TV, the XEL-1, in December 2007. In 2008, Samsung released the SDI

AMOLED (Active Matrix OLED, 31-in. diagonal with 1980×1080 pixel resolution) and Sony released an 11-in. OLED TV of 70 000 hours life.

The introduction of Active Matrix is further boosting the performance in terms of image quality and power consumption reduction. There is a growing confidence that OLED Display developers are mastering the technology, and it is time to compete with LCD technology, by leveraging on creative design. OLED displays are expected to compete with the LCD in practically all current applications. See the related "Further Reading" section at the end of this appendix for more details.

As far as the driving technique is concerned, the substantial difference between LCDs and OLEDs is that the former are voltage driven and the latter are *current driven.* Figure A.1 shows the simplest circuit diagram of an Active Matrix OLED pixel with the pass transistor M_1 (enabled by the Gate line) and the current-source transistor M_2 that drives the OLED with the appropriate current. The hold effect between two subsequent addressing is performed by capacitor C_S that keeps the gate-source voltage of M_2 constant.

The main problem of this scheme is due to the threshold voltage mismatches in the current source transistors that cause brightness variations for the same source voltage. At this purpose, more efficient driving schemes have been developed, see for instance [LEE2002, YYX2003].

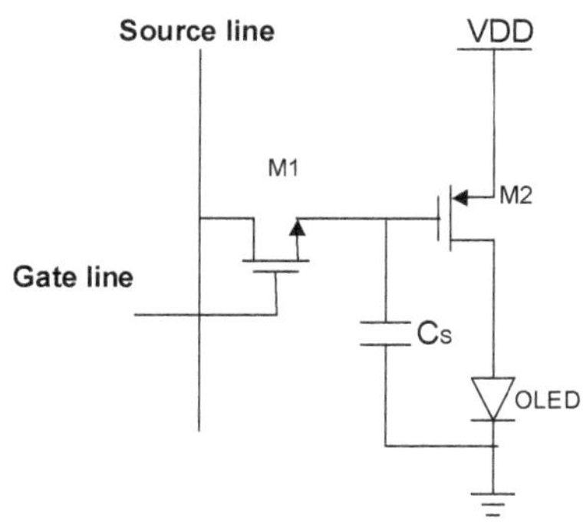

Fig. A.1 Schematic of an AMOLED pixel with addressing circuit

A.2.3 Plasma Displays

The photoluminescence is obtained by high voltage drive applied on a small cavity, as small as a RGB sub-pixel dimension, containing a gas at low pressure (neon, xenon). The gas discharge generates UV irradiation exciting the sub-pixel phosphorus layer. Therefore, the sub-pixel cell cannot be too much small, and the glass making the cell too much thin, otherwise cannot withstand with atmospheric pressure.

First market applications, display banking and military, were introduced in 1985 and in 1993 it was developed a three electrode with barrier rib isolation for color displays [SWN1993]. The first color PDP was introduced by Fujitsu in 1992 [UH2003].

The high cost of electronic drive (because of the high voltage) and a relatively large sub-pixel dimension, are characteristics compatible to large size panel for public display or home theater, where the diagonal is equal or larger than 42 in. In December 2004, Samsung demonstrated the first 102 in. PDP display with a 1920×1080 pixel resolution [S2005]. In 2008, Panasonic demonstrated a 150-in. PDP TV (4096×2160 pixel resolution).

A.3 Nonemissive Displays

Apart form LCDs the most important categories of light modulator displays are:
- DMDs (high performance for projection display)
- Bistable Displays (new developments also on plastic substrate)

A.3.1 DMDs

The DMD (digital micro mirror machine) was developed by Texas Instruments in 1987, even if the idea of such a MEM (microelectronic machine) was conceived since the beginning of the seventies. The principle is based on a metallic micro-mirror that, according to its position, can efficiently reflect (or not) the light.

Projectors have been designed and sold on the market since 1996. In 2008 the company Microvision Inc. has showed a projector prototype using three lasers as light source with a MEM scanning mirror imager to achieve a WVGA resolution.

A.3.2 Bistable Displays

The multiplex driving of PMLCD has the disadvantage to decrease the contrast ratio because of the phenomena known as "frame response". The AM technique is an artifact to transform a monostable display (the stable state is without electrical field applied) into a bistable display ("dark" and "bright" states are both stable, and electrical pulse is needed only to change the stable state).
A bistable display has therefore the great advantages:
- to give an optical stable image, without any flicker, cross talk and with contrast ratio not degraded by the frame response
- to consume electrical power only when changing the state

Several bistable technologies have been developed:
- **Electrophoretic** [OOY1973], and its evolution to micro-encapsulation electrophoretic by E-Ink Corporation and NOK [DCA1998, NKK1998].
- **Cholesteric Liquid Crystal** [GWK1973, YWC1994].
- **Surface Stabilized Ferroelectric LC** (SSFLC), [CL1980].
- **Bulk Bistable Twisted Nematic** (360° BTN), [BH1980].
- **Surface Nematic Bistable**: G. Durand of Paris Orsay University (1988), a student of the Nobel Prize professor DeGennes. Further developed by G. Durand with R. Barbieri and M. Giocondo in 1991 and 1992, University of Catanzaro [BD1991, BGD1992].
- **Surface Controlled Bistable Nematic** (BiNem) by I. Dozov, M. Nobili, G. Durand, Martinot-Lagarde (1997) of Nemoptic [DND1997, DMP1997].
- **Zenithal Bistable Display** (ZBD): G.P. Bryan-Brown (1997) [BBJ1997, WBB2000].

In the case of Cholesteric LCDs, the planar texture reflects strongly one of the circular polarizations (white state, Fig. A.2a), whilst focal conic texture (Fig A.2b), consisting of small domains separated by defect lines, transmits the light and gives a dark state due to the light absorption on the back plate

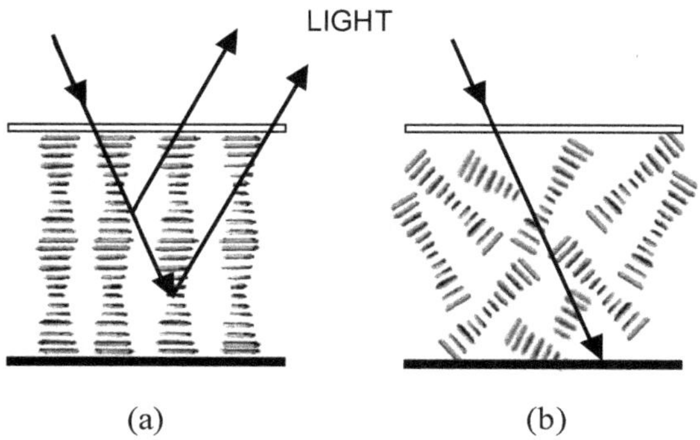

Fig. A.2 Cholesteric LCD cell: (**a**) white state and (**b**) black state

Bistable technologies are becoming the preferred ones for **flexible displays**, manufactured with a plastic substrate. The most important LCD manufacturers are also entering in competition realizing various prototypes: 14.3-in. color electrophoretic display by Samsung, 11.5-in. UXGA monochrome AMEPD (active matrix electrophoretic display) by LG Display, 127-ppi color rollable display (radius of 6 mm) by Polymer's Vision, and 13.5-in. 385-ppi electronic newspaper by Epson. Apart the inherent flexibility, the plastic substrate is used for the so called "Polymer Electronics", where the active matrix TFT (thin film transistor) and the driving circuits can be manufactured using organic material, much less expensive and easy to process, of course where circuit performances can allow it.

In 2006, Motorola used for the first time a bistable technology for a high-volume product, the Motofone F3 cellular phone. It was based on the **electrophoresis**, that is, the motion of a charged pigment particle through a liquid medium due to an applied electric field. Also for these subjects see the related "Further Reading" section at the end of this chapter.

References

[AVH2002] W.F. Aerts, S. Verlaak, and. Heremans, "Design of Organic Pixel Addressing Circuit for an Active-Matrix OLED Display," *IEEE Transaction on Electron Devices*, Vol. 49, No. 12, pp. 2124–2130, 2002.

[BBB1990] J.H. Burroughes, D.D.C. Bradley, A.R. Brown, R.N. Marks, K. Mackay, R.H. Friend, P.L. Burns, and A.B. Holmes, "Light-Emitting Diodes Based on Conjugated Polymers," *Nature*, Vol. 347, pp. 539–541, 1990.

[BBJ1997] G.P. Bryan-Brown, C.V. Brown, J.C. Jones, E.L. Wood, I.C. Sage, P. Brett, and J. Rudin, "Grating Aligned Bistable Nematic Device", *SID Digest*, Vol. 18, pp. 37–40, 1997.

[BD1991] R. Barberi and G. Durand, "Electrochirally Controlled Bistable Surface Switching in Nematic Liquid Crystal," *Applied Physics Letters*, Vol. 58, No. 25, pp. 2907–2909, 1991.

[BGD1992] R. Barberi, M. Giocondo, and G. Durand, "Flexoelectrically Controlled Surface Bistable Switching in Nematic Liquid Crystals," *Applied Physics Letters*, Vol. 60, No. 9, pp. 1085–1086, 1992.

[BH1980] D.W. Berreman and W.R. Heffner, "New Bistable Cholesteric Liquid-Crystal Display," *Applied Physics Letters*, Vol. 37, No. 1, pp. 109–111, 1980.

[CL1980] N.A. Clark and S.T. Lagerwall, "Submicrosecond Bistable Electro-Optic Switching in Liquid Crystals," *Applied Physics Letters*, Vol. 36, No. 11, pp. 899–901, 1980.

[DCA1998] P. Drzaic, B. Comiskey, J.D. Albert, L. Zhang, A. Loxley, and R. Feeney, "A Printed and Rollable Bistable Electronic Display," *SID Digest*, Vol. 28, No. 1, pp. 1131–1134, 1998.

[DMP1997] I. Dozov, P. Martinot-Lagarde, E. Polossat, I. Lelidis, M. Giocondo, and G. Durand, "Fast Bistable Nematic Display from Coupled Surface Anchoring Breaking", *Proceedings of SPIE*, Vol. 3015, pp. 61–69, 1997.

[DND1997] I. Dozov, M. Nobili, and G. Durand, "Fast Bistable Nematic Display Using Monostable Surface Switching," *Applied Physics Letters*, Vol. 70, No. 9, pp. 1179–1181, 1997.

[DKS1998a] R.M.A Dawson, M.G. Kane, Z. Shen, D.A. Furst, S. Connor, J. Hsu, R.G. Stewart, A. Ipri, C.N. King, P.J. Green, R.T. Flegal, S. Pearson, W.A. Barrow, E. Dickey, K. Ping, S. Robinson, C.W. Tang, S. Vanslyke, F. Chen, J. Shi, J.C. Sturm, and M.H. Lu, "The Impact of the Transient Response of Organic Light Emitting Diodes on the Design of Active Matrix OLED Displays," *International Electron Devices Meeting Digest*, pp. 875–878, 1998.

[DKS1998b] R.M.A. Dawson, M.G. Kane, Z. Shen, D.A. Furst, S. Connor, J. Hsu, R.G. Stewart, A. Ipri, C.N. King, P.J. Green, R.T. Flegal, S. Pearson, W.A. Barrow, E. Dickey, K. Ping, S. Robinson, C.W. Tang, S. Vanslyke, F. Chen, J. Shi, J.C. Sturm, and M.H. Lu, "Active Matrix Organic Light Emitting Diode Pixel Design Using Polysilicon Thin Film Transistors," *IEEE Lasers and Electro-Optics Society Annual Meeting*, Vol. 1, pp. 128–129, 1998.

[DKS1998c] R.M.A Dawson, M.G. Kane, Z. Shen, D.A. Furst, S. Connor, J. Hsu, R.G. Stewart, A. Ipri, C.N. King, P.J. Green, R.T. Flegal, S. Pearson, W.A. Barrow, E. Dickey, K. Ping, S. Robinson, C.W. Tang, S. Van Slyke, F. Chen, J. Shi, J.C. Sturm, and M.H. Lu, "Design of an Improved Pixel for a Polysilicon Active Matrix Organic Light Emitting Diode Displays," *SID Digest*, Vol. 29, pp. 11–14, 1998.

[GWK1973] W. Greubel, U. Wolf, and H. Kinger, "Electric Field Induced Texture Change in Certain Nematic/Cholesteric Liquid Crystal Mixtures," *Molecular Crystals and Liquid Crystals*, Vol. 24, p. 103, 1973.

[HK2000] R. Hattori and J. Kanicki, "Current-Source a-Si:H Thin-Film-Transistor Circuit for Active-Matrix Organic Light-Emitting Displays," *IEEE Electron Device Letters*, Vol. 21, No. 12, pp. 590–592, 2000.

[HKH2002] Y. Hong, J. Kanicki, and R. Hattori, "Novel Poly-Si TFT Pixel Electrode Circuits and Current Programmed Active-Matrix Driving Methods for AM-OLEDs," *SID Digest*, Vol. 36, No. 1, pp. 1–4, 2002.

[LEE2002] G.B. Levy, W. Evans, J. Ebner, P. Farrell, M. Hufford, B.H. Alison, D. Wheeler, H. Lin, O. Prache, and E. Naviasky, "An 852×600 Pixel OLED-on-Silicon Color Microdisplay Using CMOS Subthreshold-Voltage-Scaling

Current Drivers," *IEEE Journal of Solid-State Circuits*, Vol. 37, pp. 1879–1889, 2002.

[NKK1998] E. Nakamura, H. Kawai, N. Kanae, and H. Yamamoto, "Development of Electrophoretic Display Using Microcapsulated Suspension," *SID Digest*, Vol. 29, No. 1, pp. 1014–1017, 1998.

[OOY1973] I. Ota, J. Ohnishi, M. Yoshiyama, "Electrophoretic Image Display (EPID) Panel," *Proceedings of IEEE*, Vol. 61, No. 7, pp. 832–836, 1973.

[SHP1998] M. Stewart, R.S. Howell, L. Pires, M.K. Hatalis, O. Prache, and W. Howard, "Polysilicon VGA Active Matrix OLED Displays-Technology and Performance," *International Electron Devices Meeting Digest*, pp. 871–874, 1998.

[SHP2001] M. Stewart, R.S. Howell, L. Pires, and M.K. Hatalis, "Polysilicon TFT Technology for Active Matrix OLED Displays," *IEEE Transaction on Electron Devices*, Vol. 48, No. 5, pp. 845–851, 2001.

[SWN1993] T. Shinoda, M. Wakitani, T. Nanto, K. Yoshikawa, A. Otsuka, and T. Hirose, "Development of Technologies for Large-Area Color AC Plasma Displays," *SID Digest*, pp. 161–164, 1993.

[TMK2000] T.W.-B. Simon, Y. Matsueda, M. Kimura, H. Maeda, T. Shimoda, and P. Migliorato, "Improved Polysilicon TFT Drivers for Light Emitting Polymer Displays," *Proceedings of International Display Workshop*, pp. 243–246, 2000.

[TV1987] C.W. Tang and S.A. VanSlyke, "Organic electroluminescent diodes," *Applied Physics Letters*, Vol. 51, No. 12, pp. 913–915, 1987.

[WBB2000] E.L. Wood, G. Bryan-Brown, P. Brett, A. Graham, J.C. Jones, and J.R. Hughes, "Zenithal Bistable Device (ZBD) Suitable for Portable Applications," *SID Digest*, Vol. 31, No. 1, pp. 124–127, 2000.

[YWC1994] D.-K. Yang, J.L. West, L.C. Chien, and J.W. Doane, "Control of the Reflectivity and Bistability in Displays Based on Cholesteric Liquid Crystals," *Journal of Applied Physics*, Vol. 76, No. 2, pp. 1331–1333, 1994.

[YYX2003] S. Yujuan, Z. Yi, C. Xinfa, and L. Shiyong, "A Simple and Effective Ac Pixel Driving Circuit for Active Matrix OLED," *IEEE Transaction on Electron Devices*, Vol. 50, No. 4, pp. 1137–1140, 2003.

Further Reading

OLED Displays

[AVH2002] W.F. Aerts, S. Verlaak, and P. Heremans, "Design of an Organic Pixel Addressing Circuit for an Active-Matrix OLED Display," *IEEE Transactions on Electron Devices*, Vol. 49, No. 12, pp. 2124–2130, 2002.

[BBF1994] P.E. Burrows, V. Bulovic, S.R. Forrest, L.S. Sapochak, D.M. McCarty, and M.E. Thompson, "Reliability and Degradation of Organic Light Emitting Devices," *Applied Physics Letters*, Vol. 65, No. 23, pp. 2922–2924, 1994.

[KNJ2005] A. Kumar, A. Nathan, and G. Jabbour, "Does TFT Mobility Impact Pixel Size in AMOLED Backplanes?," *IEEE Transactions on Electron Devices*, Vol. 52, No. 11, pp. 2386–2394, 2005.

[NCA2005] A. Nathan, G.R. Chaji, and S.J. Ashtiani, "Driving Schemes for a-Si and LTPS AMOLED Displays," *IEEE Journal of Display Technology*, Vol. 1, No. 4, pp. 267–277, 2005.

[OK2004] S. Ono and Y. Kobayashi, "Four-Thin-Film-Transistor Pixel Circuit for Amorphous-Silicon Active-Matrix Organic Light-Emitting Diode Displays,"

Japanese Journal of Applied Physics, Vol. 43, No. 12, Part 1, pp. 7947–7952, 2004.

[PNK2005] H.S. Pae, Y.S. Na, O.K. Kwon, and H.S. Kim, "Driving Circuit of Active-Matrix Method in Display Device," *U.S. Patent #6 917 350 B2*, issued 2005.

[SSS2004] K. Sakariya, S. Sambandan, P. Servati, and A. Nathan, "Analysis and Characterization of Self-Compensating Current Programmed a-Si:H Active-Matrix Organic Light-Emitting Diode Pixel Circuits," *Journal of Vacuum Science and Technology A*, Vol. 22, No. 3, pp. 1001–1004, 2004.

[V2003] K.M. Vaeth, "OLED-Display Technology," *Information Display*, Vol. 19, No. 6, pp. 12–18, 2003.

[VSJ2008] V. Vaidya, S. Soggs, K. Jungbae, A. Haldi, J.N. Haddock, B. Kippelen, and D.M. Wilson, "Comparison of Pentacene and Amorphous Silicon AMOLED Display Driver Circuits," *IEEE Transactions on Circuits and Systems - part I*, Vol. 55, No. 5, pp. 1177–1184, 2008.

[ZPB2005] L. Zhou, S. Park, B. Bai, J. Sun, S.-C. Wu, T.N. Jackson, S. Nelson, (K.) D. Freeman, and Y. Hong, "Pentacene TFT driven AM OLED displays," *IEEE Electron Device Letters*, Vol. 26, No. 9, pp. 640–642, 2005.

[ZWW2006] L. Zhou, A. Wanga, S.-C.Wu, J. Sun, S. Park, and T.N. Jackson, "All Organic Active-Matrix Flexible Display," *Applied Physics Letters*, Vol. 88, No. 8, p. 83502(1–3), 2006.

Bistable Electrophoretic Displays

[AS2006] K. Amundson and T. Sjodin, "Achieving Graytone Images in a Microencapsulated Electrophoretic Display," *SID Digest*, Vol. 37, No. 2, pp. 1918–1921, 2006.

[BC1996] Z. Bryning and R. Cromer, "Electrophoretic Display Utilizing Phase Separation of Liquids," *U.S. Patent #5 582 700*, issued 1996.

[CAY1998] B. Comiskey, J.D. Albert, H. Yoshizawa, and J. Jacobson, "An Electrophoretic Ink for All-Printed Reflective Electronic Displays," *Nature*, Vol. 394, pp. 253–255, 1998.

[CL1980] N.A. Clark and S.T. Lagerwall, "Sub Microsecond Bistable Electrooptic Switching in Liquid Crystals," *Applied Physics Letters*, Vol. 36, pp. 899–901, 1980.

[D1977] A. Dalisa, "Electrophoretic Display Technology," *IEEE Transactions on Electron Devices*, Vol. 24, No. 7, pp. 827–834, 1977.

[DAW2005] J.H. Daniel, A.C. Arias, W.S. Wong, R. Lujan, B.S. Krusor, R.B. Apte, M.L. Chabinyc, A. Salleo, R.A. Street, N. Chopra, G. Iftime, and P.M. Kazmaier, "Flexible Electrophoretic Displays with Jet-Printed Active-Matrix Backplanes," *SID Digest*, Vol. 36, No. 1, pp. 1630–1633, 2005.

[HK2006] A. Henzen and J. van de Kamer, "The Present and Future of Electronic Paper," *Journal of the SID*, Vol. 14, No. 5, pp. 437–442, 2006.

[HN1979] M.A. Hopper and V. Novotny "An Electrophoretic Display, Its Properties, Model, and Addressing," *IEEE Transactions on Electron Devices*, Vol. 26, No. 8, pp. 1148–1152, 1979.

[KFM2003] S. Kobayashi, H. Furuta, Y. Murakami, and J. Xu, "Fabrication and Electro-optic Characteristics of Polymer-Stabilized V-Mode FLCD and Intrinsic H-V-Mode FLCD: their Application to AM LCDs," *Proceedings of SPIE*, Vol. 5003, pp. 88–95, 2003.

[KOK2004] S. Kawamoto, M. Oh-Kochi, S. Kundu, H. Hasebe, H. Takatsu, and S. Kobayashi, "Polymer-Stabilized V-Mode FLCDs and their Application to Color Sequential Fullcolor LCDs," *Displays*, Vol. 25, No. 1, pp. 45–47, 2004.

[L2004] S.T. Lagerwall, "Ferroelectric and Antiferroelectric Liquid Crystals," *Ferroelectrics*, Vol. 301, pp. 15–45, 2004.

[LMM2007] X.H. Li, A. Murauski, A. Muravsky, P.Z. Xu, H.L. Cheung, and V.G. Chigrinov, "Grayscale Generation and Stabilization in Ferroelectric Liquid Crystal Display," *IEEE Journal of Display Technology*, Vol. 3, No. 3, pp.273–279, 2007.

[NCA2005] A. Nathan, G.R. Chaji, and S.J. Ashtiani, "Driving Schemes for a-Si and LTPS AMOLED Displays," *IEEE Journal of Display Technology*, Vol. 1, No. 2, pp. 267–277, 2005.

[SAB1991] P.W.H. Surguy, P.J. Ayliffe, M.J. Birch, M.F. Bone, I. Coulson, W.A. Crossland, J.R. Hughes, P.W. Ross, F.C. Saunders, and M.J. Towler, "The 'JOERS/Alvey' Ferroelectric Multiplexing Scheme," *Ferroelectrics*, Vol. 122, pp. 63–79, 1991.

[TBN2003] A. Tagawa, P. Bonnett, and T. Numao, "Driving Scheme for Ferroelectric Liquid Crystal Display with Multithreshold Method," *Japanese Journal of Applied Physics*, Vol. 42, pp. 1675–1679, 2003.

[TBN2004] A. Tagawa, P. Bonnett, and T. Numao, "Driving Scheme for Ferroelectric Liquid Crystal Display with Multithreshold Method by Thickness Variation," *Japanese Journal of Applied Physics*, Vol. 43, pp. 7150–7154, 2004.

[W2002] R. Webber, "Image Stability in Active Matrix Microencapsulated Electrophoretic Displays," *SID Digest*, Vol. 33, No. 1, pp. 126–129, 2002.

[WWP2004] T. Whitesides, M. Wallis, R. Paolini, S. Sohn, H. Gates, M. McCreary, and J. Jacobson, "Towards Video-rate Microencapsulated Dual-Particle Electrophoretic Displays," *SID Digest*, Vol. 35, No. 1, pp. 133–135, 2004.

[ZWL2006] H.M. Zang, X. Wang, and R.C. Liang, "Composition and Process for the Sealing of Microcups in Roll-to-Roll Display Manufacturing," *U.S. Patent #7,005,468*, issued 2006.

Appendix B

Display Specifications

The performance parameters of LCDs, and FPDs in general, are related to physical, optical and electrical display characteristics. In the following Table B.1 we summarize some of the most important display specifications. The list can be extended even further to: manufacturing cost, price, lifetime, operating temperature range, viewability in sunlight, weight, volume, voltage requirements, electromagnetic emissions, electromagnetic sensitivity, and toxic material content.

Size The diagonal size of the viewable display area expressed in inches.

Table B.1 Typical display sizes for different applications

Application	Typical Size (in)
Microdisplays	<1.5
Mobile phones and PDAs	2-4
Car navigation systems	7-9
Notbook computers	8-18
Desktop computers	10-25
Direct view TVs	>25

Aspect ratio Landscape (width larger than length), equal, and portrait (length larger than width) are the three display formats. Monitors and TVs adopt the landscape one with typical aspect ratio of 4:3, 16:9 or 16:10.

Resolution The number of pixels (contraction of *picture element*, the smallest individually addressable unit of an image that can be rendered or displayed) of the entire screen expressed as the product of the number of pixels in width and the number of pixels in length. Acronyms are adopted to designate resolution. For example, SVGA means an 800×600 pixel screen which is capable of displaying 480,000 pixels. Table B.2 summarizes many of the standard resolution formats.

Table B.2 Display resolution

Name	Resolution	Aspect ratio	Full name
QCIF	176×144	4:3	Quarter CIF
QVGA	320×240	4:3	Quarter VGA
CIF	352×288	4:3	Common Intermediate Format
VGA	640×480	4:3	Video Graphics Adapter
SVGA	800×600	4:3	Super VGA
XGA	1024×768	4:3	Extended Graphics Array
XGA+	1152×864	4:3	XGA plus
SXGA	1280×1024	5:4	Super XGA
SXGA+	1400×1050	4:3	Super XGA plus
WXGA	1280×768	5:3	Wide XGA
UXGA	1600×1200	4:3	Ultra XGA
WSXGA+	1680×1050	16:10	Wide SXGA plus
HDTV	1920×1080	16:9	High Definition TV
WUXGA	1920×1200	16:10	Wide UXGA
QXGA	2048×1536	4:3	Quad XGA
WQXGA	2560×1600	16:10	
QSXGA	2560×2048	5:4	
QSXGA+	2560×2100	4:3	
WQSXGA	3200×2048	25:16	
QUXGA	3200×2400	4:3	
WQUXGA	3840×2400	16:10	
HXGA	4096×3072	4:3	Hex(adecatuple) XGA (highest-end professional and scientific cameras, no monitors)

Luminance (Brightness) Luminance is the luminous flux emitted or reflected from a display surface and is measured in candela per square meter (cd/m^2) ANSI/HFS 100–1988 prescribes that "either the character or its background, whichever is of higher luminance, shall be able to achieve a luminance of at least 35 cd/m^2 or more." As a rule, FPD brightness should be the same or slightly higher than that of the real object to be reproduced. For an LCD TV a luminance of 500–1000 cd/m^2 is usually required.

Color depth Color in a display is obtained by mixing three R, G, B, (Red, Green, and Blue) primary colors. The more saturated (i.e., pure) they are, the broader is the range of possible colors (*color gamut*). If r, g, and b is the number of bits dedicated to the R G and B subpixel, respectively, then the color depth is $2^r \times 2^g \times 2^b$. For example, by using 8 bit to generate the gray levels of each RGB subpixel we get $2^8 \times 2^8 \times 2^8 \approx 16.8$ million colors.

Contrast ratio It is defined as $CR = (L_{\mathrm{WHITE}} + R)/(L_{\mathrm{BLACK}} + R)$, where L_{WITE} are L_{BLACK} are the luminance at the white and black states, respectively, and R takes into account the reflected ambient illumination that is the same for the white of black state. To increase contrast one has to increase L_{WITE} and/or decrease reflections.

Viewing angle It is the largest angle (viewing cone) over which one is able to acceptably view an image, generally defined in terms of minimum (acceptable) contrast ratio, minimum luminance, or maximum value of color shift. In same cases the *CR* is lower than 1, meaning that gray level inversion occurs.

Response time The **rise time** is the time required to switch a pixel from OFF to ON (from 10% to 90% luminance). The **fall time** is the time required to switch a pixel from ON to OFF (from 90% to 10% luminance). **Gray-to-gray response** time is more meaningful for color displays.

Power consumption Power dissipation of reflective LCDs, in their largest versions, it is in the range of 0.5–0.75 W. The use of a backlight increases consumption to 3–10 W. This parameter is of course more important for mobile displays since it affects battery life. However, notebooks and TVs also take advantage of lower power dissipation because it simplifies thermal management. Typically, the power efficiency is expressed in lumen per Watt, lm/W, which measures the ratio of light output to the electrical input power.

Appendix C

Matrices for MLA

Rademacher, Hadamard, and Walsh matrices

C.1 Rademacher Functions

The **Rademacher** functions are square waveforms with frequencies increasing successively by factors of two. They are easy to generate since they are just the output of a binary counter.

Figure C.1 illustrates the first four Rademacher functions and associated matrix. Observe that the matrix has dimensions 4×8 and, in general, $n \times 2^{(n-1)}$, where n is the number of rows. This means that, except for n=2, Rademacher functions do not lead to a square matrix.

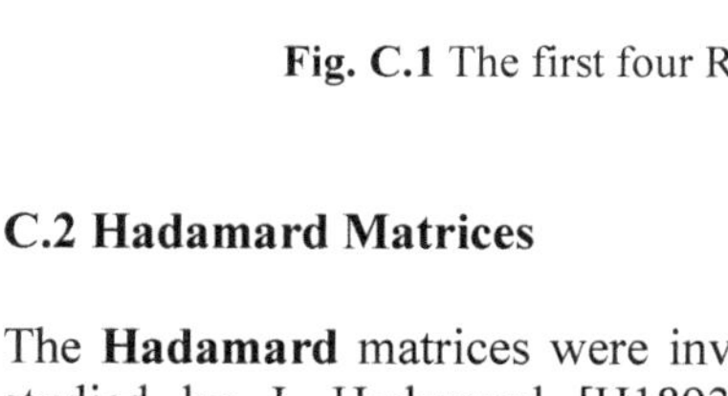

$$\mathbf{R}_4 = \begin{bmatrix} +1 & +1 & +1 & +1 & +1 & +1 & +1 & +1 \\ +1 & +1 & +1 & +1 & -1 & -1 & -1 & -1 \\ +1 & +1 & -1 & -1 & +1 & +1 & -1 & -1 \\ +1 & -1 & +1 & -1 & +1 & -1 & +1 & -1 \end{bmatrix}$$

Fig. C.1 The first four Rademacher functions and related matrix

C.2 Hadamard Matrices

The **Hadamard** matrices were invented by J.J. Sylvester in 1867 [S1867], further studied by J. Hadamard [H1893] and systematized by R. Paley [P1933]. A Hadamard matrix of order n, $\mathbf{H}_n$, is a square matrix whose entries are ±1 defined by $\mathbf{H}_n\mathbf{H}_n^{\mathrm{T}} = \mathbf{H}_n^{\mathrm{T}}\mathbf{H}_n = n\mathbf{I}_n$, where $\mathbf{H}_n^{\mathrm{T}}$ is the transpose matrix of $\mathbf{H}_n$, and $\mathbf{I}_n$ is the identity matrix of order n. A famous, but still unproved, conjecture is that $\mathbf{H}_n$ exists for all n divisible by 4. A possible recursive way to construct a Hadamard matrix was given by Sylvester

$$\mathbf{H}_2 = \begin{bmatrix} +1 & +1 \\ +1 & -1 \end{bmatrix}, \mathbf{H}_4 = \begin{bmatrix} +1 & +1 & +1 & +1 \\ +1 & -1 & +1 & -1 \\ +1 & +1 & -1 & -1 \\ +1 & -1 & -1 & +1 \end{bmatrix}, \ldots \mathbf{H}_n = \begin{bmatrix} \mathbf{H}_{n/2} & \mathbf{H}_{n/2} \\ \mathbf{H}_{n/2} & -\mathbf{H}_{n/2} \end{bmatrix}$$

A special 4×4 Hadamard matrix, widely used in LCD applications and that we can call of **second type** or **Walking 1**, $\mathbf{H}_{4,2}$, has diagonal elements being +1 while the remaining elements are −1. It is a particular case of circulant matrix, in which once defined the first row, every other row is a cyclic shift of its predecessor row.

$$\mathbf{H}_{4,2} = \begin{bmatrix} +1 & -1 & -1 & -1 \\ -1 & +1 & -1 & -1 \\ -1 & -1 & +1 & -1 \\ -1 & -1 & -1 & +1 \end{bmatrix}$$

Note that following from the definition, two distinct rows (and columns) of Hadamard matrices are orthogonal; this means that their dot product is zero, whereas the dot product of the same row (or column) gives *n*.

C.3 Walsh Matrices

The **Walsh** matrices can be derived from the Walsh functions [W1923]. They can be also obtained from the Hadamard matrix of the same dimension (Sylvester construction), by rearranging the rows so that the number of sign-changes is in increasing order.

For instance, the Walsh matrix of order 4 and its associated waveforms are shown below (Fig. C.2).

$$\mathbf{W}_4 = \begin{bmatrix} +1 & +1 & +1 & +1 \\ +1 & +1 & -1 & -1 \\ +1 & -1 & -1 & +1 \\ +1 & -1 & +1 & -1 \end{bmatrix}$$

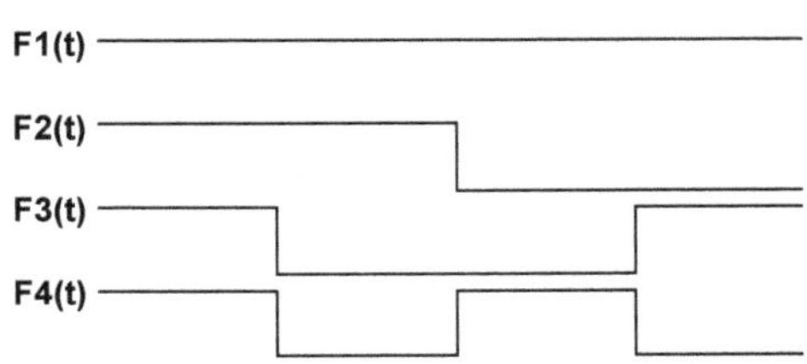

Fig. C.2 Walsh matrix of order 4 and associated waveforms

The first use of Walsh functions in LCD addressing is due to Jürgen Nehring and Allan R. Kmetz who developed an approach for theoretically analyze the performance of several different LCD addressing methods [NK1979]. Basically, they described the row and column voltages as two time-dependent vectors whereas the resultant LCD material response time is proportional to the time-averaged square of the difference of the two vectors at each pixel. The LCD panel integrates the excitation over the entire frame refresh period with a time constant characteristic of the particular liquid crystal material. One of the methods that Nehring and Kmetz described employed Walsh functions as the row vectors.

It should be noted that all the above representations lead to the same matrix of order 2×2.

In order to reduce the number of column voltages required by the MLA approach a theoretical way exploits sparse matrices. An example of sparse matrix, obtained from the Hadamard matrix $\mathbf{H}_4$ by replacing one entry in each column with a zero while maintaining orthogonality [RS2005], is shown below

$$\mathbf{S}_4 = \begin{bmatrix} 0 & +1 & +1 & +1 \\ +1 & -1 & +1 & 0 \\ +1 & +1 & 0 & -1 \\ +1 & 0 & -1 & +1 \end{bmatrix}$$

References

[S1867] J.J. Sylvester, "Thoughts on Orthogonal Matrices, Simultaneous Sign Successions, and Tessellated Pavements in Two or More Colours, with Applications to Newton's Rule, Ornamental Tile-Work, and Theory of Numbers," *Philosophical Magazine*, Vol. 34, pp. 461–475, 1867.

[H1893] J. Hadamard, "Résolution d'une Question Relative aux Determinants," (Solution of a Problem Relevant to Determinants) *Bulletin des Sciences Mathématiques*, Vol. 17, pp. 240–246, 1893.

[P1933] R.A.E.C. Paley, "On Orthogonal Matrices," *Journal of Mathematics and Physics*, Vol. 12, pp. 311–320, 1933.

[W1923] J.L. Walsh, "A Closed Set of Normal Orthogonal Functions," *American Journal of Mathematics*, Vol. 45, pp. 5–24, 1923.

[NK1979] J. Nehring and A.R. Kmetz, "Ultimate Limits for Matrix Addressing of RMS Responding Liquid Crystal Displays," *IEEE Transactions on Electron Devices*, Vol. 26, No. 5, pp. 795–802, 1979.

Appendix D

Color Perpection and Description

Among the five human senses, the view is the most important one. It has been computed that considering all the input stimuli reaching the human brain, 65% of them are coming from the eyes. The human being, in addition to the need to know, has the need to see. This is the reason of the constant and compulsory progress in the human history of both visual arts and technologies: from grapphiti, to painting, photography, cinema, and TV.

D.1 Color Perception

The human retina contains two types of photoreceptors: the **cones** and the **rods**. The rods are more numerous and are responsible for vision at low light levels (e.g., night vision). They do not mediate colors (because they are only of one type), and have a low spatial acuity. Cones have higher spatial acuity, their response times to stimuli are faster than those of rods, and are sensitive to colors. Indeed, humans have normally three types of cones that have peak sensitivity near 564–580 nm, 534–545 nm, and 420–440 nm wavelengths, respectively (see. Fig. D.1). These wavelengths are placed in the red, green, and blue region, respectively. Due to the absorption of cornea and lens, wavelengths lower than 380 nm (ultraviolet light) and higher than 780 nm (infrared) are not seen by the human eye.

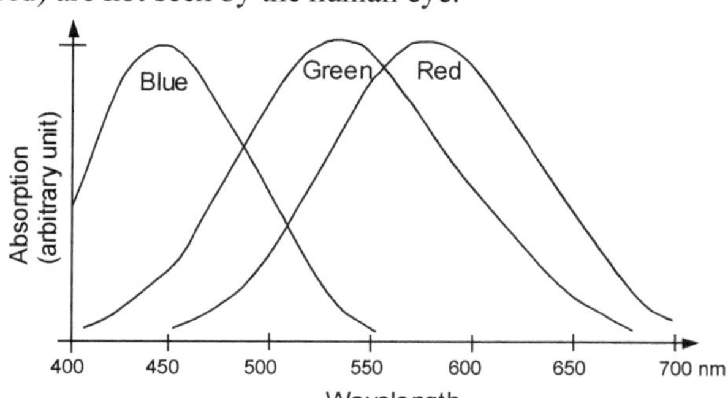

Fig. D.1 Absorption versus wavelength of the three types of cones

At the end of the 1920s, William David Wright [W1928] and John Guild [G1931] conducted independently a series of experiments on human sight which laid the foundation for both color perception science and color description. Then, in 1931 the CIE (Commission Internationale de l'Eclairage) used such experiments to define a standard set of spectral weighting functions that model the perception of color.

Color perception was measured by viewing combinations of the three standard CIE primary colors: red (at 700 nm), green (at 546.1 nm), and blue (at 435.8 nm).

Before describing the color spaces, it should be mentioned that the eye perception is nonlinear (as described in the following section with the McAdam ellipses), besides, human eyes are very sharp to detect image differences. At the same time such a comparison approach can be misleading in the perception, as can be seen from Fig. D.2. The two diagonals have the same color but being the left to right descendent diagonal surrounded by jellow, it seems lighter than the ascendant one (being surrounded by blue).

Techniques used in displays to improve their optical performance, like the error diffusion, are based upon the human vision perception system.

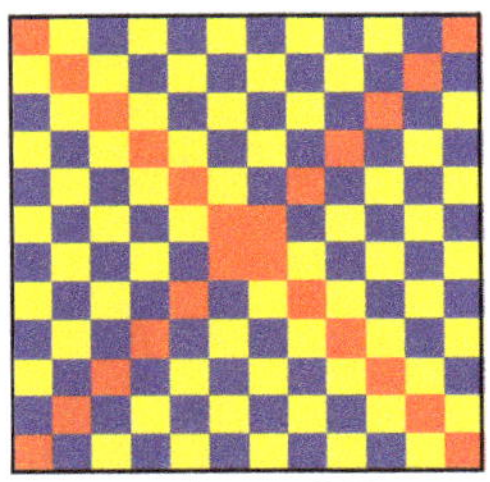

Fig. D.2 The two diagonals seem to have different colors because are differently surrounded

D.2 Color Description

One of the first mathematically defined color spaces was the **CIE 1931 XYZ** [CIE1931]. In this system, the intensities of red, green, and blue are transformed into what are called the *tristimulus values*, which are represented by the capital letters X, Y, and Z. A complete plot of all visible colors is therefore three dimensional. However, the color information can be divided into two parts: brightness and chromaticity. A derived color space, the **CIE xyY**, was then developed and used in practice [J2001]. Parameter Y was chosen so as to represent the luminance, while the chromaticity of a color was specified by the two derived parameters x and y, functions of all three tristimulus values: $x = X / (X + Y + Z)$ and $y = Y / (X + Y + Z)$. The third chromaticity value, $z = Z / (X + Y + Z)$, is redundant because can be obtained from the equation $x + y + z = 1$. The X and Z tristimulus values can be calculated back from the chromaticity and luminance. The CIE xy chromaticity diagram is shown in Fig. D.3. It represents the color **gamut** perceived by a person with normal vision (the CIE 1931 Standard Observer). In general, we refer to the gamut as the portion of the color space that can be represented.

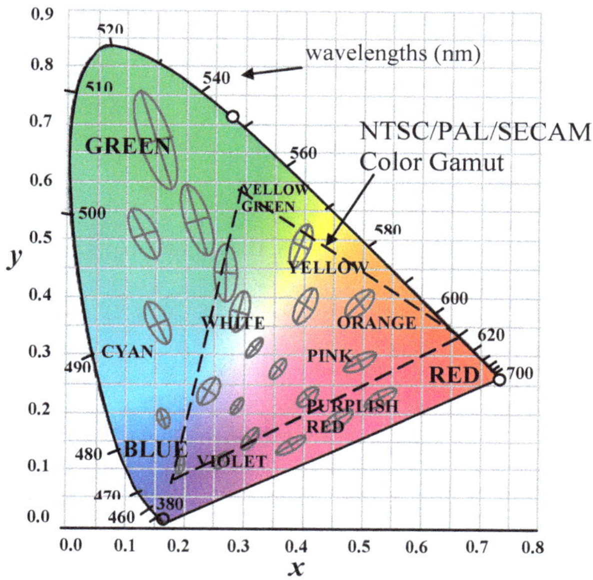

Fig. D.3 CIE 1931 Chromaticity Diagram. McAdam ellipses and NTSC/PAL/SECAM triangle of color gamut are also included

Each point on the diagram represents a unique color and is identified by its x and y coordinates.The outer curved boundary of the diagram is the *spectral locus*, with wavelengths shown in nanometers. The ends of the spectrum locus (at red and blue) are connected by a straight line that represents the purples, which are combinations of red and blue. The area within this closed boundary contains all the colors that can be generated by mixing light of different colors. The closer a color is to the boundary, the more saturated it is. Near the white center (acromatic light, $x=y=0.333$) colors are perceived as more pastel.

All colors that can be formed by mixing two color sources on the diagram lie on a straight line connecting these two points. All colors that can be formed by mixing three sources are found inside the triangle formed by the source points (and so on for multiple sources). It is seen that three sources are unable to cover the complete diagram.

Note that an equal mixture of two equally bright colors will not generally lie on the midpoint of that segment. In general, a distance on the xy chromaticity diagram does not correspond to the degree of difference between two colors.

In the early 1940s, David MacAdam studied the nature of visual sensitivity to color differences and summarized his results in the concept of a **MacAdam ellipse**. Based on this work, the CIE 1960, CIE 1964, and CIE 1976 color spaces were developed in order to achieve perceptual uniformity (a distance in the color space corresponds to equal differences in color).

D.3 Video Standards

In summary, a color space is a mathematical representation of a set of colors. The CIE 1931 standard can be immediately applicable to self-luminous sources and displays.[1] It is possible to specify a color by its chromaticity and luminance, in the form of an $x\,y\,Y$ triplet.

Many video standards use one luma and two color signals. The YUV color space is used by the PAL (Phase Alternation Line), NTSC (National Television System Committee), and SECAM (Sequentiel Couleur Avec Mémoire or Sequential Color with Memory) composite color video standards. The black-and-white system used only luma (Y) information. Alternative color spaces used in video systems are the YIQ (the "I" stands for "inphase" and the "Q" for "quadrature," which is the modulation method used to transmit the color information.) and the YCbCr.

The RGB is used in computer graphics and the CMYK is used in color printing. All of the color spaces can be derived from the RGB information provided by devices such as cameras and scanners. The RGB color space simplifies the architecture and design of the system which can take advantage of many software routines. However, transmitting and processing an image in the RGB space is not efficient because any RGB components need equal bandwidth. The frame buffer need also the same pixel depth and display resolution for each RGB component. If the system had access to an image stored directly in the brightness and color format, some processing steps would be faster.

The NTSC gamut is the triangle shown in Fig. D.3. It is seen that some colors cannot be reproduced by a TV display. The color depth (e.g., 18 bit, 24 bit, etc.) only shows how precisely we can define the color inside the triangle. To expand the triangle we will need to change the monitor luminosity spectrum thus shifting the triangle vertex coordinates. Conventional CCFL backlights provide the smallest color gamut, which is increased by the new bulbs with improved phosphors. The maximum color gamut can be achieved through light emitting diodes.

References

[CIE1931] CIE, *Commission Internationale de l'Eclairage Proceedings*, Cambridge University Press, Cambridge, 1931.

[G1931] J. Guild, "The colorimetric properties of the spectrum," *Philosophical Transactions of the Royal Society of London*, Vol. A230, pp. 149–187, 1931.

[J2001] K. Jack, *Video Demystified: A Handbook for the Digital Engineer*, LLH Technology Publishing, third edition, Eagle Rock, 2001.

[W1928] W.D. Wright, "A re-determination of the trichromatic coefficients of the spectral colours" *Transactions of the Optical Society*, Vol. 30, pp. 141–164, 1928.

---

[1]Considering reflective systems, like photography, printing or paint, the color perception is determined by the ambient light illumination. A photography not illuminated is perceived as a "black" image, whilst if the same photography is illuminated by a monochromatic light, the color will be distorted. Therefore the image must be submitted to an achromatic light, "white standard" (x=0.333, y= 0.333).

About the Authors

David Joseph Roger Cristaldi was born in Liege, Belgium, on December 18th 1978. He graduated in Electronics Engineering at the University of Catania in 2005. In his thesis, developed at STMicroelectronics, he studied low-power design methodologies for liquid crystal display drivers. In 2006 he obtained the Italian qualification for the engineer profession and in the same year he was admitted to the Ph.D. course in Electronics Engineering, Automation Engineering, and Control of Complex Systems. His current research activity concerns analysis and modeling of electromagnetic-interference effects on Sigma-Delta analog-to-digital converters. Dr. Cristaldi received the Ph.D. degree in 2009.

Salvatore Pennisi was born in Catania, Italy, in 1965. He received the laurea degree in electronic engineering in 1992 and the Ph.D. degree in electrical engineering in 1997, both from the University of Catania. He joined in 1996 the Department of Electrical Electronics and Systems Engineering (DIEES), University of Catania, as an Assistant Professor and he was appointed an Associate Professor in 2002. Since then, he has been engaged in scientific projects in collaboration with national and international academic and industrial partners. His main research interests include circuit theory and analog design with emphasis on low-voltage and current-mode techniques. In this field, he has developed several building blocks and unconventional architectures of operational amplifiers. More recently, his research activities have involved multi-stage amplifiers with related frequency compensation, data converters and the analysis of high-frequency distortion in analog circuits such as feedback amplifiers, oscillators and filters. He published over 60 papers in international journals and more than 120 contributions in conference proceedings, he is the co-author of the books *CMOS Current Amplifiers* (1999) and *Feedback Amplifiers: Theory and Design* (2001), both edited by Kluwer Academic Publishers, and has written an entry in the *Wiley Encyclopedia of Electrical and Electronics Engineering.*

Dr. Pennisi is senior member of the *IEEE* since 2004 and is a member of the IEEE CAS Analog Signal Processing Technical Committee. He currently serves as an Associate Editor of the IEEE TRANSACTIONS ON CIRCUITS AND SYSTEMS-PART II: EXPRESS BRIEFS.

Francesco Pulvirenti was born in Catania, Italy on August 8th 1964. He graduated in Electronic Engineering at the University of Pisa, Italy, in 1989. After the military service, he joined STMicroelectronics, Castelletto (Milan) site, in 1991 and then he moved to Catania site in 1993.

He started working in the Industrial and Peripheral Division developing Smart Power ICs for Industrial Applications, Linear Regulators, Switching Mode Power Supply and Power Management for Industrial, Mobile Phone, Battery Charger and Computer Applications.

In 2001 he moved to Display segment to manage the design of new Drivers dedicated to Liquid Crystal Display Applications for portable equipments and in 2004 he was appointed Design Director of the Display Division.

Since 2007 he is working in the Industrial and Power Conversion Division to define and develop new products for Photovoltaic Applications and he is currently managing a team of about 30 people as Business Unit Director.

Francesco Pulvirenti holds 32 patents on Analog and Smart Power ICs in the USA, he has 4 patents pending and he is also author or co-author of 8 papers on full custom ICs.

Index

P

R

S

T

www.ingramcontent.com/pod-product-compliance
Lightning Source LLC
Chambersburg PA
CBHW072219131224
19000CB00003B/65

9 789048 122547